MATHEMATICS-II

MATHEMATICS-II

Shadab Ahmad Khan
Lecturer in Department of Mathematics
Integral University, Lucknow

Dr. Qazi Shoeb Ahmad
Assistant Professor in Department of Mathematics
Integral University, Lucknow

Dr. Zubair Khan
Lecturer in Department of Mathematics
Integral University, Lucknow

Ane Books Pvt. Ltd.

New Delhi ♦ Chennai

Mathematics-II

Shadab Ahmad Khan, Qazi Shoeb Ahmad & Zubair Khan

Price: ₹ 350.00

Pages: xii+368=380 (PB)

Size: 6.25 x 9.50 inch

Paper Quality (GSM): 80

Published by

Ane Books Pvt. Ltd.
4821, Parwana Bhawan, 1st Floor, 24 Ansari Road,
Darya Ganj, **New Delhi** - 110 002, India
Tel.: +91(011) 23276843-44, Fax: +91(011) 23276863
e-mail: kapoor@anebooks.com, Website: www.anebooks.com

Branch Office :

Avantika Niwas, 1st Floor, 19 Doraiswamy Road, T. Nagar,
Chennai - 600 017, Tel.: +91(044) 28141554, 28141209
e-mail: anebooks_tn@airtelmail.in

ISBN : 978-81-8052-170-6

Printed at : R.K. Offset Process, New Delhi.

Preface

The authors feel great pleasure in presenting the first edition of the book **"Mathematics-II for B.C.A."**. This book is designed to meet the requirements of B.C.A. students of Integral University and its study centres (UPTEC).

The subject-matter has been discussed in such a way that the students will find no difficulty to understand it. Each chapter of this book contains complete self-explanatory theory and a large number of solved examples, followed by a collection of good exercises.

The language of the book, on the whole, is lucid and easy to understand. The authors hope that the students, teachers and other readers will find the book interesting and to the point covering the whole course. We hope that the students will receive the book warmly.

We have taken great care in eliminating the misprints, but if there are still any, we shall be highly obliged to those who will take trouble of pointing them out. Suggestions for the improvement of the book will be gratefully acknowledged.

We wish to place on record our sincere appreciation and thanks for the help, received from our respected teachers, colleagues and friends in presenting the manuscript of the book.

In the last but not the least we are also thankful to Mr. Sunil Saxena, Mr. Jai Raj Kapoor, Mr. H. Rahman and their team of **"Ane Books Pvt. Ltd."** for their kind cooperation at every stage, without which it would not have been possible to bring out this book in such a fine format.

Shadab Ahmad Khan

Dr. Qazi Shoeb Ahmad

Dr. Zubair Khan

Contents

UNIT–II : ORDINARY DIFFERENTIAL EQUATIONS

UNIT–III : PARTIAL DIFFERENTIAL EQUATION AND GEOMETRY

UNIT – I

Partial Differentiation and Its Applications

1.1 FUNCTION OF TWO VARIABLES

If $z = (x, y)$, then z is called the function of two independent variables x and y, where z is the dependent variable. The function $z = f(x, y)$ represents a surface.

1.2 PARTIAL DERIVATIVES

Let $z = f(x, y)$ be a function of two independent variables x and y. The partial derivative or partial differential coefficient of z w.r.t. x is the ordinary derivative of z w.r.t. x when y is regarded as a constant. It is usually written as $\frac{\partial z}{\partial x}$ or $\frac{\partial f}{\partial x}$ or f_x.

Thus, $\boxed{\frac{\partial z}{\partial x} = \frac{\partial f}{\partial x} = \lim_{h \to 0} \frac{f(x+h, y) - f(x, y)}{h}}$, provided the limit exist.

Similarly, the partial derivative of z w.r.t. y is the ordinary derivative of z w.r.t. y when x is regarded as a constant. It is usually written as $\frac{\partial z}{\partial y}$ or $\frac{\partial f}{\partial y}$ or f_y.

Thus, $\boxed{\frac{\partial z}{\partial y} = \frac{\partial f}{\partial y} = \lim_{k \to 0} \frac{f(x, y+k) - f(x, y)}{k}}$, provided the limit exist. $\frac{\partial z}{\partial x}$ and $\frac{\partial z}{\partial y}$ are called the first order partial derivative of z.

Note : In general if z is a function of two or more independent variables, then the partial derivative of z w.r.t. to any one of the independent variables is the ordinary derivative of z w.r.t. that independent variable, treating all other independent variable as constant.

1.3 PARTIAL DERIVATIVES OF HIGHER ORDER

The partial derivative $\frac{\partial z}{\partial x}$ and $\frac{\partial z}{\partial y}$ of z can be further differentiated partially w.r.t. x as well as y and , thus giving the second order partial derivative

$$\frac{\partial}{\partial x}\left(\frac{\partial z}{\partial x}\right) = \frac{\partial^2 z}{\partial x^2} = f_{xx} = z_{xx}$$

$$\frac{\partial}{\partial x}\left(\frac{\partial z}{\partial y}\right) = \frac{\partial^2 z}{\partial x \partial y} = f_{xy} = z_{xy}$$

$$\frac{\partial}{\partial y}\left(\frac{\partial z}{\partial y}\right) = \frac{\partial^2 z}{\partial y^2} = f_{yy} = z_{yy}$$

$$\frac{\partial}{\partial y}\left(\frac{\partial z}{\partial x}\right) = \frac{\partial^2 z}{\partial y \partial x} = f_{yx} = z_{yx}$$

Note 1 : In general, $\dfrac{\partial^2 z}{\partial x \partial y} = \dfrac{\partial^2 z}{\partial y \partial x}$

Note 2 : The rules in partial differentiation are same as that of ordinary differentiation. Hence,

(*i*) If $z = u + v$ and u, v being the functions of x and y. Then

$$\frac{\partial z}{\partial x} = \frac{\partial u}{\partial x} + \frac{\partial v}{\partial x}, \quad \frac{\partial z}{\partial y} = \frac{\partial u}{\partial y} + \frac{\partial v}{\partial y}$$

(*ii*) If $z = u(x, y).\ v(x, y)$, then

$$\frac{\partial z}{\partial x} = u.\frac{\partial v}{\partial x} + v.\frac{\partial u}{\partial x}, \quad \frac{\partial z}{\partial y} = u.\frac{\partial v}{\partial y} + v.\frac{\partial u}{\partial y}$$

(*iii*) If $z = \dfrac{u(x, y)}{v(x, y)}$, then

$$\frac{\partial z}{\partial x} = \frac{v.\dfrac{\partial u}{\partial x} - u.\dfrac{\partial v}{\partial x}}{v^2}, \quad \frac{\partial z}{\partial y} = \frac{v.\dfrac{\partial u}{\partial y} - u.\dfrac{\partial v}{\partial y}}{v^2}$$

(*iv*) If $z = f(u)$ and $u = u(x, y)$, then

$$\frac{\partial z}{\partial x} = \frac{dz}{du}.\frac{\partial u}{\partial x}, \quad \frac{\partial z}{\partial y} = \frac{dz}{du}.\frac{\partial u}{\partial y}$$

Here, $\dfrac{dz}{du}$ is used instead of $\dfrac{\partial z}{\partial u}$, since z is a function of u alone.

ILLUSTRATIVE EXAMPLES

Example1. If $u = x^2 \tan^{-1}\left(\dfrac{y}{x}\right)$, find $\dfrac{\partial u}{\partial x}$ and $\dfrac{\partial u}{\partial y}$.

Solution.

$$\frac{\partial u}{\partial x} = \frac{\partial}{\partial x}\left(x^2 \tan^{-1}\frac{y}{x}\right)$$

$$= x^2.\frac{\partial}{\partial x}\left(\tan^{-1}\frac{y}{x}\right) + \tan^{-1}\frac{y}{x}.\frac{\partial}{\partial x}\left(x^2\right)$$

$$= x^2.\frac{1}{1+\dfrac{y^2}{x^2}}.\left(\frac{-y}{x^2}\right) + \tan^{-1}\frac{y}{x}.2x = \frac{-x^2 y}{x^2+y^2}$$

And,

$$\frac{\partial u}{\partial y} = \frac{\partial}{\partial y}\left(x^2 \tan^{-1}\frac{y}{x}\right) = x^2\frac{\partial}{\partial y}\left(\tan^{-1}\frac{y}{x}\right)$$

$$= x^2.\frac{1}{1+\dfrac{y^2}{x^2}}.\left(\frac{1}{x}\right) = \frac{x^3}{x^2+y^2}$$

Example 2. If $u = x^3 y + e^{xy^2}$, prove that $\dfrac{\partial^2 u}{\partial x \partial y} = \dfrac{\partial^2 u}{\partial y \partial x}$.

Solution.

$$\frac{\partial u}{\partial x} = y\frac{\partial}{\partial x}\left(x^3\right) + \frac{\partial}{\partial x}\left(e^{xy^2}\right) = 3x^2 y + e^{xy^2}.\frac{\partial}{\partial x}\left(xy^2\right)$$

$$= 3x^2 + y^2$$

$$\frac{\partial u}{\partial y} = x^3\frac{\partial}{\partial y}(y) + \frac{\partial}{\partial y}\left(e^{xy^2}\right) = x^3 + e^{xy^2}.\frac{\partial}{\partial y}\left(xy^2\right)$$

$$= x^3 + 2xye^{xy^2}$$

$$\frac{\partial^2 u}{\partial x \partial y} = \frac{\partial}{\partial x}\left(\frac{\partial u}{\partial y}\right) = \frac{\partial}{\partial x}\left(x^3 + 2xye^{xy^2}\right)$$

$$= 3x^2 + 2y\left(xe^{xy^2}.y^2 + e^{xy^2}.1\right)$$

$\Rightarrow$

$$\frac{\partial^2 u}{\partial x \partial y} = 3x^2 + 2ye^{xy^2} + 2xy^3 e^{xy^2}$$

And

$$\frac{\partial^2 u}{\partial y \partial x} = \frac{\partial}{\partial y}\left(\frac{\partial u}{\partial x}\right) = \frac{\partial}{\partial y}\left(3x^2 y + y^2 e^{xy^2}\right)$$

$$= 3x^2.1 + \left(2y.e^{xy^2} + y^2 e^{xy^2}.2xy\right)$$

$$\Rightarrow \qquad \frac{\partial^2 u}{\partial y \partial x} = 3x^2 + 2ye^{xy^2} + 2xy^3 e^{xy^2}$$

Hence,
$$\frac{\partial^2 u}{\partial x \partial y} = \frac{\partial^2 u}{\partial y \partial x}.$$

Example 3. If $z = \dfrac{x^2+y^2}{x+y}$, verify $\left(\dfrac{\partial z}{\partial x} - \dfrac{\partial z}{\partial y}\right)^2 = 4\left(1 - \dfrac{\partial z}{\partial x} - \dfrac{\partial z}{\partial y}\right)$.

Solution.
$$z = \frac{x^2+y^2}{x+y}$$

$$\therefore \qquad \frac{\partial z}{\partial x} = \frac{(x+y).\dfrac{\partial\left(x^2+y^2\right)}{\partial x} - \left(x^2+y^2\right).\dfrac{\partial(x+y)}{\partial x}}{(x+y)^2}$$

$$\Rightarrow \qquad \frac{\partial z}{\partial x} = \frac{(x+y).2x - \left(x^2+y^2\right).1}{(x+y)^2} = \frac{x^2-y^2+2xy}{(x+y)^2}$$

And,
$$\frac{\partial z}{\partial y} = \frac{(x+y).2y - \left(x^2+y^2\right).1}{(x+y)^2} = \frac{y^2-x^2+2xy}{(x+y)^2}$$

Now,
$$\left(\frac{\partial z}{\partial x} - \frac{\partial z}{\partial y}\right)^2 = \left(\frac{x^2-y^2+2xy}{(x+y)^2} - \frac{y^2-x^2+2xy}{(x+y)^2}\right)^2$$

$$= \left[\frac{2x^2-2y^2}{(x+y)^2}\right]^2 = \left[\frac{2(x-y)(x+y)}{(x+y)^2}\right]^2 = \frac{4(x-y)^2}{(x+y)^2}$$

$$4\left(1 - \frac{\partial z}{\partial x} - \frac{\partial z}{\partial y}\right) = 4\left[1 - \frac{x^2-y^2+2xy}{(x+y)^2} - \frac{y^2-x^2+2xy}{(x+y)^2}\right]$$

$$= 4\left[\frac{x^2+y^2-2xy}{(x+y)^2}\right] = 4\frac{(x-y)^2}{(x+y)^2}$$

$$\therefore \qquad \left(\frac{\partial z}{\partial x} - \frac{\partial z}{\partial y}\right)^2 = 4\left(1 - \frac{\partial z}{\partial x} - \frac{\partial z}{\partial y}\right)$$

Example 4. If $z = f(x + ay) + g(x - ay)$, prove that $\dfrac{\partial^2 z}{\partial y^2} = a^2 \dfrac{\partial^2 z}{\partial x^2}$.

Solution.

$$z = f(x + ay) + g(x - ay)$$

$$\frac{\partial z}{\partial x} = f'(x+ay).1 + g'(x-ay).1$$

$$= f'(x+ay) + g'(x-ay)$$

$$\frac{\partial^2 z}{\partial x^2} = f''(x+ay) + g''(x-ay)$$

$$\frac{\partial z}{\partial y} = f'(x+ay).a + g'(x-ay).(-a)$$

$$= af'(x+ay) - ag'(x-ay)$$

$$\frac{\partial^2 z}{\partial y^2} = a^2 f''(x+ay) + a^2 g''(x-ay)$$

$$= a^2\left(f''(x+ay) + g''(x-ay)\right)$$

$$\Rightarrow \quad \frac{\partial^2 z}{\partial y^2} = a^2 \frac{\partial^2 z}{\partial x^2}$$

Example 5. Verify that $\dfrac{\partial^2 z}{\partial x \partial y} = \dfrac{\partial^2 z}{\partial y \partial x}$ for the function $z = x^y + y^x$.

Solution. $z = x^y + y^x$

$$\therefore \quad \frac{\partial z}{\partial x} = yx^{y-1} + y^x \log y$$

$$\Rightarrow \quad \frac{\partial^2 z}{\partial y \partial x} = \frac{\partial}{\partial y}\left(yx^{y-1} + y^x \log y\right)$$

$$= 1.x^{y-1} + yx^{y-1}\log x + xy^{x-1}\log y + y^x.\frac{1}{y}$$

$$= x^{y-1}(1 + y\log x) + y^{x-1}(1 + x\log y)$$

Again,

$$\frac{\partial z}{\partial y} = x^y \log x + xy^{x-1}$$

$$\Rightarrow \quad \frac{\partial^2 z}{\partial x \partial y} = \frac{\partial}{\partial x}\left(x^y \log x + xy^{x-1}\right)$$

$$= yx^{y-1}.\log x + x^y.\frac{1}{x} + 1.y^{x-1} + xy^{x-1}\log y$$

$$= x^{y-1}(1 + y\log x) + y^{x-1}(1 + x\log y)$$

Hence,

$$\frac{\partial^2 z}{\partial x \partial y} = \frac{\partial^2 z}{\partial y \partial x}.$$

Example 6. If $\theta = t^n e^{-r^2/4t}$, find the value of n such that $\frac{1}{r^2}\frac{\partial}{\partial r}\left(r^2\frac{\partial\theta}{\partial r}\right) = \frac{\partial\theta}{\partial t}$.

Solution. $\theta = t^n e^{-r^2/4t}$

$$\therefore \quad \frac{\partial\theta}{\partial r} = t^n e^{-r^2/4t}.\left(-\frac{2r}{4t}\right) = -\frac{1}{2}rt^{n-1}e^{-r^2/4t}$$

$$r^2\frac{\partial\theta}{\partial r} = -\frac{1}{2}r^3t^{n-1}e^{-r^2/4t}$$

$$\frac{\partial}{\partial r}\left(r^2\frac{\partial\theta}{\partial r}\right) = -\frac{1}{2}\frac{\partial}{\partial r}\left(r^3t^{n-1}e^{-r^2/4t}\right) = -\frac{1}{2}t^{n-1}\frac{\partial}{\partial r}\left(r^3e^{-r^2/4t}\right)$$

$$\frac{\partial}{\partial r}\left(r^2\frac{\partial\theta}{\partial r}\right) = -\frac{1}{2}t^{n-1}\left[r^3e^{-r^2/4t}\left(-\frac{2r}{4t}\right)+3r^2e^{-r^2/4t}\right]$$

$$= -\frac{1}{2}t^{n-1}r^2e^{-r^2/4t}\left[3-\frac{r^2}{2t}\right]$$

$$\Rightarrow \quad \frac{1}{r^2}\frac{\partial}{\partial r}\left(r^2\frac{\partial\theta}{\partial r}\right) = t^{n-1}e^{-r^2/4t}\left(\frac{r^2}{4t}-\frac{3}{2}\right)$$

Also
$$\frac{\partial\theta}{\partial t} = e^{-r^2/4t}.nt^{n-1}+t^ne^{-r^2/4t}.\frac{r^2}{4t^2} = t^{n-1}e^{-r^2/4t}\left(n+\frac{r^2}{4t}\right)$$

Now
$$\frac{1}{r^2}\frac{\partial}{\partial r}\left(r^2\frac{\partial\theta}{\partial r}\right) = \frac{\partial\theta}{\partial t} \Rightarrow \frac{r^2}{4t}-\frac{3}{2} = n+\frac{r^2}{4t}$$

$$\therefore \quad n = -\frac{3}{2}$$

Example 7. If $u = e^{xyz}$, show that $\frac{\partial^3 u}{\partial x\partial y\partial z} = (1+3xyz+x^2y^2z^2)e^{xyz}$.

Solution. Here $u = e^{xyz}$

$$\therefore \quad \frac{\partial u}{\partial z} = xy.e^{xyz}$$

Again,
$$\frac{\partial^2 u}{\partial y\partial z} = \frac{\partial}{\partial y}\left(\frac{\partial u}{\partial z}\right) = \frac{\partial}{\partial y}\left(xy.e^{xyz}\right) = x\frac{\partial}{\partial y}\left(y.e^{xyz}\right)$$

$$= x[e^{xyz}.1+y.e^{xyz}.xz] = [x+x^2yz]e^{xyz}$$

$$\frac{\partial^3 u}{\partial x\partial y\partial z} = \frac{\partial}{\partial x}\left(\frac{\partial^2 u}{\partial y\partial z}\right) = \frac{\partial}{\partial x}\left[\left(x+x^2yz\right)\right]e^{xyz}$$

$$= (x+x^2yz)e^{xyz}.yz+(1+2xyz)e^{xyz}$$

$$= e^{xyz}(1+3xyz+x^2y^2z^2).$$

Example 8. If $u = \log (x^3 + y^3 + z^3 - 3xyz)$, prove that

$$(i) \qquad \frac{\partial u}{\partial x}+\frac{\partial u}{\partial y}+\frac{\partial u}{\partial z} = \frac{3}{x+y+z} \text{ and}$$

$$(ii) \qquad \left(\frac{\partial}{\partial x}+\frac{\partial}{\partial y}+\frac{\partial}{\partial z}\right)^2 u = \frac{-9}{(x+y+z)^2}.$$

Solution. (*i*) $\qquad u = \log (x^3 + y^3 + z^3 - 3xyz)$

$$\frac{\partial u}{\partial x} = \frac{3x^2-3yz}{x^3+y^3+z^3-3xyz} \qquad \text{... (1)}$$

$$\frac{\partial u}{\partial y} = \frac{3y^2-3xz}{x^3+y^3+z^3-3xyz} \qquad \text{... (2)}$$

$$\frac{\partial u}{\partial z} = \frac{3z^2-3xy}{x^3+y^3+z^3-3xyz} \qquad \text{... (3)}$$

Adding eqns. (1), (2) and (3), we get

$$\frac{\partial u}{\partial x}+\frac{\partial u}{\partial y}+\frac{\partial u}{\partial z} = \frac{3\left(x^2+y^2+z^2-yz-xz-xy\right)}{x^3+y^3+z^3-3xyz}$$

$$\frac{\partial u}{\partial x}+\frac{\partial u}{\partial y}+\frac{\partial u}{\partial z} = \frac{3\left(x^2+y^2+z^2-yz-xz-xy\right)}{(x+y+z)\left(x^2+y^2+z^2-yz-xz-xy\right)}$$

$$= \frac{3}{(x+y+z)}$$

$$(ii) \qquad \left(\frac{\partial}{\partial x}+\frac{\partial}{\partial y}+\frac{\partial}{\partial z}\right)^2 u = \left(\frac{\partial}{\partial x}+\frac{\partial}{\partial y}+\frac{\partial}{\partial z}\right)\left(\frac{\partial u}{\partial x}+\frac{\partial u}{\partial y}+\frac{\partial u}{\partial z}\right)$$

$$= \left(\frac{\partial}{\partial x}+\frac{\partial}{\partial y}+\frac{\partial}{\partial z}\right)\left(\frac{3}{x+y+z}\right)$$

$$= -\frac{3}{(x+y+z)^2}-\frac{3}{(x+y+z)^2}-\frac{3}{(x+y+z)^2}$$

$$= -\frac{9}{(x+y+z)^2}.$$

Example 9. If $u = f(r)$, and $x = r\cos\theta$, $y = r\sin\theta$, show that

$$\frac{\partial^2 u}{\partial x^2}+\frac{\partial^2 u}{\partial y^2} = f''(r)+\frac{1}{r}f'(r).$$

Solution. Since $x = r\cos\theta$, $y = r\sin\theta$

$$r^2 = x^2 + y^2 \qquad \text{... (1)}$$

Differentiating eq. (1) partially w.r.t. x, we get

$$2r\frac{\partial r}{\partial x} = 2x \quad \Rightarrow \quad \frac{\partial r}{\partial x} = \frac{x}{r}$$

Similarly, $$\frac{\partial r}{\partial y} = \frac{y}{r}$$

Now, $$u = f(r) \Rightarrow \frac{\partial u}{\partial x} = f'(r)\frac{\partial r}{\partial x} = f'(r)\frac{x}{r}$$

$\therefore$ $$\frac{\partial^2 u}{\partial x^2} = \frac{\partial}{\partial x}\left(\frac{xf'(r)}{r}\right) = \frac{r\left[xf''(r).\frac{\partial r}{\partial x}+f'(r)\right]-xf'(r).\frac{\partial r}{\partial x}}{r^2}$$

$\Rightarrow$ $$\frac{\partial^2 u}{\partial x^2} = \frac{1}{r^2}\left[x^2 f''(r)+rf'(r)-\frac{x^2}{r}f'(r)\right] \qquad \text{... (2)}$$

Similarly, $$\frac{\partial^2 u}{\partial y^2} = \frac{1}{r^2}\left[y^2 f''(r)+rf'(r)-\frac{y^2}{r}f'(r)\right] \qquad \text{... (3)}$$

Adding eqns. (2) and (3), we get

$$\frac{\partial^2 u}{\partial x^2}+\frac{\partial^2 u}{\partial y^2} = \frac{1}{r^2}\left[\left(x^2+y^2\right)f''(r)+2rf'(r)-\frac{\left(x^2+y^2\right)}{r}f'(r)\right]$$

$$= \frac{1}{r^2}\left[r^2 f''(r)+2rf'(r)-\frac{r^2}{r}f'(r)\right]$$

$$= f''(r)+\frac{1}{r}f'(r).$$

Example 10. If $z = \log(e^x + e^y)$, show that

$$\left(\frac{\partial^2 z}{\partial x^2}\right).\left(\frac{\partial^2 z}{\partial y^2}\right) = \left(\frac{\partial^2 z}{\partial x \partial y}\right)^2$$

Solution.

$$\frac{\partial z}{\partial x} = \frac{e^x}{e^x + e^y},\ \frac{\partial z}{\partial y} = \frac{e^y}{e^x + e^y}$$

$$\frac{\partial^2 z}{\partial x^2} = \frac{(e^x + e^y)e^x - e^x.e^x}{(e^x + e^y)^2} = \frac{e^x e^y}{(e^x + e^y)^2} = \frac{e^{x+y}}{(e^x + e^y)^2}$$

$$\frac{\partial^2 z}{\partial y^2} = \frac{(e^x + e^y)e^y - e^y.e^y}{(e^x + e^y)^2} = \frac{e^x e^y}{(e^x + e^y)^2} = \frac{e^{x+y}}{(e^x + e^y)^2}$$

$$\frac{\partial^2 z}{\partial x \partial y} = \frac{\partial}{\partial x}\left(\frac{e^y}{e^x + e^y}\right) = \frac{(e^x + e^y).0 - e^y.e^x}{(e^x + e^y)^2}$$

$$= \frac{-e^x e^y}{(e^x + e^y)^2} = \frac{-e^{x+y}}{(e^x + e^y)^2}$$

$$\therefore \left(\frac{\partial^2 z}{\partial x^2}\right)\cdot\left(\frac{\partial^2 z}{\partial y^2}\right) - \left(\frac{\partial^2 z}{\partial x \partial y}\right)^2 = \frac{e^{x+y}}{(e^x + e^y)^2} \times \frac{e^{x+y}}{(e^x + e^y)^2} - \left\{-\frac{e^{x+y}}{(e^x + e^y)^2}\right\}^2 = 0$$

Example 11. If $x^2 = au + bv$, $y^2 = au - bv$, prove that

$$\left(\frac{\partial u}{\partial x}\right)_y \cdot \left(\frac{\partial x}{\partial u}\right)_v = \frac{1}{2} = \left(\frac{\partial v}{\partial y}\right)_x \cdot \left(\frac{\partial y}{\partial v}\right)_u.$$

Solution. We have $x^2 = au + bv$, $y^2 = au - bv$

which gives $$u = \frac{x^2 + y^2}{2a} \text{ and } v = \frac{x^2 - y^2}{2b}$$

Now, $\left(\frac{\partial u}{\partial x}\right)_y$ = partial derivative of u w.r.t. x keeping y constant.

$$\therefore \quad u = \frac{x^2 + y^2}{2a} \Rightarrow \left(\frac{\partial u}{\partial x}\right)_y = \frac{x}{a}$$

$\left(\frac{\partial x}{\partial u}\right)_v$ = partial derivative of x w.r.t. u keeping v constant.

$$\therefore \quad x^2 = au + bv \Rightarrow \left(\frac{\partial x}{\partial u}\right)_v = \frac{a}{2x}$$

Hence, $$\left(\frac{\partial u}{\partial x}\right)_y \cdot \left(\frac{\partial x}{\partial u}\right)_v = \frac{x}{a}.\frac{a}{2x} = \frac{1}{2}$$

Also, $\left(\dfrac{\partial v}{\partial y}\right)_x$ = partial derivative of v w.r.t. y keeping x constant.

$\therefore \qquad v = \dfrac{x^2 - y^2}{2b} \Rightarrow \left(\dfrac{\partial v}{\partial y}\right)_x = \dfrac{-y}{b}$

$\left(\dfrac{\partial y}{\partial v}\right)_u$ = partial derivative of y w.r.t. v keeping u constant.

$\therefore \qquad y^2 = au - bv \Rightarrow \left(\dfrac{\partial x}{\partial u}\right)_v = \dfrac{a}{2x}$

Hence, $\left(\dfrac{\partial v}{\partial y}\right)_x \cdot \left(\dfrac{\partial y}{\partial v}\right)_u = \dfrac{-y}{b} \cdot \dfrac{-b}{2y} = \dfrac{1}{2}$.

Example 12. If $f(x, y) = x^3y - xy^3$, find $\left(\dfrac{1}{f_x} + \dfrac{1}{f_y}\right)_{x=2, y=1}$

Solution. Here $\qquad f = x^3y - xy^3$

$\therefore \qquad f_x = \dfrac{\partial f}{\partial x} = 3x^2y - y^3$ and $f_y = \dfrac{\partial f}{\partial y} = x^3 - 3xy^2$

Putting $x = 2,\ y = 1$, we have

$f_x = 12 - 1 = 11$ and $f_y = 8 - 6 = 2$

$\therefore \qquad \left(\dfrac{1}{f_x} + \dfrac{1}{f_y}\right)_{x=2, y=1} = \dfrac{1}{11} + \dfrac{1}{2} = \dfrac{13}{22}$

Example 13. If $x^x y^y z^z = c$, then show that $\dfrac{\partial^2 z}{\partial x \partial y} = -(x \log ex)^{-1}$ at $x = y = z$.

Solution. We have $x^x y^y z^z = c$

$\Rightarrow \qquad x \log x + y \log y + z \log z = c \qquad \text{... (1)}$

where z is a function of x and y.

Differentiating eqn. (1) partially w.r.t. x, we get

$$\left(x.\frac{1}{x} + \log x.1\right) + 0 + \left(z.\frac{1}{z} + \log z.1\right)\frac{\partial z}{\partial x} = 0$$

$\Rightarrow \qquad \dfrac{\partial z}{\partial x} = -\dfrac{1 + \log x}{1 + \log z} \qquad \text{... (2)}$

Similarly, $$\frac{\partial z}{\partial y} = -\frac{1+\log y}{1+\log z} \qquad \text{... (3)}$$

Now $$\frac{\partial^2 z}{\partial x \partial y} = \frac{\partial}{\partial x}\left(\frac{\partial z}{\partial y}\right) = -\frac{\partial}{\partial x}\left(\frac{1+\log y}{1+\log z}\right)$$

$$= -(1+\log y)\frac{\partial}{\partial x}(1+\log z)^{-1}$$

$$= -(1+\log y)\left[-(1+\log z)^{-2}.\frac{1}{z}.\frac{\partial z}{\partial x}\right]$$

$$= \frac{(1+\log y)}{z(1+\log z)^2}\left\{-\frac{1+\log x}{1+\log z}\right\} \qquad \text{... (4)}$$

At $x = y = z$,

$$\frac{\partial^2 z}{\partial x \partial y} = \frac{(1+\log x)}{x(1+\log z)^2}\left\{-\frac{1+\log x}{1+\log x}\right\} = -\frac{1}{x(1+\log x)}$$

$$\frac{\partial^2 z}{\partial x \partial y} = -\frac{1}{x(\log e+\log x)} = -\frac{1}{x(\log e+\log x)}$$

Example 14. If $u = \tan^{-1}\frac{xy}{\sqrt{1+x^2+y^2}}$, show that $\frac{\partial^2 u}{\partial x \partial y} = \frac{1}{(1+x^2+y^2)^{3/2}}$.

(I.U. 2007, 08)

Solution. Given, $u = \tan^{-1}\frac{xy}{\sqrt{1+x^2+y^2}}$

$$\Rightarrow \frac{\partial u}{\partial x} = \frac{1}{1+\left(\frac{xy}{\sqrt{1+x^2+y^2}}\right)^2}\cdot\left[\frac{(1+x^2+y^2)^{1/2}.y - xy.\frac{1}{2}.(1+x^2+y^2)^{-1/2}(2x)}{(1+x^2+y^2)}\right]$$

$$= \frac{1}{1+x^2+y^2+x^2y^2}\cdot\left[y(1+x^2+y^2)^{1/2} - \frac{x^2 y}{(1+x^2+y^2)^{1/2}}\right]$$

$$= \frac{1}{1+x^2+y^2+x^2y^2}\cdot\left[\frac{y(1+x^2+y^2)-x^2 y}{(1+x^2+y^2)^{1/2}}\right]$$

$$= \frac{y(1+y^2)}{(1+x^2+y^2+x^2y^2)(1+x^2+y^2)^{1/2}}$$

$$\Rightarrow \quad \frac{\partial u}{\partial x} = \frac{y(1+y^2)}{(1+x^2)(1+y^2)(1+x^2+y^2)^{1/2}} = \frac{y}{(1+x^2)(1+x^2+y^2)^{1/2}}$$

$$\therefore \quad \frac{\partial^2 u}{\partial x \partial y} = \frac{1}{(1+x^2)}.\frac{\partial}{\partial y}\left[\frac{y}{(1+x^2+y^2)^{1/2}}\right]$$

$$= \frac{1}{(1+x^2)}\left[\frac{(1+x^2+y^2)^{1/2}.1 - y.\frac{1}{2}(1+x^2+y^2)^{-1/2}.2y}{(1+x^2+y^2)}\right]$$

$$= \frac{1}{(1+x^2)}\left[\frac{1+x^2+y^2-y^2}{(1+x^2+y^2)(1+x^2+y^2)^{1/2}}\right]$$

$$\Rightarrow \quad \frac{\partial^2 u}{\partial x \partial y} = \frac{1}{(1+x^2+y^2)^{3/2}}$$

EXERCISE 1.1

1. If $z^3 - 3yz - 3x = 0$, then show that $\left[\frac{\partial^2 z}{\partial x \partial y} + \left(\frac{\partial z}{\partial x}\right)^2\right] = \frac{1}{z}.\frac{\partial^2 z}{\partial y^2}$.

2. Verify that $\frac{\partial^2 u}{\partial x \partial y} = \frac{\partial^2 u}{\partial y \partial x}$, for

(*i*) $\log_e \frac{x^2+y^2}{x+y}$ (*ii*) $u = x \cos y + y \cos x$ (*iii*) $\log \tan \frac{y}{x}$.

3. If $u = e^{ax+by} f(ax - by)$, prove that $b\frac{\partial u}{\partial x} + a\frac{\partial u}{\partial y} = 2abu$.

4. If $u = (x^2 + y^2 + z^2)^{1/2}$, then prove that $\frac{\partial^2 u}{\partial x^2} + \frac{\partial^2 u}{\partial y^2} + \frac{\partial^2 u}{\partial z^2} = \frac{z}{u}$.

5. If $f(x, y) = \frac{1}{\sqrt{y}} e^{-\frac{(x-a)^2}{4y}}$, verify that $f_{xy} = f_{yx}$.

6. If $x = r \cos \theta$, $y = r \sin \theta$, prove that

(*i*) $\frac{\partial r}{\partial x} = \frac{\partial x}{\partial r}$ (*ii*) $r.\frac{\partial \theta}{\partial x} = \frac{1}{r}.\frac{\partial x}{\partial \theta}$ (*iii*) $\frac{\partial^2 r}{\partial x^2} + \frac{\partial^2 r}{\partial y^2} = \frac{1}{r}\left[\left(\frac{\partial r}{\partial x}\right)^2 + \left(\frac{\partial r}{\partial y}\right)^2\right]$

(*iv*) $\frac{\partial^2 \theta}{\partial x^2} + \frac{\partial^2 \theta}{\partial y^2} = 0$

7. If $x = \frac{r}{2}(e^{\theta} + e^{-\theta}), \quad y = \frac{r}{2}(e^{\theta} - e^{-\theta})$, prove that $\frac{\partial r}{\partial x} = \frac{\partial x}{\partial r}$.

8. If $u = (1 - 2xy + y^2)^{-1/2}$, prove that $\frac{\partial}{\partial x}\left\{(1 - x^2)\frac{\partial u}{\partial x}\right\} + \frac{\partial}{\partial y}\left\{y^2 \frac{\partial u}{\partial y}\right\} = 0$.

9. If $u = \log_e \sqrt{x^2 + y^2 + z^2}$, prove that $(x^2 + y^2 + z^2)\left(\frac{\partial^2 u}{\partial x^2} + \frac{\partial^2 u}{\partial y^2} + \frac{\partial^2 u}{\partial z^2}\right) = 1$.

10. If $u = f(ax^2 + by^2 + 2hxy)$, $v = \phi\,(ax^2 + by^2 + 2hxy)$, prove that $\frac{\partial}{\partial y}\left(u\frac{\partial v}{\partial x}\right) = \frac{\partial}{\partial x}\left(u\frac{\partial v}{\partial y}\right)$.

11. If $u = r^m$, $r^2 = x^2 + y^2 + z^2$, then prove that $\frac{\partial^2 u}{\partial x^2} + \frac{\partial^2 u}{\partial y^2} + \frac{\partial^2 u}{\partial z^2} = m(m+1)r^{n-2}$.

12. If $ux + vy = 0$ and $\frac{u}{x} + \frac{v}{y} = 1$, show that $\frac{\partial u}{\partial x} - \frac{\partial v}{\partial y} = \frac{x^2 + y^2}{y^2 - x^2}$.

1.4 HOMOGENEOUS FUNCTIONS

A function in which every term is of the same degree is called a homogeneous function.

Thus, $f(x, y) = a_0 x^n + a_1 x^{n-1} y + a_2 x^{n-2} y^2 + \ldots\ldots\ldots + a_{n-1} x y^{n-1} + a_n y^n$

be a homogeneous function of x and y of degree n.

This can also be written as:

$$f(x, y) = x^n\left[a_0 + a_1\left(\frac{y}{x}\right) + a_2\left(\frac{y}{x}\right)^2 + \ldots\ldots\ldots + a_{n-1}\left(\frac{y}{x}\right)^{n-1} + a_n\left(\frac{y}{x}\right)^n\right]$$

$$= x^n \phi\left(\frac{y}{x}\right)$$

Thus, every homogeneous function of x and y of degree n can be expressed in the form $x^n \phi\left(\frac{y}{x}\right)$.

1.5 TEST OF HOMOGENEITY OF A FUNCTION OF *x* AND *y*

Let $f(x, y)$ be any function of x and y. Then $f(x, y)$ is said to be homogeneous function of x and y of degree n, if

$$f(tx, ty) = t^n f(x, y)$$

Example 1. If $f(x, y) = \dfrac{\sqrt{x}+\sqrt{y}}{x^2+y^2}$,

then $$f(x, y) = \frac{\sqrt{x}+\sqrt{y}}{x^2+y^2} = \frac{\sqrt{x}\left(1+\sqrt{\dfrac{y}{x}}\right)}{x^2\left(1+\dfrac{y^2}{x^2}\right)} = x^{-\frac{3}{2}}\phi\left(\frac{y}{x}\right)$$

or $$f(tx, ty) = \frac{\sqrt{tx}+\sqrt{ty}}{(tx)^2+(ty)^2} = t^{-\frac{3}{2}}\left[\frac{\sqrt{x}+\sqrt{y}}{(x)^2+(y)^2}\right] = t^{-\frac{3}{2}}f(x,y)$$

Thus, $f(x, y)$ is a homogeneous function of x and y of degree (–3/2).

Example 2. If $f(x, y) = x^3 \log\left(\dfrac{y}{x}\right)$,

then $$f(tx, ty) = (tx)^3 \log\left(\frac{ty}{tx}\right) = t^3 f(x, y)$$

Thus, $f(x, y)$ is a homogeneous function of x and y of degree 3.

1.6 EULER'S THEOREM ON HOMOGENEOUS FUNCTIONS

If $u(x, y)$ is a homogeneous function of x and y of degree n, then

$$x\frac{\partial u}{\partial x} + y\frac{\partial u}{\partial y} = nu$$

Proof: Since $u = u(x, y)$ is a homogeneous function of x and y of degree n, therefore we can write

$$u(x, y) = x^n F\left(\frac{y}{x}\right) \qquad \text{... (1)}$$

$$\therefore \quad \frac{\partial u}{\partial x} = nx^{n-1}F\left(\frac{y}{x}\right) + x^n F'\left(\frac{y}{x}\right)\left(-\frac{y}{x^2}\right)$$

$$\Rightarrow \quad x\frac{\partial u}{\partial x} = nx^n F\left(\frac{y}{x}\right) - x^{n-1}yF'\left(\frac{y}{x}\right) \qquad \text{... (2)}$$

Again $$\frac{\partial u}{\partial y} = x^n F'\left(\frac{y}{x}\right)\left(\frac{1}{x}\right)$$

$$\Rightarrow \quad y\frac{\partial u}{\partial y} = x^{n-1}yF'\left(\frac{y}{x}\right) \qquad \text{... (3)}$$

Adding eqns. (2) and (3), we get

$$x\frac{\partial u}{\partial x}+y\frac{\partial u}{\partial y} = nx^n F\left(\frac{y}{x}\right) \Rightarrow x\frac{\partial u}{\partial x}+y\frac{\partial u}{\partial y}=nu$$

Note : Euler's Theorem can be extended to a homogeneous function of any number of variables. Thus, if f is a homogeneous function of three variables x, y and z of degree n, then

$$x\frac{\partial u}{\partial x}+y\frac{\partial u}{\partial y}+z\frac{\partial u}{\partial z}=nu$$

1.7 EULER'S THEOREM FOR SECOND ORDER DERIVATIVES

If u is a homogeneous function of x and y of degree, then

(*i*) $$x\frac{\partial^2 u}{\partial x^2}+y\frac{\partial^2 u}{\partial x\partial y} = (n-1)\frac{\partial u}{\partial x}$$

(*ii*) $$x\frac{\partial^2 u}{\partial x\partial y}+y\frac{\partial^2 u}{\partial y^2} = (n-1)\frac{\partial u}{\partial y}$$

(*iii*) $$x^2\frac{\partial^2 u}{\partial x^2}+2xy\frac{\partial^2 u}{\partial x\partial y}+y^2\frac{\partial^2 u}{\partial y^2} = n(n-1)u$$

Proof : Since u is a homogeneous function of x and y of degree n, therefore by Euler's Theorem we have

$$x\frac{\partial u}{\partial x}+y\frac{\partial u}{\partial y} = nu \qquad \text{... (1)}$$

Differentiating eqn. (1) partially w.r.t. x, we get

$$x\frac{\partial^2 u}{\partial x^2}+\frac{\partial u}{\partial x}+y\frac{\partial^2 u}{\partial x\partial y} = n\frac{\partial u}{\partial x}$$

$$x\frac{\partial^2 u}{\partial x^2}+y\frac{\partial^2 u}{\partial x\partial y} = (n-1)\frac{\partial u}{\partial x} \qquad \text{... (2)}$$

Again differentiating eqn. (1) partially w.r.t. y, we get

$$x\frac{\partial^2 u}{\partial y\partial x}+\frac{\partial u}{\partial y}+y\frac{\partial^2 u}{\partial y^2} = n\frac{\partial u}{\partial y}$$

$$x\frac{\partial^2 u}{\partial x\partial y}+y\frac{\partial^2 u}{\partial y^2} = (n-1)\frac{\partial u}{\partial y} \qquad \left(\text{फ } \frac{\partial^2 u}{\partial x\partial y}=\frac{\partial^2 u}{\partial y\partial x}\right) \qquad \text{... (3)}$$

Multiplying eqn. (2) by x and eqn. (3) by y and adding, we get

$$x^2\frac{\partial^2 u}{\partial x^2}+2xy\frac{\partial^2 u}{\partial x\partial y}+y^2\frac{\partial^2 u}{\partial y^2} = n(n-1)\left(x\frac{\partial u}{\partial x}+y\frac{\partial u}{\partial y}\right)=n(n-1)u \qquad \text{[using (1)]}$$

1.8 DEDUCTIONS FROM EULER'S THEOREM

If $u = u(x, y)$ is a homogeneous function of three variables x, y and z of degree n and $f = f(u)$ is function of u, then

(*i*) $$x\frac{\partial u}{\partial x}+y\frac{\partial u}{\partial y} = n\frac{F(u)}{F'(u)}$$

(*ii*) $$x^2\frac{\partial^2 u}{\partial x^2}+2xy\frac{\partial^2 u}{\partial x\partial y}+y^2\frac{\partial^2 u}{\partial y^2} = g(u)[g'(u)-1] \text{ where } g(u)=n\frac{F(u)}{F'(u)}$$

ILLUSTRATIVE EXAMPLES

Example1. Verify Euler's theorem for $u = x^4-3x^3y+5x^2y^2+4xy^3-2y^4$.

Solution. Let $u = f(x, y) = x^4-3x^3y+5x^2y^2+4xy^3-2y^4$

then $f(tx, ty) = (tx)^4-3(tx)^3(ty)+5(tx)^2(ty)^2+4(tx)(ty)^3-2(ty)^4 = t^4 f(x, y)$

thus u is a homogeneous function of x and y of degree 4.

Now, $$\frac{\partial u}{\partial x} = 4x^3 - 9x^2y + 10xy^2 + 4y^3$$

$\Rightarrow$ $$x\frac{\partial u}{\partial x} = 4x^4 - 9x^3y + 10x^2y^2 + 4xy^3$$

and $$\frac{\partial u}{\partial y} = -3x^3 + 10x^2y + 12xy^2 - 8y^3$$

$\Rightarrow$ $$y\frac{\partial u}{\partial y} = -3x^3y + 10x^2y^2 + 12xy^3 - 8y^4$$

$$x\frac{\partial u}{\partial x}+y\frac{\partial u}{\partial y} = 4x^4 - 12x^3y + 20x^2y^2 + 16xy^3 - 8y^4$$
$$= 4(x^4 - 3x^3y + 5x^2y^2 + 4xy^3 - 2y^4) = 4\,u$$

This verifies Euler's theorem.

Example 2. Verify Euler's theorem for $u = \dfrac{x^{\frac{1}{4}}+y^{\frac{1}{4}}}{x^{\frac{1}{5}}+y^{\frac{1}{5}}}$

Solution. Let $$u = f(x, y) = \frac{x^{\frac{1}{4}}+y^{\frac{1}{4}}}{x^{\frac{1}{5}}+y^{\frac{1}{5}}}$$

then $$u = f(tx, ty) = \frac{(tx)^{\frac{1}{4}} + (ty)^{\frac{1}{4}}}{(tx)^{\frac{1}{5}} + (ty)^{\frac{1}{5}}} = t^{\frac{1}{4}-\frac{1}{5}} \frac{x^{\frac{1}{4}} + y^{\frac{1}{4}}}{x^{\frac{1}{5}} + y^{\frac{1}{5}}}$$

$$= t^{\frac{1}{20}} f(x, y)$$

thus u is a homogeneous function of x and y of degree (1/20).

Now, $$\frac{\partial u}{\partial x} = \frac{\left(x^{\frac{1}{5}} + y^{\frac{1}{5}}\right)\left(\frac{1}{4}x^{-\frac{3}{4}}\right) - \left(x^{\frac{1}{4}} + y^{\frac{1}{4}}\right)\left(\frac{1}{5}x^{-\frac{4}{5}}\right)}{\left(x^{\frac{1}{5}} + y^{\frac{1}{5}}\right)^2}$$

$\Rightarrow$ $$x\frac{\partial u}{\partial x} = \frac{\left(x^{\frac{1}{5}} + y^{\frac{1}{5}}\right)\left(\frac{1}{4}x^{\frac{1}{4}}\right) - \left(x^{\frac{1}{4}} + y^{\frac{1}{4}}\right)\left(\frac{1}{5}x^{\frac{1}{5}}\right)}{\left(x^{\frac{1}{5}} + y^{\frac{1}{5}}\right)^2}$$

and $$\frac{\partial u}{\partial y} = \frac{\left(x^{\frac{1}{5}} + y^{\frac{1}{5}}\right)\left(\frac{1}{4}y^{-\frac{3}{4}}\right) - \left(x^{\frac{1}{4}} + y^{\frac{1}{4}}\right)\left(\frac{1}{5}y^{-\frac{4}{5}}\right)}{\left(x^{\frac{1}{5}} + y^{\frac{1}{5}}\right)^2}$$

$\Rightarrow$ $$y\frac{\partial u}{\partial y} = \frac{\left(x^{\frac{1}{5}} + y^{\frac{1}{5}}\right)\left(\frac{1}{4}y^{\frac{1}{4}}\right) - \left(x^{\frac{1}{4}} + y^{\frac{1}{4}}\right)\left(\frac{1}{5}y^{\frac{1}{5}}\right)}{\left(x^{\frac{1}{5}} + y^{\frac{1}{5}}\right)^2}$$

Therefore, $$x\frac{\partial u}{\partial x} + y\frac{\partial u}{\partial y} =$$

$$\frac{\left(x^{\frac{1}{5}} + y^{\frac{1}{5}}\right)\left(\frac{1}{4}x^{\frac{1}{4}}\right) \left(x^{\frac{1}{4}} + y^{\frac{1}{4}}\right)\left(\frac{1}{5}x^{\frac{1}{5}}\right) + \left(x^{\frac{1}{5}} + y^{\frac{1}{5}}\right)\left(\frac{1}{4}y^{\frac{1}{4}}\right) - \left(x^{\frac{1}{4}} + y^{\frac{1}{4}}\right)\left(\frac{1}{5}y^{\frac{1}{5}}\right)}{\left(x^{\frac{1}{5}} + y^{\frac{1}{5}}\right)^2}$$

$$= \frac{\frac{1}{4}\left(x^{\frac{1}{5}} + y^{\frac{1}{5}}\right)\left(x^{\frac{1}{4}} + y^{\frac{1}{4}}\right) - \frac{1}{5}\left(x^{\frac{1}{4}} + y^{\frac{1}{4}}\right)\left(x^{\frac{1}{5}} + x^{\frac{1}{5}}\right)}{\left(x^{\frac{1}{5}} + y^{\frac{1}{5}}\right)^2}$$

$$= \frac{\frac{1}{20}\left(x^{\frac{1}{5}} + y^{\frac{1}{5}}\right)\left(x^{\frac{1}{4}} + y^{\frac{1}{4}}\right)}{\left(x^{\frac{1}{5}} + y^{\frac{1}{5}}\right)^2} = \frac{1}{20}\frac{\left(x^{\frac{1}{4}} + y^{\frac{1}{4}}\right)}{\left(x^{\frac{1}{5}} + y^{\frac{1}{5}}\right)} = \frac{1}{20}u$$

This verifies Euler's theorem.

Example 3. Verify Euler's theorem for $u = \tan^{-1}\left(\frac{y^2}{x^2}\right)$.

Solution. Let $u = f(x, y) = \tan^{-1}\left(\frac{(ty)^2}{(tx)^2}\right)$

then $$u = f(tx, ty) = \tan^{-1}\left(\frac{(ty)^2}{(tx)^2}\right) = t^0 \tan^{-1}\left(\frac{y^2}{x^2}\right) = t^0 f(x, y)$$

thus u is a homogeneous function of x and y of degree 0.

Now, $$\frac{\partial u}{\partial x} = \frac{1}{\sqrt{1+\left(\frac{y^2}{x^2}\right)^2}}\left(\frac{-2y^2}{x^3}\right) = \frac{-2y^2}{x\sqrt{x^2+y^2}}$$

$\Rightarrow$ $$x\frac{\partial u}{\partial x} = \frac{-2y^2}{\sqrt{x^2+y^2}}$$

and $$\frac{\partial u}{\partial y} = \frac{1}{\sqrt{1+\left(\frac{y^2}{x^2}\right)^2}}\left(\frac{2y}{x^2}\right) = \frac{2y}{\sqrt{x^2+y^2}}$$

$\Rightarrow$ $$y\frac{\partial u}{\partial y} = \frac{2y^2}{\sqrt{x^2+y^2}}$$

$\therefore$ $$x\frac{\partial u}{\partial x} + y\frac{\partial u}{\partial y} = \frac{2y^2}{\sqrt{x^2+y^2}}(1-1) = 0.u = 0$$

This verifies Euler's theorem.

Example 4. If $u = \frac{x^2y^2}{x+y}$, prove that (*i*) $x\frac{\partial u}{\partial x} + y\frac{\partial u}{\partial y} = 3u$ and

(*ii*) $x^2\frac{\partial^2 u}{\partial x^2} + 2xy\frac{\partial^2 u}{\partial x \partial y} + y^2\frac{\partial^2 u}{\partial y^2} = 6u$

Solution. $u(tx, ty) = \dfrac{(tx)^2 (ty)^2}{tx+ty} = t^3\left(\dfrac{x^2y^2}{x+y}\right) = t^3u(x,y)$

thus u is a homogeneous function of x and y of degree 3.

$\therefore$ By Euler's theorem, we have

(*i*) $$x\frac{\partial u}{\partial x} + y\frac{\partial u}{\partial y} = 3u.$$

(*ii*) $$x^2\frac{\partial^2 u}{\partial x^2} + 2xy\frac{\partial^2 u}{\partial x \partial y} + y^2\frac{\partial^2 u}{\partial y^2} = 3(3-1)u = 6u$$

Example 5. If $u = \dfrac{x^3+y^3+z^3}{xy+yz+zx}$, prove that $x\dfrac{\partial u}{\partial x} + y\dfrac{\partial u}{\partial y} + z\dfrac{\partial u}{\partial z} = u.$

Solution. $u(tx, ty, tz) = \dfrac{(tx)^3+(ty)^3+(tz)^3}{t^2xy+t^2yz+t^2zx} = t\left(\dfrac{x^3+y^3+y^3}{xy+yz+zx}\right) = tu(x,y,z)$

thus u is a homogeneous function of x and y of degree 1.

$\therefore$ By Euler's theorem, we have

$$x\frac{\partial u}{\partial x} + y\frac{\partial u}{\partial y} + z\frac{\partial u}{\partial z} = u.$$

Example 6. If $u = \dfrac{1}{x^2} + \dfrac{1}{xy} + \dfrac{\log x - \log y}{x^2+y^2}$, prove that $x\dfrac{\partial u}{\partial x} + y\dfrac{\partial u}{\partial y} + 2u = 0$

Solution. We have, $$u = \frac{1}{x^2} + \frac{1}{xy} + \frac{\log\left(\dfrac{x}{y}\right)}{x^2+y^2}$$

$\Rightarrow$ $$u(tx, ty) = t^{-2}u(x, y)$$

thus u is a homogeneous function of x and y of degree (-2).

By Euler's theorem, we have

$$x\frac{\partial u}{\partial x} + y\frac{\partial u}{\partial y} = -2u \Rightarrow x\frac{\partial u}{\partial x} + y\frac{\partial u}{\partial y} + 2u = 0.$$

Example 7. If $u = \log\dfrac{x^2+y^2}{x+y}$, prove that (*i*) $x\dfrac{\partial u}{\partial x} + y\dfrac{\partial u}{\partial y} = 1$

(*ii*) $x^2\dfrac{\partial^2 u}{\partial x^2} + 2xy\dfrac{\partial^2 u}{\partial x \partial y} + y^2\dfrac{\partial^2 u}{\partial y^2} = -1$

Solution. We have, $u = \log\dfrac{x^2+y^2}{x+y}$, which is not a homogeneous function.

$$\Rightarrow \qquad e^u = \frac{x^2+y^2}{x+y} = \frac{x^2\left(1+\dfrac{y^2}{x^2}\right)}{x\left(1+\dfrac{y}{x}\right)} = x\phi\left(\frac{y}{x}\right)$$

thus $F(u) = e^u$ is a homogeneous function of x and y of degree 1.

By Euler's theorem, we have

$$(i) \qquad x\frac{\partial u}{\partial x} + y\frac{\partial u}{\partial y} = n\frac{F(u)}{F'(u)} = 1\frac{e^u}{e^u} = 1$$

$$(ii) \text{ Here} \qquad g(u) = n\frac{F(u)}{F'(u)} = 1\frac{e^u}{e^u} = 1 \Rightarrow g'(u) = 0$$

$$x^2\frac{\partial^2 u}{\partial x^2} + 2xy\frac{\partial^2 u}{\partial x\partial y} + y^2\frac{\partial^2 u}{\partial y^2} = g(u)[g'(u)-1]$$

$$= 1(0-1) = -1$$

Example 8. If $u = \tan^{-1}\left(\dfrac{x^3+y^3}{x-y}\right)$, prove that

(i) $x\dfrac{\partial u}{\partial x} + y\dfrac{\partial u}{\partial y} = \sin 2u$ $\quad (ii)$ $x^2\dfrac{\partial^2 u}{\partial x^2} + 2xy\dfrac{\partial^2 u}{\partial x\partial y} + y^2\dfrac{\partial^2 u}{\partial y^2} = 2\cos 3u \sin u$

Solution. We have, $u = \tan^{-1}\left(\dfrac{x^3+y^3}{x-y}\right)$, which is not a homogeneous function.

$$\Rightarrow \qquad \tan u = \frac{x^3+y^3}{x-y} = x^2\phi\left(\frac{y}{x}\right)$$

thus $F(u) = \tan u$ is a homogeneous function of degree 2.

By Euler's theorem, we have

$$(i) \qquad x\frac{\partial u}{\partial x} + y\frac{\partial u}{\partial y} = n\frac{F(u)}{F'(u)} = 2\frac{\tan u}{\sec^2 u} = 2\left(\frac{\sin u/\cos u}{1/\cos^2 u}\right)$$

$$= 2\sin u\cos u = \sin 2u$$

$$(ii) \text{ Here} \qquad g(u) = n\frac{F(u)}{F'(u)} = \sin 2u \Rightarrow g'(u) = 2\cos 2u$$

$$\therefore \quad x^2\frac{\partial^2 u}{\partial x^2}+2xy\frac{\partial^2 u}{\partial x\partial y}+y^2\frac{\partial^2 u}{\partial y^2} = g(u)[g'(u)-1]$$
$$= \sin 2u\,(2\cos 2u - 1)$$
$$= 2\sin 2u\cos 2u - \sin 2u$$
$$= \sin 4u - \sin 2u = 2\cos 3u \sin u$$

Example 9. If $u = \sin^{-1}\left(\dfrac{x^2+y^2}{x+y}\right)$, prove that

(*i*) $x\dfrac{\partial u}{\partial x}+y\dfrac{\partial u}{\partial y}=\tan u$ (*ii*) $x^2\dfrac{\partial^2 u}{\partial x^2}+2xy\dfrac{\partial^2 u}{\partial x\partial y}+y^2\dfrac{\partial^2 u}{\partial y^2}=\tan^3 u.$

Solution. We have, $u = \sin^{-1}\left(\dfrac{x^2+y^2}{x+y}\right)$,which is not a homogeneous function.

$$\Rightarrow \quad \sin u = \frac{x^2+y^2}{x+y} = x\phi\left(\frac{y}{x}\right)$$

thus $F(u) = \sin u$ is a homogeneous function of degree 1.

(*i*) $$x\frac{\partial u}{\partial x}+y\frac{\partial u}{\partial y} = n\frac{F(u)}{F'(u)} = \frac{\sin u}{\cos u} = \tan u$$

(*ii*) Here $$g(u) = n\frac{F(u)}{F'(u)} = \tan u \Rightarrow g'(u) = \sec^2 u$$

$$\therefore \quad x^2\frac{\partial^2 u}{\partial x^2}+2xy\frac{\partial^2 u}{\partial x\partial y}+y^2\frac{\partial^2 u}{\partial y^2} = g(u)[g'(u)-1]$$
$$= \tan u(\sec^2 u - 1)$$
$$= \frac{\sin u}{\cos u}\left(\frac{1}{\cos^2 u}-1\right)$$
$$= \frac{\sin u}{\cos u}\left(\frac{1-\cos^2 u}{\cos^2 u}\right) = \frac{\sin u}{\cos u}\left(\frac{\sin^2 u}{\cos^2 u}\right)$$
$$= \tan^3 u.$$

Example 10. If $u = (x^3+y^3)^{\frac{1}{2}}$, find the value of $x^2\dfrac{\partial^2 u}{\partial x^2}+2xy\dfrac{\partial^2 u}{\partial x\partial y}+y^2\dfrac{\partial^2 u}{\partial y^2}$.

Solution. We have, $u = (x^3+y^3)^{\frac{1}{2}}$

which is a homogeneous function of degree (3/2).

∴ By Euler's theorem, we have

$$x^2\frac{\partial^2 u}{\partial x^2}+2xy\frac{\partial^2 u}{\partial x\partial y}+y^2\frac{\partial^2 u}{\partial y^2} = n(n-1)u$$

$$= \frac{3}{2}\left(\frac{3}{2}-1\right)u = \frac{3}{4}u.$$

Example 11. If $u = \operatorname{cosec}^{-1}\left(\frac{x^{\frac{1}{2}}+y^{\frac{1}{2}}}{x^{\frac{1}{3}}+y^{\frac{1}{3}}}\right)^{\frac{1}{2}}$, prove that

$$x^2\frac{\partial^2 u}{\partial x^2}+2xy\frac{\partial^2 u}{\partial x\partial y}+y^2\frac{\partial^2 u}{\partial y^2} = \frac{\tan u}{144}(13+\tan^2 u).$$

Solution. Here u is not a homogeneous function but $F(u) = \operatorname{cosec} u$ is a homogeneous function of degree (1/12).

Here
$$g(u) = n\frac{F(u)}{F'(u)} = \frac{1}{12}\left(\frac{\operatorname{cosec} u}{-\operatorname{cosec} u\cot u}\right) = -\frac{1}{12}\tan u$$

$$g'(u) = -\frac{1}{12}\sec^2 u$$

$$\therefore \quad x^2\frac{\partial^2 u}{\partial x^2}+2xy\frac{\partial^2 u}{\partial x\partial y}+y^2\frac{\partial^2 u}{\partial y^2} = g(u)[g'(u)-1]$$

$$= -\frac{1}{12}\tan u\left(-\frac{1}{12}\sec^2 u-1\right)$$

$$= \frac{1}{144}\tan u(\sec^2 u+12)$$

$$= \frac{1}{144}\tan u(1+\tan^2 u+12)$$

$$= \frac{1}{144}\tan u(13+\tan^2 u)$$

EXERCISE 1.2

Verify Euler's theorem for the following functions:

1. $u = \frac{x^2\left(x^2-y^2\right)^3}{\left(x^2+y^2\right)^3}$.

2. $u = \log\left(\frac{x^2+y^2}{xy}\right)$

3. $u = 3x^2yz + 5xy^2z + 4z^4$

4. $u = \sin^{-1}\left(\frac{x}{y}\right) + \tan^{-1}\left(\frac{y}{x}\right)$

5. If $u = \frac{x^3y^3}{x^3+y^3}$, prove that

(*i*) $x\frac{\partial u}{\partial x} + y\frac{\partial u}{\partial y} = 3u$ and (*ii*) $x^2\frac{\partial^2 u}{\partial x^2} + 2xy\frac{\partial^2 u}{\partial x\partial y} + y^2\frac{\partial^2 u}{\partial y^2} = 6u.$

6. If $u = x\,\mathrm{cosec}^{-1}\left(\frac{x+y}{x-y}\right) + \sqrt{x^2-y^2}$, find the value of

(*i*) $x\frac{\partial u}{\partial x} + y\frac{\partial u}{\partial y}$ (*ii*) $x^2\frac{\partial^2 u}{\partial x^2} + 2xy\frac{\partial^2 u}{\partial x\partial y} + y^2\frac{\partial^2 u}{\partial y^2}.$

7. If $u = \sin^{-1}\left(\frac{x+y}{\sqrt{x}+\sqrt{y}}\right)$, prove that

(*i*) $x\frac{\partial u}{\partial x} + y\frac{\partial u}{\partial y} = \frac{1}{2}\tan u$ (*ii*) $x^2\frac{\partial^2 u}{\partial x^2} + 2xy\frac{\partial^2 u}{\partial x\partial y} + y^2\frac{\partial^2 u}{\partial y^2} = -\frac{\sin u\cos 2u}{4\cos^3 u}.$

8. If $u = \cos^{-1}\left(\frac{x+y}{\sqrt{x}+\sqrt{y}}\right)$, find the value of $x\frac{\partial u}{\partial x} + y\frac{\partial u}{\partial y}.$

9. If $u = \tan^{-1}\left(\frac{y^2}{x}\right)$, find the value of $x^2\frac{\partial^2 u}{\partial x^2} + 2xy\frac{\partial^2 u}{\partial x\partial y} + y^2\frac{\partial^2 u}{\partial y^2}.$

10. If $u = \sin^{-1}\left(\frac{x^4+y^4+z^4}{\sqrt{ax}+\sqrt{by}+\sqrt{cz}}\right)$, prove that $x\frac{\partial u}{\partial x} + y\frac{\partial u}{\partial y} + z\frac{\partial u}{\partial z} = \frac{7}{2}\tan u.$

11. If $u = \sin^{-1}\left(x^3+y^3\right)^{\frac{2}{5}}$, prove that

$$x^2\frac{\partial^2 u}{\partial x^2} + 2xy\frac{\partial^2 u}{\partial x\partial y} + y^2\frac{\partial^2 u}{\partial y^2} = \frac{6}{25}\tan u(6\sec^2 u - 5).$$

12. If $u = \log\left(\frac{2xy}{x^2+y^2}\right)$, find the value of $x\frac{\partial u}{\partial x} + y\frac{\partial u}{\partial y}.$

ANSWERS

6. (*i*) u (*ii*) 0 **8.** $-\frac{1}{2}\cot u$ **9.** $-\sin 2u\sin^2 u$ **12.** 0

1.9 TOTAL DIFFERENTIAL COEFFICIENT

Let $$z = f(x, y) \quad \text{... (1)}$$

If δx , δy be the increments in x and y respectively, let δz be the corresponding increment of z.

Then $$z + \delta z = f(x+\delta x, y+\delta y) \quad \text{... (2)}$$

Subtracting eqn. (1) from (2), we have

$$\delta z = f(x+\delta x, y+\delta y) - f(x, y) \quad \text{... (3)}$$

Adding and subtracting $f(x, y+\delta y)$ on R.H.S. of (3), we have

$$\delta z = f(x+\delta x, y+\delta y) - f(x, y+\delta y) + f(x, y+\delta y) - f(x, y)$$

$$\delta z = \frac{f(x+\delta x, y+\delta y) - f(x, y+\delta y)}{\delta x}\delta x + \frac{f(x, y+\delta y) - f(x, y)}{\delta y}\delta y$$

On taking limit $\delta x \to 0$ and $\delta y \to 0$

$$dz = \frac{\partial f}{\partial x}dx + \frac{\partial f}{\partial y}dy \quad \text{... (4)}$$

dz is called the total differential coefficient of z.

Corollary 1 : If $z = f(x, y)$

where $x = \varphi(t)$

and $y = \psi(t)$

Here z is a composite function of t .

Dividing eqn. (4) by dt, we have

$$\frac{dz}{dt} = \frac{\partial z}{\partial x}\frac{dx}{dt} + \frac{\partial z}{\partial y}\frac{dy}{dt} \quad (5)$$

Then dz is called the total differential coefficient of z.

Corollary 2 : If $z = f(x, y)$

where $x = \varphi(u, v)$

and $y = \psi(u, v)$

Then from (5), we have

$$\frac{\partial z}{\partial u} = \frac{\partial f}{\partial x}\cdot\frac{\partial x}{\partial u} + \frac{\partial f}{\partial y}\cdot\frac{\partial y}{\partial u}$$

and $$\frac{\partial z}{\partial v} = \frac{\partial f}{\partial x}\cdot\frac{\partial x}{\partial v} + \frac{\partial f}{\partial y}\cdot\frac{\partial y}{\partial v}$$

ILLUSTRATIVE EXAMPLES

Example 1. If $u = \sin^{-1}(x-y)$; $x = 3t$, $y = 4t^3$, show that $\dfrac{\partial u}{\partial t} = \dfrac{3}{\sqrt{1-t^2}}$.

Solution. Here $u = f(x, y)$; $x = \varphi(t)$, $y = \psi(t)$

$$\therefore \quad \frac{du}{dt} = \frac{\partial u}{\partial x}\frac{dx}{dt} + \frac{\partial u}{\partial y}\frac{dy}{dt}$$

Now, $\dfrac{\partial u}{\partial x} = \dfrac{1}{\sqrt{1-(x-y)^2}}$, $\dfrac{\partial u}{\partial y} = \dfrac{-1}{\sqrt{1-(x-y)^2}}$, $\dfrac{dx}{dt} = 3$, $\dfrac{dy}{dt} = 12t^2$

$$\therefore \quad \frac{du}{dt} = \frac{1}{\sqrt{1-(x-y)^2}}.3 - \frac{12t^2}{\sqrt{1-(x-y)^2}}$$

$$= \frac{3-12t^2}{\sqrt{1-(x-y)^2}} = \frac{3}{\sqrt{1-t^2}}$$

(after putting values of x and y).

Example 2. If $z = f(x, y)$; $x = e^u \cos v$, $y = e^u \sin v$, show that

$$y\frac{\partial z}{\partial u} + x\frac{\partial z}{\partial v} = e^{2u}\frac{\partial z}{\partial y}$$

Solution. We have, $x = e^u \cos v$, $\quad y = e^u \sin v$

$$\therefore \quad \frac{\partial z}{\partial u} = \frac{\partial z}{\partial x}\frac{\partial x}{\partial u} + \frac{\partial z}{\partial y}\frac{\partial y}{\partial u}$$

$$= \frac{\partial z}{\partial x}e^u \cos v + \frac{\partial z}{\partial y}e^u \sin v = x\frac{\partial z}{\partial x} + y\frac{\partial z}{\partial y}$$

$$y\frac{\partial z}{\partial u} = xy\frac{\partial z}{\partial x} + y^2\frac{\partial z}{\partial y} \qquad \text{... (1)}$$

And

$$\frac{\partial z}{\partial v} = \frac{\partial z}{\partial x}\frac{\partial x}{\partial v} + \frac{\partial z}{\partial y}\frac{\partial y}{\partial v}$$

$$= \frac{\partial z}{\partial x}(-e^u \sin v) + \frac{\partial z}{\partial y}(e^u \cos v)$$

$$= -y\frac{\partial z}{\partial x} + x\frac{\partial z}{\partial y}$$

$$x\frac{\partial z}{\partial v} = -xy\frac{\partial z}{\partial x} + x^2\frac{\partial z}{\partial y} \quad \text{... (2)}$$

On adding eqns. (1) and (2), we have

$$y\frac{\partial z}{\partial u} + x\frac{\partial z}{\partial v} = (x^2 + y^2)\frac{\partial z}{\partial y}$$

$$= (e^{2u}\cos^2 v + e^{2u}\sin^2 v)\frac{\partial z}{\partial y}$$

$$= e^{2u}\,(\cos^2 v + \sin^2 v)\frac{\partial z}{\partial y} = e^{2u}\frac{\partial z}{\partial y}$$

Example 3. If $w = f(u, v)$ where, $u = x + y,\quad v = x - y$ show that

$$\frac{\partial w}{\partial x} + \frac{\partial w}{\partial y} = 2\frac{\partial w}{\partial u}.$$

Solution.
$$\frac{\partial w}{\partial x} = \frac{\partial w}{\partial u}\frac{\partial u}{\partial x} + \frac{\partial w}{\partial v}\frac{\partial v}{\partial x}$$

$$\Rightarrow \quad \frac{\partial w}{\partial x} = \frac{\partial w}{\partial u}.1 + \frac{\partial w}{\partial v}.1 = \frac{\partial w}{\partial u} + \frac{\partial w}{\partial v} \quad \text{... (1)}$$

$$\frac{\partial w}{\partial y} = \frac{\partial w}{\partial u}\frac{\partial u}{\partial y} + \frac{\partial w}{\partial v}\frac{\partial v}{\partial y}$$

$$\Rightarrow \quad \frac{\partial w}{\partial y} = \frac{\partial w}{\partial u}.1 + \frac{\partial w}{\partial v}.(-1) = \frac{\partial w}{\partial u} - \frac{\partial w}{\partial v} \quad \text{... (2)}$$

Adding eqns. (1) and (2), we have

$$\frac{\partial w}{\partial x} + \frac{\partial w}{\partial y} = 2\frac{\partial w}{\partial u}$$

Example 4. If $V = f(y - z, z - x, x - y)$, prove that $\frac{\partial V}{\partial x} + \frac{\partial V}{\partial y} + \frac{\partial V}{\partial z} = 0$.

Solution. Let $X = y - z$, $Y = z - x$, $Z = x - y$; then $V = f(X, Y, Z)$

$$\therefore \quad \frac{\partial V}{\partial x} = \frac{\partial V}{\partial X}\frac{\partial X}{\partial x} + \frac{\partial V}{\partial Y}\frac{\partial Y}{\partial x} + \frac{\partial V}{\partial Z}\frac{\partial Z}{\partial x}$$

$$= \frac{\partial V}{\partial X}.(0) + \frac{\partial V}{\partial Y}.(-1) + \frac{\partial V}{\partial Z}.(1)$$

$$\Rightarrow \quad \frac{\partial V}{\partial x} = -\frac{\partial V}{\partial Y} + \frac{\partial V}{\partial Z}$$

By symmetry we have,

$$\frac{\partial V}{\partial y} = -\frac{\partial V}{\partial Z}+\frac{\partial V}{\partial X}$$

And
$$\frac{\partial V}{\partial z} = -\frac{\partial V}{\partial X}+\frac{\partial V}{\partial Y}$$

$$\therefore \quad \frac{\partial V}{\partial x}+\frac{\partial V}{\partial y}+\frac{\partial V}{\partial z} = 0$$

Example 5. If $u = f(r)$ and $x = r\cos\theta, y = r\sin\theta$, prove that

$$\frac{\partial^2 u}{\partial x^2}+\frac{\partial^2 u}{\partial y^2} = f''(r)+\frac{1}{r}f'(r).$$

Solution. फ $x = r\cos\theta, y = r\sin\theta$

$$r^2 = x^2 + y^2$$

So that
$$\frac{\partial r}{\partial x} = \frac{x}{r} \text{ and } \frac{\partial r}{\partial y} = \frac{y}{r}$$

$$\frac{\partial u}{\partial x} = \frac{df}{dr}\frac{\partial r}{\partial x} = \frac{df}{dr}\frac{x}{r}$$

Differentiating again w.r.t. x, we get

$$\frac{\partial^2 u}{\partial x^2} = \left(\frac{d^2 f}{dr^2}\frac{\partial r}{\partial x}\right).\frac{x}{r}+\frac{df}{dr}.\left[\frac{r.1-x\dfrac{\partial r}{\partial x}}{r^2}\right]$$

$$= \left(\frac{d^2 f}{dr^2}\frac{x}{r}\right).\frac{x}{r}+\frac{df}{dr}.\left[\frac{r.1-x.\dfrac{x}{r}}{r^2}\right]$$

$$= \frac{d^2 f}{dr^2}\frac{x^2}{r^2}+\frac{df}{dr}\frac{r^2-x^2}{r^3}$$

$$\Rightarrow \quad \frac{\partial^2 u}{\partial x^2} = \frac{d^2 f}{dr^2}\frac{x^2}{r^2}+\frac{df}{dr}\frac{y^2}{r^3} \quad \text{... (1)}$$

Similarly,
$$\frac{\partial^2 u}{\partial y^2} = \frac{d^2 f}{dr^2}\frac{y^2}{r^2}+\frac{df}{dr}\frac{x^2}{r^3} \quad \text{... (2)}$$

Adding eqns. (1) and (2), we get

$$\frac{\partial^2 u}{\partial x^2}+\frac{\partial^2 u}{\partial y^2} = \frac{d^2 f}{dr^2}\frac{x^2+y^2}{r^2}+\frac{df}{dr}\frac{x^2+y^2}{r^3}$$

$$= \frac{d^2 f}{dr^2}\frac{r^2}{r^2}+\frac{df}{dr}\frac{r^2}{r^3}$$

$$\Rightarrow \qquad \frac{\partial^2 u}{\partial x^2}+\frac{\partial^2 u}{\partial y^2} = f''(r)+\frac{1}{r}f'(r).$$

Example 6. If $u = x \log xy$ where $x^3 + y^3 + 3xy = 1$, find $\frac{du}{dx}$.

Solution. $\quad u = x \log xy \Rightarrow \frac{\partial u}{\partial x} = 1.\log xy + x.\frac{1}{xy}(y) = \log xy + 1$

and $\quad \frac{\partial u}{\partial y} = x.\frac{1}{xy}(x) = \frac{x}{y}$

Also, $\quad x^3+y^3+3xy = 1 \Rightarrow 3x^2+3y^2\frac{dy}{dx}+3y+3y\frac{dy}{dx}=0$

$$\Rightarrow \qquad \frac{dy}{dx} = -\frac{x^2+y}{x+y^2}$$

Now, $\quad \frac{du}{dx} = \frac{\partial u}{\partial x}\frac{dx}{dx}+\frac{\partial u}{\partial y}\frac{dy}{dx}$

$$= (1+\log xy).1-\frac{x}{y}\left(\frac{x^2+y}{x+y^2}\right)$$

$$= 1+\log xy-\frac{x}{y}\left(\frac{x^2+y}{x+y^2}\right).$$

Example 7. If z is a function of x and y and $x = e^u + e^{-v}, \quad y = e^{-u} - e^v$, prove that

$$\frac{\partial z}{\partial u}-\frac{\partial z}{\partial v} = x\frac{\partial z}{\partial x}-y\frac{\partial z}{\partial y}.$$

Solution. $z = f(x, y)$; $x = e^u + e^{-v}, \quad y = e^{-u} - e^v$

$$\therefore \qquad \frac{\partial z}{\partial u} = \frac{\partial z}{\partial x}\frac{\partial x}{\partial u}+\frac{\partial z}{\partial y}\frac{\partial y}{\partial u} = \frac{\partial z}{\partial x}.e^u+\frac{\partial z}{\partial y}(-e^{-u})$$

$$= e^u\frac{\partial z}{\partial x}-e^{-u}\frac{\partial z}{\partial y}$$

and $$\frac{\partial z}{\partial v} = \frac{\partial z}{\partial x}\frac{\partial x}{\partial v}+\frac{\partial z}{\partial y}\frac{\partial y}{\partial v} = \frac{\partial z}{\partial x}(-e^{v})+\frac{\partial z}{\partial y}(-e^{v})$$

$$= -e^{-v}\frac{\partial z}{\partial x}-e^{v}\frac{\partial z}{\partial y}$$

Hence, $$\frac{\partial z}{\partial u}-\frac{\partial z}{\partial v} = (e^{u}+e^{-v})\frac{\partial z}{\partial x}-(e^{-u}-e^{v})\frac{\partial z}{\partial y}$$

$$= x\frac{\partial z}{\partial x}-y\frac{\partial z}{\partial y}$$

Example 8. If $x = r\cos\theta,\ y = r\sin\theta$, prove that $\frac{\partial x}{\partial r}=\frac{\partial r}{\partial x}$, $\frac{1}{r}\frac{\partial x}{\partial\theta}=r\frac{\partial\theta}{\partial x}$

Solution. $x = r\cos\theta$, $\therefore\ \frac{\partial x}{\partial r}=\cos\theta$ and $\frac{\partial x}{\partial\theta}=-r\sin\theta$

Also, $$r^2 = x^2+y^2 \text{ and } \theta=\tan^{-1}\frac{y}{x}$$

$\therefore$ $$2r\frac{\partial r}{\partial x} = 2x \Rightarrow \frac{\partial r}{\partial x}=\frac{x}{r}=\cos\theta$$

and $$\frac{\partial\theta}{\partial x} = \frac{1}{1+\frac{y^2}{x^2}}\left(-\frac{y}{x^2}\right)$$

and $$\frac{\partial\theta}{\partial x} = \frac{1}{1+\frac{y^2}{x^2}}\left(-\frac{y}{x^2}\right)=-\frac{y}{y^2+x^2}$$

$$= -\frac{r\sin\theta}{r^2}=-\frac{\sin\theta}{r}$$

$\Rightarrow$ $$\frac{\partial x}{\partial r} = \frac{\partial r}{\partial x}=\cos\theta \text{ and } \frac{1}{r}\frac{\partial x}{\partial\theta}=r\frac{\partial\theta}{\partial x}=-\sin\theta.$$

1.10 IMPORTANT RESULTS

Let $$z = f(x, y)$$

then, $$dz = \frac{\partial f}{\partial x}dx+\frac{\partial f}{\partial y}dy$$

If $z = 0$, $dz = 0$

$\therefore$ $$0 = \frac{\partial f}{\partial x}dx+\frac{\partial f}{\partial y}dy \Rightarrow \frac{\partial f}{\partial y}\frac{dy}{dx} = -\frac{\partial f}{\partial x}$$

$$\Rightarrow \qquad \boxed{\frac{dy}{dx} = -\frac{\frac{\partial f}{\partial x}}{\frac{\partial f}{\partial y}}}$$

Now, by differentiating the above result, we can get

$$\boxed{\frac{d^2y}{dx^2} = -\frac{q^2r - 2pqs + p^2t}{q^3}}$$

where, $p = \frac{\partial z}{\partial x}$, $q = \frac{\partial z}{\partial y}$, $r = \frac{\partial^2 z}{\partial x^2}$, $s = \frac{\partial^2 z}{\partial x \partial y}$, $t = \frac{\partial^2 z}{\partial y^2}$

ILLUSTRATIVE EXAMPLES

Example 1. Find $\frac{dy}{dx}$ if $ax^2 + by^2 + 2hxy = 1$.

Solution. Let $f(x, y) = ax^2 + by^2 + 2hxy - 1 = 0$

$$\therefore \qquad \frac{\partial f}{\partial x} = 2ax + 2hy \text{ and } \frac{\partial f}{\partial y} = 2by + 2hx$$

$$\Rightarrow \qquad \frac{dy}{dx} = -\frac{\frac{\partial f}{\partial x}}{\frac{\partial f}{\partial y}} = -\frac{2ax + 2hy}{2by + 2hx} = -\frac{ax + hy}{by + hx}$$

Example 2. Find $\frac{dy}{dx}$ and $\frac{d^2y}{dx^2}$ if $x^3 + y^3 - 3axy = 0$.

Solution. Let $\qquad f(x, y) = x^3 + y^3 - 3axy = 0$

$$\therefore \qquad p = \frac{\partial f}{\partial x} = 3x^2 - 3ay\,;\ q = \frac{\partial f}{\partial y} = 3y^2 - 3ax\,;\ r = \frac{\partial^2 f}{\partial x^2} = 6x\,;\ s = \frac{\partial^2 f}{\partial x \partial y} = -3a\,;$$

$$t = \frac{\partial^2 f}{\partial y^2} = 6y$$

$$\Rightarrow \qquad \frac{dy}{dx} = -\frac{\frac{\partial f}{\partial x}}{\frac{\partial f}{\partial y}} = -\frac{3x^2 - 3ay}{3y^2 - 3ax} = -\frac{x^2 - ay}{y^2 - ax} = \frac{ay - x^2}{y^2 - ax}$$

Also,

$$\frac{d^2y}{dx^2} = -\frac{q^2r - 2pqs + p^2t}{q^3}$$

$$\Rightarrow \quad \frac{d^2y}{dx^2} = -\frac{(3y^2-3ax)^2 6x - 2(3x^2-3ay)(3y^2-3ax)(-3a) + (3x^2-3ay)^2 6y}{(3y^2-3ax)^3}$$

$$= -\frac{2x(y^2-ax)^2 + 2a(x^2-ay)(y^2-ax) + 2y(x^2-ay)^2}{(y^2-ax)^3}$$

$$= \frac{2a^3xy}{(ax-y^2)^3}$$

Example 3. Find $\frac{dy}{dx}$ if $x^y = y^x$.

Solution. Let $f(x, y) = x^y - y^x = 0$

$$\therefore \quad p = \frac{\partial f}{\partial x} = yx^{y-1} - y^x \log y \; ; \; q = \frac{\partial f}{\partial y} = x^y \log x - xy^{x-1}$$

$$\Rightarrow \quad \frac{dy}{dx} = -\frac{\frac{\partial f}{\partial x}}{\frac{\partial f}{\partial y}} = -\frac{yx^{y-1} - y^x \log y}{x^y \log x - xy^{x-1}} = -\frac{\frac{y}{x}x^y - y^x \log y}{x^y \log x - \frac{x}{y}y^x}$$

$$= -\frac{\frac{y}{x}x^y - x^y \log y}{x^y \log x - \frac{x}{y}x^y} = -\frac{\frac{y}{x} - \log y}{\log x - \frac{x}{y}}$$

$$= -\frac{y(y - x\log y)}{x(y\log x - x)} = \frac{y(y - x\log y)}{x(x - y\log x)}$$

(Putting $x^y = y^x$)

EXERCISE 1.3

1. If $u = x^3 + y^3$; $x = a\cos t$, $y = b\sin t$, find $\frac{du}{dt}$.
2. If $z = f(x, y)$; $x = r\cos\theta$, $y = r\sin\theta$, prove that

$$\left(\frac{\partial z}{\partial r}\right)^2 + \frac{1}{r^2}\left(\frac{\partial z}{\partial \theta}\right)^2 = \left(\frac{\partial f}{\partial x}\right)^2 + \left(\frac{\partial f}{\partial y}\right)^2$$

3. If $z = u^2 + v^2$; $u = at^2$, $v = 2at$, find $\frac{dz}{dt}$.
4. If $u = f(x^2 + 2yz,\; y^2 + 2zx)$, prove that

$$(y^2 - zx)\frac{\partial u}{\partial x} + (x^2 - yz)\frac{\partial u}{\partial y} + (z^2 - xy)\frac{\partial u}{\partial z} = 0$$

5. If $z = f(x, y)$; $x = uv$, $y = \dfrac{u+v}{u-v}$ show that $2x\dfrac{\partial z}{\partial x} = u\dfrac{\partial z}{\partial u} + v\dfrac{\partial z}{\partial v}$.

6. If $u = f(r, s, t)$ where $r = \dfrac{x}{y}$, $s = \dfrac{y}{z}$, $t = \dfrac{z}{x}$, show that

$$x\frac{\partial u}{\partial x} + y\frac{\partial u}{\partial y} + z\frac{\partial u}{\partial z} = 0.$$

7. Find $\dfrac{dy}{dx}$ if $x^3 + y^3 + 3x^2y + 6xy^2 = 1$.

8. If $x^3 + y^3 = 3ax^2$ find $\dfrac{d^2y}{dx^2}$.

9. Find $\dfrac{dy}{dx}$ when $(\cos x)^y = (\sin y)^x$.

10. If $x^4 + y^4 = 4a^2xy$, show that $(a^2x - y^3)^3\dfrac{d^2y}{dx^2} = 2a^2xy(3a^4 + x^2y^2)$.

ANSWERS

1. $-3a^3\cos^2 t\sin t + 3b^3\sin^2 t\cos t$

3. $4a^2t(t^2+2)$

7. $\dfrac{dy}{dx} = -\dfrac{x^2+2xy+2y^2}{x^2+4xy+y^2}$

8. $-\dfrac{2a^2x^2}{y^5}$

9. $\dfrac{y\tan x + \log\sin y}{\log\cos x - x\cot y}$

1.11 ERRORS

If $y = f(x)$, then we have $\dfrac{dy}{dx} = \lim\limits_{\delta x \to 0}\dfrac{\delta y}{\delta x}$

i.e., $\dfrac{dy}{dx}$ is approximately equal to $\dfrac{\delta y}{\delta x}$

i.e., $\delta y = \left(\dfrac{dy}{dx}\right)\delta x$ approximately.

Note :

1. δx is called the absolute error in the value of x.
2. $\dfrac{\delta x}{x}$ is called the relative error in x.
3. $\dfrac{\delta x}{x}\times 100$ is called the percentage error in x.

Again, if $f = f(x, y)$ is a function of two variables, then we have

$$\boxed{\delta f = \frac{\partial f}{\partial x}\delta x + \frac{\partial f}{\partial y}\delta y} \text{ approximately.}$$

this gives small error δf in the value of f due to errors δx and δy in the value of x and respectively.

ILLUSTRATIVE EXAMPLES

Example 1. Find the percentage error in the area of a rectangle when an error of +1% is made in measuring its length and breadth.

Solution. Let the length and breadth of the rectangle be x and y and A be the area.

Given, $\frac{\delta x}{x} \times 100 = 1$ and $\frac{\delta y}{y} \times 100 = 1$.

$\therefore$ Area, $\quad A = xy \Rightarrow \log A = \log x + \log y$

On differentiating, we have

$$\frac{\delta A}{A} = \frac{\delta x}{x} + \frac{\delta y}{y} \Rightarrow \frac{\delta A}{A} \times 100 = \frac{\delta x}{x} \times 100 + \frac{\delta y}{y} \times 100$$

$$\Rightarrow \quad \frac{\delta A}{A} \times 100 = 1 + 1 = 2$$

Error in the area = 2%.

Example 2. If the kinetic energy is given by, $T = \frac{mv^2}{2}$. Find approximately the change in T as m changes from 49 to 49.5 and v changes from 1600 to 1590.

Solution. We have, $T = \frac{mv^2}{2} \Rightarrow \log T = \log m + 2\log \ v - \log 2$

On differentiating, we have

$$\frac{\delta T}{T} = \frac{\delta m}{m} + 2\frac{\delta v}{v} - 0 \Rightarrow \frac{\delta T}{T} = \frac{\delta m}{m} + 2\frac{\delta v}{v} \qquad \ldots (1)$$

Given, error in m, $\delta m = 0.5$ and $m = 49$

error in v , $\delta v = -10$ and $v = 1600$

$$\therefore \quad T = \frac{mv^2}{2} = \frac{49 \times (1600)^2}{2} = 62720000$$

By eqn. (1) we get, $\quad \frac{\delta T}{T} = \frac{\delta m}{m} + 2\frac{\delta v}{v} \Rightarrow \delta T = 62720000\left[\frac{0.5}{49} + 2 \times \frac{-10}{1600}\right]$

$$\Rightarrow \qquad \delta T = 62720000\left[\frac{0.5}{49} - 2\times\frac{10}{1600}\right]$$

$$= 62720000\times\frac{-9}{3920} = -144000$$

Hence, change in $T = -144000$.

Example 3. Find the possible percentage error in computing the parallel resistance r of two resistances r_1 and r_2 from the formula $\frac{1}{r} = \frac{1}{r_1} + \frac{1}{r_2}$ where r_1 and r_2 are both in error by +2% each.

Solution. We have, $\qquad \frac{1}{r} = \frac{1}{r_1} + \frac{1}{r_2}$

Differentiating, we get

$$-\frac{1}{r^2}\delta r = -\frac{1}{r_1^2}\delta r_1 - \frac{1}{r_2^2}\delta r_2$$

$$\Rightarrow \qquad \frac{1}{r}\left(\frac{\delta r}{r}\times 100\right) = \frac{1}{r_1}\left(\frac{\delta r_1}{r_1}\times 100\right) + \frac{1}{r_2}\left(\frac{\delta r_2}{r_2}\times 100\right)$$

$$\Rightarrow \qquad \frac{1}{r}\left(\frac{\delta r}{r}\times 100\right) = \frac{1}{r_1}\left(\frac{\delta r_1}{r_1}\times 100\right) + \frac{1}{r_2}\left(\frac{\delta r_2}{r_2}\times 100\right)$$

$$\Rightarrow \qquad \frac{1}{r}\left(\frac{\delta r}{r}\times 100\right) = \frac{1}{r_1}.2 + \frac{1}{r_2}.2 = 2\left(\frac{1}{r_1} + \frac{1}{r_2}\right) = 2\left(\frac{1}{r}\right)$$

$\therefore$ Percentage error in $r = 2\%$

Example 4. The period of a simple pendulum is given by $T = 2\pi\sqrt{\frac{l}{g}}$. Find the maximum error in T due to possible errors up to 1% in l and 2.5% in g.

Solution. We have, $\qquad T = 2\pi\sqrt{\frac{l}{g}}$

$$\Rightarrow \qquad \log T = \log 2\pi + \frac{1}{2}\log\ l - \frac{1}{2}\log\ g$$

On differentiating, we have

$$\frac{\delta T}{T} = 0 + \frac{1}{2}\frac{\delta l}{l} - \frac{1}{2}\frac{\delta g}{g}$$

$$\Rightarrow \quad \frac{\delta T}{T}\times 100 = \frac{1}{2}\left[\frac{\delta l}{l}\times 100 - \frac{\delta g}{g}\times 100\right]$$

$$\Rightarrow \quad \frac{\delta T}{T}\times 100 = \frac{1}{2}[1\pm 2.5] = 1.75\ \% \quad or \quad 0.75\%$$

$\therefore$ Maximum error in T = 1.75%.

Example 5. A balloon is in the form of a right circular cylinder of radius 1.5 m and length 4 m and is surmounted by hemispherical ends. If the radius is increased by 0.01 m and the length by 0.05 m , find the percentage change in the volume of the balloon.

Solution. Let the radius and height (length) of the cylinder be r and h.

Given, $r = 1.5$ m, $h = 4$ m, $\delta r = 0.01$m, $\delta h = 0.05$ m.

Let V be the volume of the cylinder, then

$$V = \pi r^2 h + \frac{2}{3}\pi r^3 + \frac{2}{3}\pi r^3 = \pi r^2 h + \frac{4}{3}\pi r^3$$

$$\Rightarrow \quad \delta V = \pi(2rh\delta r + r^2\delta h) + \frac{4}{3}.3\pi r^2\delta r$$

$$\Rightarrow \quad \delta V = \pi r\left[2h\delta r + r\delta h + 4r\delta r\right]$$

$$\Rightarrow \quad \frac{\delta V}{V} = \frac{\pi r\left[2h\delta r + r\delta h + 4r\delta r\right]}{\pi r^2 h + \frac{4}{3}\pi r^3} = \frac{2h\delta r + r\delta h + 4r\delta r}{rh + \frac{4}{3}r^2}$$

$$\Rightarrow \quad \frac{\delta V}{V} = \frac{2\times 4\times 0.01 + 1.5\times 0.05 + 4\times 1.5\times 0.01}{1.5\times 4 + \frac{4}{3}\times(1.5)^2}$$

$$\Rightarrow \quad \frac{\delta V}{V} = \frac{0.08 + 0.075 + 0.06}{6+3} = \frac{0.215}{9} = 0.02389$$

$$\Rightarrow \quad \frac{\delta V}{V}\times 100 = 2.389\%$$

Hence, per cent change in the volume = 2.389%.

Example 6. The torsional rigidity of a length of a wire is obtained from the formula $N = \frac{8\pi Il}{t^2 r^4}$ .If l is decreased by 2%, r is increased by 2%, t increased by 1.5%. Show that the value of N is diminished by 13% approximately.

Solution. Given, $\frac{\delta l}{l}\times 100 = -2$; $\frac{\delta r}{r}\times 100 = 2$; $\frac{\delta t}{t}\times 100 = 1.5$.

Now, $$N = \frac{8\pi Il}{t^2 r^4}$$

$$\Rightarrow \quad \log N = \log 8\pi + \log I + \log l - 2\log t - 4\log r$$

$$\Rightarrow \quad \frac{\delta N}{N} = 0+0+\frac{\delta l}{l}-2\frac{\delta t}{t}-4\frac{\delta r}{r} = \frac{\delta l}{l}-2\frac{\delta t}{t}-4\frac{\delta r}{r}$$

$$\Rightarrow \quad \frac{\delta N}{N}\times 100 = \frac{\delta l}{l}\times 100-2\frac{\delta t}{t}\times 100-4\frac{\delta r}{r}\times 100$$

$$\Rightarrow \quad \frac{\delta N}{N}\times 100 = -2-2\times 1.5-4\times 2 = -2-3-8 = -13$$

Hence, N is diminished by 13%.

Example 7. In estimating the cost of a pile of bricks measured as $2m\times 15m\times 1.2m$, the tape is stretched 1% beyond the standard length. If the count is 450 bricks in 1 $m^3$and bricks costs Rs.1.30 per 1000.Find the approximate error in the cost.

Solution. Let x, y, z be the dimensions and V be the volume of the pile of bricks, then

$$V = xyz \Rightarrow \log V = \log x+\log y+\log z$$

Differentiating, we get

$$\frac{\delta V}{V} = \frac{\delta x}{x}+\frac{\delta y}{y}+\frac{\delta z}{z}$$

$$\Rightarrow \quad \frac{\delta V}{V}\times 100 = \frac{\delta x}{x}\times 100+\frac{\delta y}{y}\times 100+\frac{\delta z}{z}\times 100$$

$$\Rightarrow \quad \frac{\delta V}{V}\times 100 = 1+1+1 = 3$$

$$\therefore \quad \delta V = \frac{3\times V}{100} = \frac{3(2\times 15\times 1.2)}{100} = \frac{108}{100} = 1.08m^3$$

$\therefore$ Number of bricks in $\delta V = 450\times 1.08 = 486$

Thus, error in the cost $= \dfrac{1.30\times 486}{1000}$ = Rs. 0.6318.

Example 8. The radius of a sphere is found to be 20 cm with a possible error of 0.03 cm. Find the relative and percentage error in computing the volume.

Solution. Let r be the radius and V be the volume of the sphere, then we have

$$V = \frac{4}{3}\pi r^3 \Rightarrow \log V = \log\left(\frac{4\pi}{3}\right)+3\log r$$

$$\Rightarrow \quad \frac{\delta V}{V} = 0+3\frac{\delta r}{r} = 3\frac{\delta r}{r}$$

$$\Rightarrow \quad \frac{\delta V}{V} = 3\times\frac{0.03}{20} = 0.0045$$

$\therefore$ Relative error in volume = 0.0045

Hence, percentage error in volume = 0.0045 × 100 = 0.45%

Example 9. If in a triangle *ABC*, the side *c* and angle *C* remain constant, while the remaining elements are changed slightly, show that $\frac{\delta a}{\cos A}+\frac{\delta b}{\cos B}=0$.

Solution. In a triangle, we have

$$\frac{a}{\sin A} = \frac{b}{\sin B}=\frac{c}{\sin C} \qquad \text{(By sine-formula)}$$

Given, *c* and *C* are constants, so $\frac{c}{\sin C}=k$ (say).

We get,

$$\frac{a}{\sin A} = \frac{b}{\sin B}=k \Rightarrow a=k\sin A,\quad b=k\sin B$$

$$\Rightarrow \quad \delta a = k\cos A\delta A,\ \delta b=k\cos B\delta B$$

$$\Rightarrow \quad \frac{\delta a}{\cos A}+\frac{\delta b}{\cos B} = k(\delta A+\delta B)=k\delta(A+B)=k\delta(\pi-C)=0$$

Hence, $$\frac{\delta a}{\cos A}+\frac{\delta b}{\cos B} = 0$$

Example 10. If in a triangle *ABC*, the angles and sides *a* and *b* are made to vary in such a way that the area remains constant, the side *c* also remain constant. Show that if *a* and *b* vary by small amounts δa and δb respectively, then $\cos A\delta a+\cos B\delta b=0$.

Solution. Let Δ be the area of the triangle, then we have

$$\Delta = \frac{1}{2}bc\sin A$$

$$\Rightarrow \quad \delta\Delta = \frac{1}{2}c(\delta b.\sin A+b\cos A.\delta A) \qquad [\because c \text{ is constant}]$$

$$\Rightarrow \quad 0 = \frac{1}{2}c(\delta b.\sin A+b\cos A.\delta A) \ [\because \Delta \text{ is constant } \delta\Delta=0]$$

$$\Rightarrow \quad \delta A = -\delta b.\frac{\tan A}{b} \qquad ...(1)$$

Also, we know that

$$a^2 = b^2+c^2-2bc\cos A$$

$$\Rightarrow \quad 2a\delta a = 2b\delta b+0-2c(\delta b.\cos A-b\sin A.\delta A)$$

$$[\because c \text{ is constant}]$$

$$\Rightarrow \quad a\delta a = b\delta b-c\cos A\delta b+bc\sin A\left(-\delta b.\frac{\tan A}{b}\right) \quad \text{[using (1)]}$$

$$\Rightarrow \quad a\delta a.\cos A = b\delta b.\cos A-c\cos^2 A\delta b-c\sin^2 A.\delta b$$

$$\Rightarrow \quad a\delta a.\cos A = b\delta b.\cos A-c(\cos^2 A+\sin^2 A).\delta b$$

$$\Rightarrow \quad a\delta a.\cos A = b\delta b.\cos A-c.\delta b$$

But, $$c = a\cos B+b\cos A$$

$$\therefore \quad a\delta a.\cos A = b\delta b.\cos A-(a\cos B+b\cos A).\delta b$$

$$\Rightarrow \quad a\delta a.\cos A = -a\cos B.\delta b$$

$$\Rightarrow \quad \cos A\delta a+\cos B\delta b = 0$$

Example 11. Find the fourth root of 83 approximately.

Solution. Let $f(x) = x^{1/4}$

Now, $\delta f(x) = f(x+\delta x) - f(x)$

$$\Rightarrow \quad \frac{1}{4}x^{-3/4}\delta x = f(x+\delta x) - f(x)$$

$$\Rightarrow \quad f(x+\delta x) = \frac{1}{4}x^{-3/4}\delta x + f(x)$$

Consider, $x = 81$, $\delta x = 2$, we get

$$\Rightarrow \quad f(83) = \frac{1}{4}(81)^{-3/4}.2 + (81)^{1/4}$$

$$\Rightarrow \quad (83)^{1/4} = \frac{1}{2}\cdot\frac{1}{(81)^{3/4}} + (81)^{1/4} = \frac{1}{2}\cdot\frac{1}{3^3} + 3 = \frac{1}{54} + 3$$

$$\Rightarrow \quad (83)^{1/4} = 0.0185 + 3 = 3.0185.$$

Example 12. Evaluate $\sqrt{(102)^2 + (99)^2}$ approximately.

Solution. Let $f(x, y) = \sqrt{x^2 + y^2}$

$$\delta f(x, y) = f(x+\delta x, y+\delta y) - f(x, y)$$

$$\Rightarrow \quad \frac{(x\delta x + y\delta y)}{\sqrt{x^2 + y^2}} = f(x+\delta x, y+\delta y) - f(x, y)$$

$$\Rightarrow \quad f(x+\delta x, y+\delta y) = \frac{(x\delta x + y\delta y)}{\sqrt{x^2 + y^2}} + f(x, y)$$

Consider, $x = 100, \delta x = 2, y = 100, \delta y = -1$, we get

$$\Rightarrow \quad f(100+2,\ 100-1) = \frac{(100\times 2 + 100\times -1)}{\sqrt{100^2 + 100^2}} + \sqrt{100^2 + 100^2}$$

$$\Rightarrow \quad \sqrt{102^2 + 99^2} = \frac{100}{100\sqrt{2}} + 100\sqrt{2} = \frac{1}{\sqrt{2}} + 100\sqrt{2} = \frac{201}{\sqrt{2}}$$

$$\Rightarrow \quad \sqrt{102^2 + 99^2} = \frac{201}{1.414} = 142.1499\cdot$$

Example 13. If a, b, c be the sides of a triangle and Δ be its area, show that the error in Δ due to a small error δc in c is given by

$$\delta\Delta = \frac{\Delta}{4}\left[\frac{1}{s} + \frac{1}{s-a} + \frac{1}{s-b} - \frac{1}{s-c}\right]\delta c\text{, where } s = \frac{a+b+c}{2}.$$

Solution. We know that

$$\Delta = \sqrt{s(s-a)(s-b)(s-c)}$$

$$\therefore \quad \log\Delta = \frac{1}{2}[\log s + \log(s-a) + \log(s-b) + \log(s-c)]$$

$$\Rightarrow \quad \frac{1}{\Delta}\delta\Delta = \frac{1}{2}\left[\frac{\delta s}{s} + \frac{\delta s}{s-a} + \frac{\delta s}{s-b} + \frac{\delta s - \delta c}{s-c}\right] \quad \ldots (1)$$

Now, $\quad 2s = a+b+c \quad \Rightarrow \quad 2\delta s = \delta c \quad \Rightarrow \delta s = \frac{\delta c}{2}$

By eqn. (1), we get

$$\delta\Delta = \frac{\Delta}{2}\left[\frac{\delta c}{2s} + \frac{\delta c}{2(s-a)} + \frac{\delta c}{2(s-b)} + \frac{\frac{\delta s}{2} - \delta c}{s-c}\right]$$

$$\Rightarrow \quad \delta\Delta = \frac{\Delta}{4}\left[\frac{1}{s} + \frac{1}{s-a} + \frac{1}{s-b} - \frac{1}{s-c}\right]\delta c.$$

EXERCISE 1.4

1. The density of the atmosphere at the height h metres is given by $\rho = \rho_0 e^{-h/800}$, where ρ_0 is the density at the sea level. Find the approximate increase in height corresponding to a fall of 1% in the density.
2. The dimensions of a cone are, radius 4 cm, height 6 cm. What is the error in its volume if the scale used in taking the measurement is short by 0.01 cm per cm?
3. Find the percentage error in the area of an ellipse when an error of +1% is made in measuring the major and minor axis. (Area of ellipse = πxy, where x and y are the length of the semi-major and semi-minor axis respectively).
4. The work that must be done to propel a ship of displacement D for a distance s in time t is given by $W = \frac{s^2 D^{2/3}}{t^2}$.

   Find approximately the percentage increase of work necessary when the displacement is increased by 1%, the time is diminished by 1% and the distance is increased by 3%.
5. Evaluate the cube root of 127 approximately.
6. Find the approximate value of $\sqrt{(2.01)^2 + (1.94)^2}$.
7. Find the possible percentage error in computing the parallel resistance r of three resistances r_1, r_2 and r_3 from the formula $\frac{1}{r} = \frac{1}{r_1} + \frac{1}{r_2} + \frac{1}{r_3}$ where r_1, r_2, r_3 are in error by +1.2% each.

8. If in a triangle ABC, the sides and angles vary in such a way that its circum radius remains constant, prove that $\frac{\delta a}{\cos A}+\frac{\delta b}{\cos B}+\frac{\delta c}{\cos C}=0$.

9. In estimating the cost of a pile of bricks measured as $5m\times 10m\times 5m$, the tape is stretched 2% beyond the standard length. If the count is 100 bricks per m^3 and bricks costs Rs.2000 per 1000. Find the approximate error in the cost.

10. Find the value of $\log_{10} 10.1$, where $\log_{10} e = 0.4343$.

11. In determining the specific gravity by the formula $S=\frac{A}{A-w}$, where A is the weight in air and w is the weight in water. A can be read within 0.01 gm and w within 0.02 gm. Find approximately the maximum error in S if the readings are A =1.1 gm, w = 0.6 gm. Find also the maximum relative error.

12. The radius of a circle is found to be 100 cm with a possible error of 1 mm. Find the relative and percentage error in the area of the circle.

ANSWERS

1. 8 m	**2.** 0.96π cubic cms.	**3.** 2%	**4.** $-\frac{4}{3}\%$
5. 5.027	**6.** 2.7935	**7.** 1.2%	**9.** Rs.3000
10. 1.004343	**11.** 0.112, 0.05091	**12.** 0.002 sq.cm, 0.2%.	

1.12 JACOBIANS

If u and v are functions of two independent variables x and y, then the determinant

$$\begin{vmatrix} \frac{\partial u}{\partial x} & \frac{\partial u}{\partial y} \\ \frac{\partial v}{\partial x} & \frac{\partial v}{\partial y} \end{vmatrix}$$

is called the Jacobian of u, v with respect to x, y and is denoted by $\frac{\partial(u,v)}{\partial(x,y)}$ or $J(u, v)$.

Similarly, the Jacobian of three independent variables u, v, w with respect to x, y, z is given by

$$J(u,v,w)=\frac{\partial(u,v,w)}{\partial(x,y,z)}=\begin{vmatrix} \frac{\partial u}{\partial x} & \frac{\partial u}{\partial y} & \frac{\partial u}{\partial z} \\ \frac{\partial v}{\partial x} & \frac{\partial v}{\partial y} & \frac{\partial v}{\partial z} \\ \frac{\partial w}{\partial x} & \frac{\partial w}{\partial y} & \frac{\partial w}{\partial z} \end{vmatrix}$$

Similarly, if $u_1, u_2, u_3, \ldots\ldots., u_n$ are functions of independent variables $x_1, x_2, x_3, \ldots\ldots., x_n$, then the Jacobian of $u_1, u_2, u_3, \ldots\ldots., u_n$ with respect to $x_1, x_2, x_3, \ldots\ldots., x_n$ is given by

$$J(u_1, u_2, \ldots., u_n) = \frac{\partial(u_1, u_2, \ldots., u_n)}{\partial(x_1, x_2, \ldots., x_n)} = \begin{vmatrix} \frac{\partial u_1}{\partial x_1} & \frac{\partial u_1}{\partial x_2} & \frac{\partial u_1}{\partial x_3} & \cdots & \frac{\partial u_1}{\partial x_n} \\ \frac{\partial u_2}{\partial x_1} & \frac{\partial u_2}{\partial x_2} & \frac{\partial u_2}{\partial x_3} & \cdots & \frac{\partial u_2}{\partial x_n} \\ \cdots & \cdots & \cdots & \cdots & \cdots \\ \cdots & \cdots & \cdots & \cdots & \cdots \\ \frac{\partial u_n}{\partial x_1} & \frac{\partial u_n}{\partial x_2} & \frac{\partial u_n}{\partial x_3} & \cdots & \frac{\partial u_n}{\partial x_n} \end{vmatrix}$$

1.13 PROPERTIES OF JACOBIANS

(1) If u and v are functions of x and y, then

$$\boxed{\frac{\partial(u, v)}{\partial(x, y)} \times \frac{\partial(x, y)}{\partial(u, v)} = 1} \quad \text{or} \quad J.J' = 1.$$

Proof : Let $u = u(x, y)$ and $v = v(x, y)$ be functions of x and y.
Differentiating these functions partially w.r.t. u and v, we get

$$\frac{\partial u}{\partial u} = 1 = \frac{\partial u}{\partial x}\cdot\frac{\partial x}{\partial u} + \frac{\partial u}{\partial y}\cdot\frac{\partial y}{\partial u}$$

$$\frac{\partial u}{\partial v} = 0 = \frac{\partial u}{\partial x}\cdot\frac{\partial x}{\partial v} + \frac{\partial u}{\partial y}\cdot\frac{\partial y}{\partial v}$$

$$\frac{\partial v}{\partial u} = 0 = \frac{\partial v}{\partial x}\cdot\frac{\partial x}{\partial u} + \frac{\partial v}{\partial y}\cdot\frac{\partial y}{\partial u}$$

$$\frac{\partial v}{\partial v} = 1 = \frac{\partial v}{\partial x}\cdot\frac{\partial x}{\partial v} + \frac{\partial v}{\partial y}\cdot\frac{\partial y}{\partial v}$$

Now $$\frac{\partial(u, v)}{\partial(x, y)} \times \frac{\partial(x, y)}{\partial(u, v)} = \begin{vmatrix} \frac{\partial u}{\partial x} & \frac{\partial u}{\partial y} \\ \frac{\partial v}{\partial x} & \frac{\partial v}{\partial y} \end{vmatrix} \times \begin{vmatrix} \frac{\partial x}{\partial u} & \frac{\partial x}{\partial v} \\ \frac{\partial y}{\partial u} & \frac{\partial y}{\partial v} \end{vmatrix}$$

$$= \begin{vmatrix} \frac{\partial u}{\partial x}\cdot\frac{\partial x}{\partial u} + \frac{\partial u}{\partial y}\cdot\frac{\partial y}{\partial u} & \frac{\partial u}{\partial x}\cdot\frac{\partial x}{\partial v} + \frac{\partial u}{\partial y}\cdot\frac{\partial y}{\partial v} \\ \frac{\partial v}{\partial x}\cdot\frac{\partial x}{\partial u} + \frac{\partial v}{\partial y}\cdot\frac{\partial y}{\partial u} & \frac{\partial v}{\partial x}\cdot\frac{\partial x}{\partial v} + \frac{\partial v}{\partial y}\cdot\frac{\partial y}{\partial v} \end{vmatrix} = \begin{vmatrix} 1 & 0 \\ 0 & 1 \end{vmatrix} = 1$$

(2) If u, v are functions of r, s and r, s are functions of x, y, then

$$\boxed{\frac{\partial(u,v)}{\partial(r,s)}\times\frac{\partial(r,s)}{\partial(x,y)}=\frac{\partial(u,v)}{\partial(x,y)}}$$

Proof: We have

$$\frac{\partial(u,v)}{\partial(r,s)}\times\frac{\partial(r,s)}{\partial(x,y)}=\begin{vmatrix}\frac{\partial u}{\partial r} & \frac{\partial u}{\partial s}\\ \frac{\partial v}{\partial r} & \frac{\partial v}{\partial s}\end{vmatrix}\begin{vmatrix}\frac{\partial r}{\partial x} & \frac{\partial r}{\partial y}\\ \frac{\partial s}{\partial x} & \frac{\partial s}{\partial y}\end{vmatrix}$$

$$=\begin{vmatrix}\frac{\partial u}{\partial r}.\frac{\partial r}{\partial x}+\frac{\partial u}{\partial s}.\frac{\partial s}{\partial x} & \frac{\partial u}{\partial r}.\frac{\partial r}{\partial y}+\frac{\partial u}{\partial s}.\frac{\partial s}{\partial y}\\ \frac{\partial v}{\partial r}.\frac{\partial r}{\partial x}+\frac{\partial v}{\partial s}.\frac{\partial s}{\partial x} & \frac{\partial v}{\partial r}.\frac{\partial r}{\partial y}+\frac{\partial v}{\partial s}.\frac{\partial s}{\partial y}\end{vmatrix}$$

But

$$\frac{\partial u}{\partial x}=\frac{\partial u}{\partial r}.\frac{\partial r}{\partial x}+\frac{\partial u}{\partial s}.\frac{\partial s}{\partial x}$$

$$\frac{\partial u}{\partial y}=\frac{\partial u}{\partial r}.\frac{\partial r}{\partial y}+\frac{\partial u}{\partial s}.\frac{\partial s}{\partial y}$$

$$\frac{\partial v}{\partial x}=\frac{\partial v}{\partial r}.\frac{\partial r}{\partial x}+\frac{\partial v}{\partial s}.\frac{\partial s}{\partial x}$$

$$\frac{\partial v}{\partial y}=\frac{\partial v}{\partial r}.\frac{\partial r}{\partial y}+\frac{\partial v}{\partial s}.\frac{\partial s}{\partial y}$$

$$\therefore \quad \frac{\partial(u,v)}{\partial(r,s)}\times\frac{\partial(r,s)}{\partial(x,y)}=\begin{vmatrix}\frac{\partial u}{\partial x} & \frac{\partial u}{\partial y}\\ \frac{\partial v}{\partial x} & \frac{\partial v}{\partial y}\end{vmatrix}=\frac{\partial(u,v)}{\partial(x,y)}$$

(3) If functions u, v, w of three independent variables x, y, z are not independent then the Jacobian of u, v, w with respect to x, y, z vanishes.

Proof: If u, v, w are not independent, then there exists a relation $F(u, v, w) = 0$ connecting them.

Differentiating this relation w.r.t. x, y and z, we get

$$\frac{\partial F}{\partial u}.\frac{\partial u}{\partial x}+\frac{\partial F}{\partial v}.\frac{\partial v}{\partial x}+\frac{\partial F}{\partial w}.\frac{\partial w}{\partial x}=0$$

$$\frac{\partial F}{\partial u}.\frac{\partial u}{\partial y}+\frac{\partial F}{\partial v}.\frac{\partial v}{\partial y}+\frac{\partial F}{\partial w}.\frac{\partial w}{\partial y}=0$$

$$\frac{\partial F}{\partial u}.\frac{\partial u}{\partial z}+\frac{\partial F}{\partial v}.\frac{\partial v}{\partial z}+\frac{\partial F}{\partial w}.\frac{\partial w}{\partial z}=0$$

Eliminating $\frac{\partial F}{\partial u}, \frac{\partial F}{\partial v}, \frac{\partial F}{\partial w}$ from the above equations, we get

$$\begin{vmatrix} \frac{\partial u}{\partial x} & \frac{\partial u}{\partial y} & \frac{\partial u}{\partial z} \\ \frac{\partial v}{\partial x} & \frac{\partial v}{\partial y} & \frac{\partial v}{\partial z} \\ \frac{\partial w}{\partial x} & \frac{\partial w}{\partial y} & \frac{\partial w}{\partial z} \end{vmatrix} = 0 \Rightarrow \frac{\partial(u,v,w)}{\partial(x,y,z)} = 0$$

ILLUSTRATIVE EXAMPLES

Example 1. If $x = r\cos\theta$, $y = r\sin\theta$, evaluate $\frac{\partial(x,y)}{\partial(r,\theta)}$ and $\frac{\partial(r,\theta)}{\partial(x,y)}$.

Solution. We have $\frac{\partial(x,y)}{\partial(r,\theta)} = \begin{vmatrix} \frac{\partial x}{\partial r} & \frac{\partial x}{\partial \theta} \\ \frac{\partial y}{\partial r} & \frac{\partial y}{\partial \theta} \end{vmatrix} = \begin{vmatrix} \cos\theta & -r\sin\theta \\ \sin\theta & r\cos\theta \end{vmatrix}$

$$= r(\cos^2\theta + \sin^2\theta) = r.1 = r$$

Now, since $\frac{\partial(x,y)}{\partial(r,\theta)} \times \frac{\partial(r,\theta)}{\partial(x,y)} = 1 \Rightarrow \frac{\partial(r,\theta)}{\partial(x,y)} = \frac{1}{\frac{\partial(x,y)}{\partial(r,\theta)}} = \frac{1}{r}.$

Example 2. If $u = \frac{yz}{x}$, $v = \frac{xz}{y}$, $w = \frac{xy}{z}$, prove that $J(u,v,w) = 4$.

Solution. We have $J(u,v,\ w) = \frac{\partial(u,v,w)}{\partial(x,y,z)}$

$$= \begin{vmatrix} \frac{\partial u}{\partial x} & \frac{\partial u}{\partial y} & \frac{\partial u}{\partial z} \\ \frac{\partial v}{\partial x} & \frac{\partial v}{\partial y} & \frac{\partial v}{\partial z} \\ \frac{\partial w}{\partial x} & \frac{\partial w}{\partial y} & \frac{\partial w}{\partial z} \end{vmatrix} = \begin{vmatrix} -\frac{yz}{x^2} & \frac{z}{x} & \frac{y}{x} \\ \frac{z}{y} & -\frac{xz}{y^2} & \frac{x}{y} \\ \frac{y}{z} & \frac{x}{z} & \frac{-xy}{z^2} \end{vmatrix}$$

$$= \frac{1}{x^2y^2z^2}\begin{vmatrix} -yz & xz & xy \\ yz & -xz & xy \\ yz & xz & -xy \end{vmatrix}$$

$$= \frac{(yz)(xz)(xy)}{x^2y^2z^2}\begin{vmatrix} -1 & 1 & 1 \\ 1 & -1 & 1 \\ 1 & 1 & -1 \end{vmatrix}$$

$$= \begin{vmatrix} -1 & 1 & 1 \\ 1 & -1 & 1 \\ 1 & 1 & -1 \end{vmatrix} = -1(1-1)-1(-1-1)+1(1+1)$$

$$= 0+2+2 = 4.$$

Example 3. If $x = r\sin\theta\cos\phi,\ y = r\sin\theta\sin\phi,\ z = r\cos\theta$, show that $\frac{\partial(r,\theta,\phi)}{\partial(x,y,z)} = \frac{1}{r^2\sin\theta}$.

Solution. We have
$$\frac{\partial x}{\partial r} = \sin\theta\cos\phi,\ \frac{\partial x}{\partial\theta} = r\cos\theta\cos\phi,\ \frac{\partial x}{\partial\phi} = -r\sin\theta\sin\phi;$$

$$\frac{\partial y}{\partial r} = \sin\theta\sin\phi,\ \frac{\partial y}{\partial\theta} = r\cos\theta\sin\phi,\ \frac{\partial y}{\partial\phi} = r\sin\theta\cos\phi;$$

$$\frac{\partial z}{\partial r} = \cos\theta,\ \frac{\partial z}{\partial\theta} = -r\sin\theta,\ \frac{\partial z}{\partial\phi} = 0.$$

$$\therefore \quad \frac{\partial(x,y,z)}{\partial(r,\theta,\phi)} = \begin{vmatrix} \frac{\partial x}{\partial r} & \frac{\partial x}{\partial\theta} & \frac{\partial x}{\partial\phi} \\ \frac{\partial y}{\partial r} & \frac{\partial y}{\partial\theta} & \frac{\partial y}{\partial\phi} \\ \frac{\partial z}{\partial r} & \frac{\partial z}{\partial\theta} & \frac{\partial z}{\partial\phi} \end{vmatrix} = \begin{vmatrix} \sin\theta\cos\phi & r\cos\theta\cos\phi & -r\sin\theta\sin\phi \\ \sin\theta\sin\phi & r\cos\theta\sin\phi & r\sin\theta\cos\phi \\ \cos\theta & -r\sin\theta & 0 \end{vmatrix}$$

$$= r^2\sin\theta\begin{vmatrix} \sin\theta\cos\phi & \cos\theta\cos\phi & -\sin\phi \\ \sin\theta\sin\phi & \cos\theta\sin\phi & \cos\phi \\ \cos\theta & -\sin\theta & 0 \end{vmatrix}$$

$$= r^2\sin\theta\left\{\begin{matrix} \cos\theta(\cos\theta\cos^2\phi + \cos\theta\sin^2\phi) \\ +\sin\theta(\sin\theta\cos^2\phi + \cos\theta\sin^2\phi) \end{matrix}\right\}$$

$$= r^2 \sin\theta\left\{\cos^2\theta(\cos^2\phi + \sin^2\phi) + \sin^2\theta(\cos^2\phi + \sin^2\phi)\right\}$$
$$= r^2 \sin\theta\left\{\cos^2\theta.1 + \sin^2\theta.1\right\} = r^2\sin\theta\left\{\cos^2\theta + \sin^2\theta\right\}$$
$$= r^2\sin\theta.1 = r^2\sin\theta.$$

$$\therefore \quad \frac{\partial(x,y,z)}{\partial(r,\theta,\phi)} \times \frac{\partial(r,\theta,\phi)}{\partial(x,y,z)} = 1$$

$$\Rightarrow \quad \frac{\partial(r,\theta,\phi)}{\partial(x,y,z)} = \frac{1}{\dfrac{\partial(x,y,z)}{\partial(r,\theta,\phi)}} = \frac{1}{r^2\sin\theta}.$$

Example 4. If $u = \dfrac{y^2}{2z}$, $v = \dfrac{x^2+z^2}{2x}, w = \dfrac{x}{2}$, find $\dfrac{\partial(u,v,w)}{\partial(x,y,z)}$.

Solution. We have,

$$\frac{\partial(u,v,w)}{\partial(x,y,z)} = \begin{vmatrix} \frac{\partial u}{\partial x} & \frac{\partial u}{\partial y} & \frac{\partial u}{\partial z} \\ \frac{\partial v}{\partial x} & \frac{\partial v}{\partial y} & \frac{\partial v}{\partial z} \\ \frac{\partial w}{\partial x} & \frac{\partial w}{\partial y} & \frac{\partial w}{\partial z} \end{vmatrix} = \begin{vmatrix} 0 & \frac{y}{z} & \frac{-y^2}{2z^2} \\ \frac{2x^2-y^2}{4x^2} & \frac{y}{x} & 0 \\ \frac{1}{2} & 0 & 0 \end{vmatrix}$$

$$= 0 - \left(\frac{2x^2 2xz^2 - y^2}{4x^2}\right)(0-0) + \frac{1}{2}\left(0 + \frac{y^3}{2xz^2}\right)$$

$$= 0 + \frac{1}{2}\left(0 + \frac{y^3}{2xz^2}\right) = \frac{y^3}{4xz^2}.$$

Example 5. If $u = xyz$, $v = x^2 + y^2 + z^2, w = x + y + z$, find $\dfrac{\partial(x,y,z)}{\partial(u,v,w)}$.

Solution. We have,

$$\therefore \quad \frac{\partial(u,v,w)}{\partial(x,y,z)} = \begin{vmatrix} \frac{\partial u}{\partial x} & \frac{\partial u}{\partial y} & \frac{\partial u}{\partial z} \\ \frac{\partial v}{\partial x} & \frac{\partial v}{\partial y} & \frac{\partial v}{\partial z} \\ \frac{\partial w}{\partial x} & \frac{\partial w}{\partial y} & \frac{\partial w}{\partial z} \end{vmatrix} = \begin{vmatrix} yz & xz & xy \\ 2x & 2y & 2z \\ 1 & 1 & 1 \end{vmatrix}$$

$$= 2\begin{vmatrix} yz & xz & xy \\ x & y & z \\ 1 & 1 & 1 \end{vmatrix} = 2\begin{vmatrix} -z(x-y) & -x(y-z) & xy \\ x-y & y-z & z \\ 0 & 0 & 1 \end{vmatrix}$$

$$[C_1 \to C_1 - C_2; C_2 \to C_2 - C_3]$$

$$= 2(x-y)(y-z)\begin{vmatrix} -z & -x & xy \\ 1 & 1 & z \\ 0 & 0 & 1 \end{vmatrix}$$

$$= 2(x-y)(y-z)\{0-0+1(-z+x)\}$$

$$= -2(x-y)(y-z)(z-x)$$

$$\Rightarrow \quad \frac{\partial(x,y,z)}{\partial(u,v,w)} = \frac{1}{\dfrac{\partial(u,v,w)}{\partial(x,y,z)}} = \frac{-1}{2(x-y)(y-z)(z-x)}$$

Example 6. If $u = x^2 - y^2$, $v = 2xy$; $x = r\cos\theta, y = r\sin\theta$, find $\dfrac{\partial(u,v)}{\partial(r,\theta)}$.

Solution. We have,

$$\therefore \quad \frac{\partial(u,v)}{\partial(x,y)} = \begin{vmatrix} \dfrac{\partial u}{\partial x} & \dfrac{\partial u}{\partial y} \\ \dfrac{\partial v}{\partial x} & \dfrac{\partial v}{\partial y} \end{vmatrix} = \begin{vmatrix} 2x & -2y \\ 2y & 2x \end{vmatrix}$$

$$= 4\begin{vmatrix} x & -y \\ y & x \end{vmatrix} = 4(x^2+y^2)$$

$$\text{and} \quad \frac{\partial(x,y)}{\partial(r,\theta)} = \begin{vmatrix} \dfrac{\partial x}{\partial r} & \dfrac{\partial x}{\partial \theta} \\ \dfrac{\partial y}{\partial r} & \dfrac{\partial y}{\partial \theta} \end{vmatrix} = \begin{vmatrix} \cos\theta & -r\sin\theta \\ \sin\theta & r\cos\theta \end{vmatrix}$$

$$= r\cos^2\theta + r\sin^2\theta = r$$

$$\text{Hence,} \quad \frac{\partial(u,v)}{\partial(r,\theta)} = \frac{\partial(u,v)}{\partial(x,y)} \times \frac{\partial(x,y)}{\partial(r,\theta)}$$

$$= 4(x^2+y^2)r = 4r^2 . r = 4r^3.$$

$(\because x^2 + y^2 = r^2)$

Example 7. If $x = r\cos\theta, y = r\sin\theta, z = z$, find the value of $\frac{\partial(r,\theta,z)}{\partial(x,y,z)}$.

Solution. We have,

$$\frac{\partial(x,y,z)}{\partial(r,\theta,z)} = \begin{vmatrix} \frac{\partial x}{\partial r} & \frac{\partial x}{\partial \theta} & \frac{\partial x}{\partial z} \\ \frac{\partial y}{\partial r} & \frac{\partial y}{\partial \theta} & \frac{\partial y}{\partial z} \\ \frac{\partial z}{\partial r} & \frac{\partial z}{\partial \theta} & \frac{\partial z}{\partial z} \end{vmatrix} = \begin{vmatrix} \cos\theta & -r\sin\theta & 0 \\ \sin\theta & r\cos\theta & 0 \\ 0 & 0 & 1 \end{vmatrix}$$

$$= 0 - 0 + 1(r\cos^2\theta + r\sin^2\theta) = r$$

$$\therefore \quad \frac{\partial(r,\theta,z)}{\partial(x,y,z)} = \frac{1}{\frac{\partial(x,y,z)}{\partial(r,\theta,z)}} = \frac{1}{r}$$

Example 8. Show that the functions $u = x^2 + y^2 + z^2$, $v = x + y + z$, $w = xy + yz + zx$ are not independent of each other.

Solution. We have,

$$\therefore \quad \frac{\partial(u,v,w)}{\partial(x,y,z)} = \begin{vmatrix} \frac{\partial u}{\partial x} & \frac{\partial u}{\partial y} & \frac{\partial u}{\partial z} \\ \frac{\partial v}{\partial x} & \frac{\partial v}{\partial y} & \frac{\partial v}{\partial z} \\ \frac{\partial w}{\partial x} & \frac{\partial w}{\partial y} & \frac{\partial w}{\partial z} \end{vmatrix} = \begin{vmatrix} 2x & 2y & 2z \\ 1 & 1 & 1 \\ y+z & x+z & x+y \end{vmatrix}$$

$$= 2\begin{vmatrix} x & y & z \\ 1 & 1 & 1 \\ y+z & x+z & x+y \end{vmatrix} = 2\begin{vmatrix} x-y & y-z & z \\ 0 & 0 & 1 \\ y-x & z-y & x+y \end{vmatrix}$$

$$[C_1 \to C_1 - C_2; C_2 \to C_2 - C_3]$$

$$= 2[(x-y)(z-y) - (y-z)(y-x)]$$

$$= 2[(x-y)(z-y) - (z-y)(x-y)] = 0$$

$$\Rightarrow \quad \frac{\partial(u,v,w)}{\partial(x,y,z)} = 0$$

Hence, the three functions are not independent of each other.

1.14 JACOBIAN OF IMPLICIT FUNCTIONS

If u, v are implicit functions of x, y and they are connected by the relations

$$F_1(u,v,x,y) = 0$$

and
$$F_2(u,v,x,y) = 0$$

then
$$\frac{\partial(u,v)}{\partial(x,y)} = (-1)^2\left[\frac{\partial(F_1,F_2)}{\partial(x,y)}\bigg/\frac{\partial(F_1,F_2)}{\partial(u,v)}\right]$$

Similarly, if u, v, w are connected to x, y, z by the relations

$$F_1(u,v,w,x,y,z) = 0$$

$$F_2(u,v,w,x,y,z) = 0$$

and
$$F_3(u,v,w,x,y,z) = 0$$

then
$$\boxed{\frac{\partial(u,v.w)}{\partial(x,y,z)} = (-1)^3\left[\frac{\partial(F_1,F_2,F_3)}{\partial(x,y,z)}\bigg/\frac{\partial(F_1,F_2,F_3)}{\partial(u,v,w)}\right]}$$

ILLUSTRATIVE EXAMPLES

Example 1. If $u^3+v^3=x+y, u^2+v^2=x^3+y^3$, then show that

$$\frac{\partial(u,v)}{\partial(x,y)} = \frac{y^2-x^2}{2uv(u-v)}.$$

Solution. Let
$$F_1 = u^3+v^3-x-y$$

$$F_2 = u^2+v^2-x^3-y^3$$

We have,
$$\frac{\partial(u,v)}{\partial(x,y)} = (-1)^2\left[\frac{\partial(F_1,F_2)}{\partial(x,y)}\bigg/\frac{\partial(F_1,F_2)}{\partial(u,v)}\right]$$

Now,
$$\frac{\partial(F_1,F_2)}{\partial(x,y)} = \begin{vmatrix}\frac{\partial F_1}{\partial x} & \frac{\partial F_1}{\partial y}\\ \frac{\partial F_2}{\partial x} & \frac{\partial F_2}{\partial y}\end{vmatrix} = \begin{vmatrix}-1 & -1\\ -3x^2 & -3y^2\end{vmatrix} = 3(y^2-x^2)$$

and,
$$\frac{\partial(F_1,F_2)}{\partial(u,v)} = \begin{vmatrix}\frac{\partial F_1}{\partial u} & \frac{\partial F_1}{\partial v}\\ \frac{\partial F_2}{\partial u} & \frac{\partial F_2}{\partial v}\end{vmatrix} = \begin{vmatrix}3u^2 & 3v^2\\ 2u & 2v\end{vmatrix} = 6uv(u-v)$$

Hence,
$$\frac{\partial(u,v)}{\partial(x,y)} = \frac{3(y^2-x^2)}{6uv(u-v)} = \frac{y^2-x^2}{uv(u-v)}.$$

Example 2. If
$$u^3+v+w=x+y^2+z^2,$$
$$u+v^3+w=x^2+y+z^2,$$
$$u+v+w^3=x^2+y^2+z,$$

then show that
$$\frac{\partial(u,v,w)}{\partial(x,y,z)} = \frac{1-4(xy+yz+zx)+16xyz}{2-3(u^2+v^2+w^2)+27u^2v^2w^2}.$$

Solution. Let

$$F_1 = u^3 + v + w - x - y^2 - z^2,$$
$$F_2 = u + v^3 + w - x^2 - y - z^2,$$
$$F_3 = u + v + w^3 - x^2 - y^2 - z,$$

We have,

$$\frac{\partial(u,v,w)}{\partial(x,y,z)} = (-1)^3 \left[\frac{\partial(F_1,F_2,F_3)}{\partial(x,y,z)} \Big/ \frac{\partial(F_1,F_2,F_3)}{\partial(u,v,w)} \right]$$

Now,

$$\frac{\partial(F_1,F_2,F_3)}{\partial(x,y,z)} = \begin{vmatrix} \frac{\partial F_1}{\partial x} & \frac{\partial F_1}{\partial y} & \frac{\partial F_1}{\partial z} \\ \frac{\partial F_2}{\partial x} & \frac{\partial F_2}{\partial y} & \frac{\partial F_2}{\partial z} \\ \frac{\partial F_3}{\partial x} & \frac{\partial F_3}{\partial y} & \frac{\partial F_3}{\partial z} \end{vmatrix} = \begin{vmatrix} -1 & -2y & -2z \\ -2x & -1 & -2z \\ -2x & -2y & -1 \end{vmatrix}$$

$$= -1 \begin{vmatrix} 1 & 2y & 2z \\ 2x & 1 & 2z \\ 2x & 2y & 1 \end{vmatrix}$$

$$= -\left[1(1-4yz) - 2y(2x-4xz) + 2z(4xy-2x)\right]$$
$$= -\left[1 - 4(xy+yz+zx) + 16xyz\right]$$

and,

$$\frac{\partial(F_1,F_2,F_3)}{\partial(u,v,w)} = \begin{vmatrix} \frac{\partial F_1}{\partial u} & \frac{\partial F_1}{\partial v} & \frac{\partial F_1}{\partial w} \\ \frac{\partial F_2}{\partial u} & \frac{\partial F_2}{\partial v} & \frac{\partial F_2}{\partial w} \\ \frac{\partial F_3}{\partial u} & \frac{\partial F_3}{\partial v} & \frac{\partial F_3}{\partial w} \end{vmatrix} = \begin{vmatrix} 3u^2 & 1 & 1 \\ 1 & 3v^2 & 1 \\ 1 & 1 & 3w^2 \end{vmatrix}$$

$$= 3u^2(9v^2w^2 - 1) - 1(3w^2 - 1) + 1(1 - 3v^2)$$
$$= 2 - 3(u^2 + v^2 + w^2) + 27u^2v^2w^2$$

Hence,

$$\frac{\partial(u,v,w)}{\partial(x,y,z)} = (-1)^3 \frac{-\left[1 - 4(xy+yz+zx) + 16xyz\right]}{2 - 3(u^2+v^2+w^2) + 27u^2v^2w^2}$$

$$= \frac{\left[1 - 4\left(xy+yz+zx\right) + 16xyz\right]}{2 - 3(u^2+v^2+w^2) + 27u^2v^2w^2}$$

Example 3. If $u = \frac{x}{\sqrt{1-r^2}}, v = \frac{y}{\sqrt{1-r^2}}, w = \frac{z}{\sqrt{1-r^2}}$, then show that

$$\frac{\partial(u,v,w)}{\partial(x,y,z)} = \frac{1}{(1-r^2)^{\frac{5}{2}}}, \text{ where } r^2 = x^2 + y^2 + z^2.$$

Solution. $u = \dfrac{x}{\sqrt{1-r^2}} \Rightarrow u^2 = \dfrac{x^2}{1-r^2} \Rightarrow u^2(1-r^2) = x^2$

$\Rightarrow \quad u^2(1-x^2-y^2-z^2) = x^2$

$\Rightarrow \quad F_1 = x^2 - u^2(1-x^2-y^2-z^2)$

Similarly, $F_2 = y^2 - v^2(1-x^2-y^2-z^2)$

$F_3 = z^2 - w^2(1-x^2-y^2-z^2)$

We have, $\dfrac{\partial(u,v.w)}{\partial(x,y,z)} = (-1)^3\left[\dfrac{\partial(F_1,F_2,F_3)}{\partial(x,y,z)} \Big/ \dfrac{\partial(F_1,F_2,F_3)}{\partial(u,v,w)}\right]$

Now, $$\frac{\partial(F_1,F_2,F_3)}{\partial(x,y,z)} = \begin{vmatrix} \frac{\partial F_1}{\partial x} & \frac{\partial F_1}{\partial y} & \frac{\partial F_1}{\partial z} \\ \frac{\partial F_2}{\partial x} & \frac{\partial F_2}{\partial y} & \frac{\partial F_2}{\partial z} \\ \frac{\partial F_3}{\partial x} & \frac{\partial F_3}{\partial y} & \frac{\partial F_3}{\partial z} \end{vmatrix}$$

$$= \begin{vmatrix} 2x(1+u^2) & 2u^2y & 2u^2z \\ 2v^2x & 2y(1+v^2) & 2v^2z \\ 2w^2x & 2w^2z & 2z(1+w^2) \end{vmatrix}$$

$$= 8xyz\begin{vmatrix} 1+u^2 & u^2 & u^2 \\ v^2 & 1+v^2 & v^2 \\ w^2 & w^2 & 1+w^2 \end{vmatrix}$$

$$= 8xyz\begin{vmatrix} 1+u^2+v^2+w^2 & 1+u^2+v^2+w^2 & 1+u^2+v^2+w^2 \\ v^2 & 1+v^2 & v^2 \\ w^2 & w^2 & 1+w^2 \end{vmatrix}$$

$[R_1 \to R_1 + (R_2 + R_3)]$

$$= 8xyz(1+u^2+v^2+w^2)\begin{vmatrix} 1 & 1 & 1 \\ v^2 & 1+v^2 & v^2 \\ w^2 & w^2 & 1+w^2 \end{vmatrix}$$

$$= 8xyz\left(1+u^2+v^2+w^2\right)\begin{vmatrix} 1 & 0 & 0 \\ v^2 & 1 & 0 \\ w^2 & 0 & 1 \end{vmatrix} \begin{bmatrix} C_2 \to C_2 - C_1; \\ C_3 \to C_3 - C_1 \end{bmatrix}$$

$$= 8xyz\left(1+u^2+v^2+w^2\right)$$

$$= 8xyz\left(1+\frac{x^2}{1-r^2}+\frac{y^2}{1-r^2}+\frac{z^2}{1-r^2}\right)$$

$$= 8xyz\left(1+\frac{x^2+y^2+z^2}{1-r^2}\right) = 8xyz\left(1+\frac{r^2}{1-r^2}\right) = \frac{8xyz}{1-r^2}$$

$$= \frac{8uvw\left(1-r^2\right)^{\frac{3}{2}}}{1-r^2} = 8uvw\left(1-r^2\right)^{\frac{1}{2}}$$

and,
$$\frac{\partial\left(F_1,F_2,F_3\right)}{\partial(u,v,w)} = \begin{vmatrix} \frac{\partial F_1}{\partial u} & \frac{\partial F_1}{\partial v} & \frac{\partial F_1}{\partial w} \\ \frac{\partial F_2}{\partial u} & \frac{\partial F_2}{\partial v} & \frac{\partial F_2}{\partial w} \\ \frac{\partial F_3}{\partial u} & \frac{\partial F_3}{\partial v} & \frac{\partial F_3}{\partial w} \end{vmatrix}$$

$$= \begin{vmatrix} -2u\left(1-x^2-y^2-z^2\right) & 0 & 0 \\ 0 & -2v\left(1-x^2-y^2-z^2\right) & 0 \\ 0 & 0 & -2w\left(1-x^2-y^2-z^2\right) \end{vmatrix}$$

$$= -8uvw\left(1-x^2-y^2-z^2\right)^3 \begin{vmatrix} 1 & 0 & 0 \\ 0 & 1 & 0 \\ 0 & 0 & 1 \end{vmatrix}$$

$$= -8uvw\left(1-x^2-y^2-z^2\right)^3 = -8uvw\left(1-r^2\right)^3$$

Hence, $\dfrac{\partial(u,v.w)}{\partial(x,y,z)} = (-1)^3 8xyz(1-r^2)^{\frac{1}{2}}.\dfrac{1}{-8uvw(1-r^2)^3} = \dfrac{1}{(1-r^2)^{\frac{5}{2}}}.$

Example 4. If u, v, w are the roots of the equation $(a-x)^3+(a-y)^3+(a-z)^3=0$, evaluate $\dfrac{\partial(x,y,z)}{\partial(u,v,w)}$.

Solution. $(a-x)^3+(a-y)^3+(a-z)^3=0$

$\Rightarrow \quad 3a^3-3a^2(x+y+z)+3a(x^2+y^2+z^2)-(x^3+y^3+z^3)=0$

$\therefore$ Sum of the roots $\equiv u+v+w=x+y+z$

Product of the roots $\equiv uv+vw+wu=x^2+y^2+z^2$

$$uvw=\frac{1}{3}\left(x^3+y^3+z^3\right)$$

Let

$$F_1=u+v+w-x-y-z,$$

$$F_2=uv+vw+wu-x^2-y^2-z^2,$$

$$F_3=uvw-\frac{1}{3}\left(x^3+y^3+z^3\right)$$

We have,

$$\frac{\partial(x,y,z)}{\partial(u,v.w)}=(-1)^3\left[\frac{\partial(F_1,F_2,F_3)}{\partial(u,v.w)}\Big/\frac{\partial(F_1,F_2,F_3)}{\partial(x,y,z)}\right]$$

Now,

$$\frac{\partial(F_1,F_2,F_3)}{\partial(x,y,z)}=\begin{vmatrix}\frac{\partial F_1}{\partial x} & \frac{\partial F_1}{\partial y} & \frac{\partial F_1}{\partial z}\\ \frac{\partial F_2}{\partial x} & \frac{\partial F_2}{\partial y} & \frac{\partial F_2}{\partial z}\\ \frac{\partial F_3}{\partial x} & \frac{\partial F_3}{\partial y} & \frac{\partial F_3}{\partial z}\end{vmatrix}=\begin{vmatrix}-1 & -1 & -1\\ -2x & -2y & -2z\\ -x^2 & -y^2 & -z^2\end{vmatrix}$$

$$=-2\begin{vmatrix}1 & 1 & 1\\ x & y & z\\ x^2 & y^2 & z^2\end{vmatrix}=-2\begin{vmatrix}0 & 0 & 1\\ x-y & y-z & z\\ (x-y)(x+y) & (y-z)(y+z) & z^2\end{vmatrix}$$

$[C_1\to C_1-C_2;C_2\to C_2-C_3]$

$$=-2(x-y)(y-z)\begin{vmatrix}0 & 0 & 1\\ 1 & 1 & z\\ x+y & y+z & z^2\end{vmatrix}$$

$$=-2(x-y)(y-z)(y+z-x-y)=-2(x-y)(y-z)(z-x)$$

and,

$$\frac{\partial(F_1,F_2,F_3)}{\partial(u,v,w)}=\begin{vmatrix}\frac{\partial F_1}{\partial u} & \frac{\partial F_1}{\partial v} & \frac{\partial F_1}{\partial w}\\ \frac{\partial F_2}{\partial u} & \frac{\partial F_2}{\partial v} & \frac{\partial F_2}{\partial w}\\ \frac{\partial F_3}{\partial u} & \frac{\partial F_3}{\partial v} & \frac{\partial F_3}{\partial w}\end{vmatrix}=\begin{vmatrix}1 & 1 & 1\\ v+w & w+u & u+v\\ vw & uw & uv\end{vmatrix}$$

$$=\begin{vmatrix}0 & 0 & 1\\ v-u & w-v & u+v\\ w(v-u) & u(w-v) & uv\end{vmatrix}\quad\begin{bmatrix}C_1\to C_1-C_2;\\ C_2\to C_2-C_3\end{bmatrix}$$

$$= (v-u)(w-v)\begin{vmatrix} 0 & 0 & 1 \\ 1 & 1 & u+v \\ w & u & uv \end{vmatrix} = (v-u)(w-v)(u-w)$$

$$= -(u-v)(v-w)(w-u)$$

Hence, $$\frac{\partial(x,y,z)}{\partial(u,v,w)} = -\left[\frac{-(u-v)(v-w)(w-u)}{-2(x-y)(y-z)(z-x)}\right] = -\frac{(u-v)(v-w)(w-u)}{2(x-y)(y-z)(z-x)}.$$

1.15 FUNCTIONAL RELATIONSHIP

Let $u_1, u_2, u_3, ..., u_n$ are n functions of n independent variables $x_1, x_2, x_3, ..., x_n$, then there exists an identical relation of the form $F(u_1, u_2, ..., u_n) = 0$ if and only if the Jacobian $\dfrac{\partial(u_1, u_2, ..., u_n)}{\partial(x_1, x_2, ..., x_n)}$ vanish identically *i.e.*, $\dfrac{\partial(u_1, u_2, ..., u_n)}{\partial(x_1, x_2, ..., x_n)} = 0.$

ILLUSTRATIVE EXAMPLES

Example 1. If $u = \dfrac{x-y}{x+y}$, $v = \dfrac{x+y}{x}$; verify whether u and v are functionally dependent, and if so, find the relation between them.

Solution.

$$\frac{\partial u}{\partial x} = \frac{(x+y).1-(x-y).1}{(x+y)^2} = \frac{2y}{(x+y)^2};$$

$$\frac{\partial u}{\partial y} = \frac{(x+y).(-1)-(x-y).1}{(x+y)^2} = \frac{-2x}{(x+y)^2}$$

$$\frac{\partial v}{\partial x} = \frac{x.1-(x+y).1}{x^2} = \frac{-y}{x^2};$$

$$\frac{\partial v}{\partial y} = \frac{x.1-(x+y).0}{x^2} = \frac{1}{x}$$

$\therefore$ $$\frac{\partial(u,v)}{\partial(x,y)} = \begin{vmatrix} \dfrac{\partial u}{\partial x} & \dfrac{\partial u}{\partial y} \\ \dfrac{\partial v}{\partial x} & \dfrac{\partial v}{\partial y} \end{vmatrix} = \begin{vmatrix} \dfrac{2y}{(x+y)^2} & \dfrac{-2x}{(x+y)^2} \\ \dfrac{-y}{x^2} & \dfrac{1}{x} \end{vmatrix}$$

$$= \frac{2}{x^2(x+y)^2}\begin{vmatrix} y & -x \\ -y & x \end{vmatrix} = \frac{2}{x^2(x+y)^2}(xy-xy) = 0$$

Hence u and v are functionally related.

Now, $$uv = \frac{x-y}{x} = \frac{2x-x-y}{x} = \frac{2x-(x+y)}{x}$$

$$= 2 - \frac{(x+y)}{x} = 2 - v$$

$\Rightarrow$ $$uv = 2 - v \Rightarrow (u+1)v = 2$$

$\Rightarrow$ $$u+1 = \frac{2}{v}$$

which is the required relation.

Example 2. Prove that $u = 3x+2y-z,\ v = x-2y+z,\ w = x(x+2y-z)$ are not independent. Also find the relation between them.

Solution. $$\frac{\partial(u,v,w)}{\partial(x,y,z)} = \begin{vmatrix} \frac{\partial u}{\partial x} & \frac{\partial u}{\partial y} & \frac{\partial u}{\partial z} \\ \frac{\partial v}{\partial x} & \frac{\partial v}{\partial y} & \frac{\partial v}{\partial z} \\ \frac{\partial w}{\partial x} & \frac{\partial w}{\partial y} & \frac{\partial w}{\partial z} \end{vmatrix} = \begin{vmatrix} 3 & 2 & -1 \\ 1 & -2 & 1 \\ 2x & 2x & -x \end{vmatrix}$$

$$= 3(2x-2x) - 2(-x-2x) - 1(2x+4x)$$

$$= 0 + 6x - 6x = 0$$

Hence u and v are not independent.

Now, $$u - v = 2x+4y-2z \text{ and } u+v = 4x$$

$\therefore$ $$u^2 - v^2 = (u+v)(u-v) = 8x(x+2y-z) = 8w$$

$\Rightarrow$ $$u^2 - v^2 = 8w$$

which is the required relation.

Example 3. If $u = \frac{y-x}{1+xy}$, $v = \tan^{-1} y - \tan^{-1} x$. Find whether and are functionally related. If so, find the relation between them.

Solution. $$\frac{\partial(u,v)}{\partial(x,y)} = \begin{vmatrix} \frac{\partial u}{\partial x} & \frac{\partial u}{\partial y} \\ \frac{\partial v}{\partial x} & \frac{\partial v}{\partial y} \end{vmatrix} = \begin{vmatrix} \frac{-(1+y^2)}{(1+xy)^2} & \frac{1+x^2}{(1+xy)^2} \\ \frac{-1}{1+x^2} & \frac{1}{1+y^2} \end{vmatrix}$$

$$= -\frac{1}{(1+xy)^2} + \frac{1}{(1+xy)^2} = 0$$

Hence u and v are not independent.

Now, $\tan^{-1} y - \tan^{-1} x = \tan^{-1} \dfrac{y-x}{1+xy} = \tan^{-1} u$

$\Rightarrow$ $v = \tan^{-1} u \;\Rightarrow\; u = \tan v$

which is the required relation.

Example 4. If $u = \dfrac{x-y}{x+y}, v = \dfrac{xy}{(x+y)^2}$;verify whether u and v are functionally dependent, and if so, find the relation between them.

Solution.

$$\frac{\partial u}{\partial x} = \frac{(x+y).1-(x-y).1}{(x+y)^2} = \frac{2y}{(x+y)^2};$$

$$\frac{\partial u}{\partial y} = \frac{(x+y).(-1)-(x-y).1}{(x+y)^2} = \frac{-2x}{(x+y)^2}$$

$$\frac{\partial v}{\partial x} = \frac{(x+y)^2.y - xy.2(x+y).1}{(x+y)^4} = \frac{y^2-xy}{(x+y)^3};$$

$$\frac{\partial v}{\partial y} = \frac{(x+y)^2.x - xy.2(x+y).1}{(x+y)^4} = \frac{x^2-xy}{(x+y)^3}$$

$$\therefore \quad \frac{\partial(u,v)}{\partial(x,y)} = \begin{vmatrix} \frac{\partial u}{\partial x} & \frac{\partial u}{\partial y} \\ \frac{\partial v}{\partial x} & \frac{\partial v}{\partial y} \end{vmatrix} = \begin{vmatrix} \frac{2y}{(x+y)^2} & \frac{-2x}{(x+y)^2} \\ \frac{y^2-xy}{(x+y)^3} & \frac{x^2-xy}{(x+y)^3} \end{vmatrix}$$

$$= \frac{2}{(x+y)^5} \begin{vmatrix} y & -x \\ y^2-xy & x^2-xy \end{vmatrix}$$

$$= \frac{2}{(x+y)^5}(x^2y - xy^2 + xy^2 - x^2y) = 0$$

Hence u and v are functionally dependent.

Now, $$u^2 = \left(\frac{x-y}{x+y}\right)^2 = \frac{(x+y)^2-4xy}{(x+y)^2} = 1 - \frac{4xy}{(x+y)^2} = 1-4v$$

$\Rightarrow$ $u^2 + 4v = 1$

which is the required relation.

Example 5. If $y_1 = \frac{x_2 y_3}{x_1}$, $y_2 = \frac{x_3 x_1}{x_2}$, $y_3 = \frac{x_1 x_2}{x_3}$, show that the Jacobian of y_1, y_2, y_3 with respect to x_1, x_2, x_3 is 4. *(I.U. 2007, 08)*

Solution.
$$\frac{\partial(y_1, y_2, y_3)}{\partial(x_1, x_2, x_3)} = \begin{vmatrix} \frac{\partial y_1}{\partial x_1} & \frac{\partial y_1}{\partial x_2} & \frac{\partial y_1}{\partial x_3} \\ \frac{\partial y_2}{\partial x_1} & \frac{\partial y_2}{\partial x_2} & \frac{\partial y_2}{\partial x_3} \\ \frac{\partial y_3}{\partial x_1} & \frac{\partial y_3}{\partial x_2} & \frac{\partial y_3}{\partial x_3} \end{vmatrix} = \begin{vmatrix} -\frac{x_2 x_3}{x_1^2} & \frac{x_3}{x_1} & \frac{x_2}{x_1} \\ \frac{x_3}{x_2} & \frac{-x_3 x_1}{x_2} & \frac{x_1}{x_2} \\ \frac{x_2}{x_3} & \frac{x_1}{x_3} & \frac{-x_1 x_2}{x_3^2} \end{vmatrix}$$

$$= \frac{1}{x_1^2 x_2^2 x_3^2} \begin{vmatrix} -x_2 x_3 & x_3 x_1 & x_2 x_1 \\ x_3 x_2 & -x_3 x_1 & x_1 x_2 \\ x_2 x_3 & x_1 x_3 & -x_1 x_2 \end{vmatrix}$$

$$= \frac{x_1^2 x_2^2 x_3^2}{x_1^2 x_2^2 x_3^2} \begin{vmatrix} -1 & 1 & 1 \\ 1 & -1 & 1 \\ 1 & 1 & -1 \end{vmatrix}$$

$$= \begin{vmatrix} -1 & 1 & 1 \\ 1 & -1 & 1 \\ 1 & 1 & -1 \end{vmatrix} = -1(1 - 1) - 1(-1-1) + 1(1 + 1)$$

$$= 0 + 2 + 2$$

$$\Rightarrow \quad \frac{\partial(y_1, y_2, y_3)}{\partial(x_1, x_2, x_3)} = 4.$$

EXERCISE 1.5

1. If $u = ax + by, v = cx + dy$, evaluate $\frac{\partial(u, v)}{\partial(x, y)}$.

2. If $u = \frac{1}{x}, v = \frac{x^2}{y}, w = x + y + y^2 z$, find $\frac{\partial(x, y, z)}{\partial(u, v, w)}$.

3. If $u = \frac{x}{y - z}, v = \frac{y}{z - x}, w = \frac{z}{x - y}$, prove that $\frac{\partial(u, v, w)}{\partial(x, y, z)} = 0$.

4. If $u = xyz, v = x^2 + y^2 + z^2, w = x + y + z$, show that

$$\frac{\partial(x, y, z)}{\partial(u, v, w)} = \frac{-1}{2(y-z)(z-x)(x-y)}.$$

5. If $u = x + y, v = x - y$ where $x = r\cos\theta, y = r\sin\theta$, find $\frac{\partial(u, v)}{\partial(r, \theta)}$.

6. If $u = e^x \cos y, v = e^x \sin y$ where $x = lr + sm, y = mr - sl$, verify

$$\frac{\partial(u, v)}{\partial(r, s)} = \frac{\partial(u, v)}{\partial(x, y)} \times \frac{\partial(x, y)}{\partial(r, s)}.$$

7. If $u = x + y + z, uv = y + z, uvw = z$, show that $\frac{\partial(u, v, w)}{\partial(x, y, z)} = u^2 v$.

8. If
$$u^3 + v^3 + w^3 = x + y + z,$$
$$u^2 + v^2 + w^2 = x^3 + y^3 + z^3,$$
$$u + v + w = x^2 + y^2 + z^2,$$

show that $\frac{\partial(u, v, w)}{\partial(x, y, z)} = \frac{(x-y)(y-z)(z-x)}{(u-v)(v-w)(w-u)}$.

9. If $u^3 = xyz, \frac{1}{v} = \frac{1}{x} + \frac{1}{y} + \frac{1}{z}, w^2 = x^2 + y^2 + z^2$, evaluate $\frac{\partial(u, v, w)}{\partial(x, y, z)}$.

10. If u, v, w are the roots of the equation $\frac{x}{a+\lambda} + \frac{y}{b+\lambda} + \frac{z}{c+\lambda} = 1$, prove that

$$\frac{\partial(x, y, z)}{\partial(u, v, w)} = -\frac{(u-v)(v-w)(w-u)}{(a-b)(b-c)(c-a)}.$$

11. Prove that the functions $u = x(y+z),\ v = y(x+z),\ w = z(x-y)$ are not independent. Also find the relation between them.

12. If $u = x + y + z,\ v = xy + yz + zx,\ w = x^3 + y^3 + z^3 - 3xyz$, show that u, v, w are not independent and find the relation between them.

13. Prove whether the functions $u = x + y - z,\ v = x - y + z, w = x^2 + y^2 + z^2 - 2yz$ are functionally dependent or not. If yes, find the relation between them.

14. If $u = \sin^{-1} x + \sin^{-1} y, v = x\sqrt{1-y^2} + y\sqrt{1-x^2}$. Find whether u and v are functionally related. If so, find the relation between them.

15. If $u = x\sqrt{1-y^2} + y\sqrt{1-x^2}, v = \sqrt{(1-x^2)(1-y^2)} - xy$, then prove that u and v are not independent and find a relation between them.

ANSWERS

1. $ad - bc$ **2.** 1 **5.** $-2\sqrt{x^2 + y^2}$

9. $\dfrac{v(x-y)(y-z)(z-x)(x+y+z)}{3u^2v(xy+yz+zx)}$ **11.** $u - v = w$

12. $w = u^3 - 3uv$ **13.** Yes, $u^2 + v^2 = 2w$ **14.** $v = \sin u$

15. $u^2 + v^2 = 1$.

1.16 EXPANSION OF FUNCTIONS OF ONE VARIABLE

Taylor's Theorem (Taylor's Series): If $f(x)$ possesses continuous derivatives of all orders in the interval $(a, a+h)$, then for all positive integer *n*, we have

$$f(a+h) = f(a) + hf'(a) + \frac{h^2}{2!}f''(a) + \frac{h^3}{3!}f'''(a) + \dots\dots\dots + \frac{h^n}{n!}f^n(a) + \dots.. \quad \dots (1)$$

Corollary 1: Putting $a = x$ in (1), we get

$$f(x+h) = f(x) + hf'(x) + \frac{h^2}{2!}f''(x) + \frac{h^3}{3!}f'''(x) + \dots\dots\dots + \frac{h^n}{n!}f^n(x) + \dots.. \quad \dots (2)$$

Corollary 2: Putting $h = x - a$ in (1), we get

$$f(x) = f(a) + (x-a)f'(a) + \frac{(x-a)^2}{2!}f''(a) + \dots\dots\dots + \frac{(x-a)^n}{n!}f^n(a) + \dots.. \quad \dots (3)$$

It is called ***Taylor's series*** of $f(x)$ about the point $x = a$.

Corollary 3: Putting $a = 0$ in (3), we get

$$f(x) = f(0) + xf'(0) + \frac{x^2}{2!}f''(0) + \dots\dots\dots + \frac{x^n}{n!}f^n(0) + \dots.. \quad \dots (4)$$

It is Taylor's series of $f(x)$ about the point $x = 0$ and is called ***Maclaurin's Theorem***.

ILLUSTRATIVE EXAMPLES

Example 1. Expand the following by Maclaurin's Theorem :

(*i*) $\sin x$ (*ii*) e^x.

Solution. (*i*) Let

$f(x) = \sin x$ $f(0) = \sin 0 = 0$

$f'(x) = \cos x$ $f'(0) = \cos 0 = 1$

$f''(x) = -\sin x$ $f''(0) = -\sin 0 = 0$

$f'''(x) = -\cos x$ $f'''(0) = -\cos 0 = -1$

$f^{iv}(x) = \sin x$ $f^{iv}(0) = \sin 0 = 0$

$f^{v}(x) = \cos x$ $f^{v}(0) = \cos 0 = 1$ and so on.

$\therefore$ By Maclaurin's Theorem,

$$f(x) = f(0)+xf'(0)+\frac{x^2}{2!}f''(0)+.........+\frac{x^n}{n!}f^n(0)+.....$$

$$\sin x = 0+x.1+\frac{x^2}{2!}.0+\frac{x^3}{3!}.(-1)+\frac{x^4}{4!}.0+\frac{x^5}{5!}.1+....$$

$$\sin x = x-\frac{x^3}{3!}+\frac{x^5}{5!}+.....$$

(*ii*) Let

$$\begin{array}{ll} f(x)=e^x & f(0)=e^0=1 \\ f'(x)=e^x & f'(0)=e^0=1 \\ f''(x)=e^x & f''(0)=e^0=1 \\ f'''(x)=e^x & f'''(0)=e^0=1 \\ f^{iv}(x)=e^x & f^{iv}(0)=e^0=1 \text{ and so on.} \end{array}$$

$\therefore$ By Maclaurin's Theorem,

$$f(x) = f(0)+xf'(0)+\frac{x^2}{2!}f''(0)+.....+\frac{x^n}{n!}f^n(0)+.....$$

$$e^x = 1+x.1+\frac{x^2}{2!}.1+\frac{x^3}{3!}.1+\frac{x^4}{4!}.1+.....$$

$$e^x = 1+x+\frac{x^2}{2!}+\frac{x^3}{3!}+\frac{x^4}{4!}+.....$$

Example 2. Expand the following by Maclaurin's Theorem :

(*i*) $e^x \cos x$ (*ii*) $\log(1+x)$.

Solution. (*i*) Let

$$\begin{array}{ll} f(x) = e^x \cos x & f(0) = e^0 \cos 0 = 1.1 = 1 \\ f'(x) = e^x \cos x + e^x(-\sin x) & f'(0) = e^0 \cos 0 + e^0(-\sin 0) \\ & = 1.1 \quad 1.0 = 1 \\ f''(x) = -2e^x \sin x & f''(0) = -2e^0 \sin 0 = -2.2.0 = 0 \\ f'''(x) = -2e^x \sin x - 2e^x \cos x & f'''(0) = -2e^0 \sin 0 \quad 2e^0 \cos 0 \\ & = -2.1.0 - 2.1.1 = -2 \\ f^{iv}(x) = -4e^x \cos x & f^{iv}(0) = -4e^0 \cos 0 \\ & = -4.1.1 = -4 = -2^2 \\ f^{v}(x) = -4e^x \cos x + 4e^x \sin x & f^{v}(0) = -4e^0 \cos 0 + 4e^0 \sin 0 \\ & = -4.1.1 + 4.1.0 = -4 = -2^2 \\ f^{vi}(x) = -8e^x \sin x & f^{vi}(0) = -8e^0 \sin 0 = -8.1.0 = 0 \\ f^{vii}(x) = -8e^x \sin x - 8e^x \cos x & f^{vii}(0) = -8 = -2^3 \text{ and so on.} \end{array}$$

∴ By Maclaurin's Theorem, we get

$$e^x \cos x = 1 + x.1 + \frac{x^2}{2!}.0 + \frac{x^3}{3!}.(-2) + \frac{x^4}{4!}.\left(-2^2\right) + \frac{x^5}{5!}.\left(-2^2\right) + \frac{x^6}{6!}.0 + \frac{x^7}{7!}.\left(-2^3\right) + ...$$

$$e^x \cos x = 1 + x - \frac{2x^3}{3!} - \frac{2^2 x^4}{4!} - \frac{2^2 x^5}{5!} - \frac{2^3 x^7}{7!} + ...$$

(*ii*) Let $f(x) = \log(1+x)$ $\qquad f(0) = \log(1+0) = 0$

$$f'(x) = \frac{1}{1+x} \qquad f'(0) = 1$$

$$f''(x) = \frac{1}{(1+x)^2} \qquad f''(0) = -1$$

$$f'''(x) = \frac{2}{(1+x)^3} \qquad f'''(0) = 2$$

$$f^{iv}(x) = -\frac{6}{(1+x)^4} \qquad f^{iv}(0) = -6$$

∴ By Maclaurin's Theorem, we get

$$\log(1+x) = 0 + x.1 + \frac{x^2}{2!}.(-1) + \frac{x^3}{3!}.2 + \frac{x^4}{4!}.(-6) + ...$$

$$\log(1+x) = x - \frac{x^2}{2} + \frac{x^3}{3} - \frac{x^4}{4} + ...$$

Example 3. Expand $\dfrac{e^x}{e^x+1}$ in powers of x upto third degree term.

Solution. Let

$$f(x) = \frac{e^x}{e^x+1} \qquad f(0) = \frac{e^0}{e^0+1} = \frac{1}{2}$$

$$f'(x) = \frac{(e^x+1)e^x - e^x.e^x}{(e^x+1)^2} = \frac{e^x}{(e^x+1)^2} \qquad f'(0) = \frac{1}{4}$$

$$f''(x) = \frac{(e^x+1)^2 e^x - e^x.2(e^x+1).e^x}{(e^x+1)^4} = \frac{e^x - e^{2x}}{(e^x+1)^3} \qquad f''(0) = \frac{e^0 - e^0}{(e^0+1)^3} = 0$$

$$f'''(x) = \frac{(e^x+1)^3(e^x - 2e^x) - 3(e^x - e^{2x})(e^x+1)^2.e^x}{(e^x+1)^6}$$

$$= \frac{e^x - 4e^{2x} + e^{3x}}{(e^x+1)^4} \qquad f'''(0) = -\frac{1}{8} \text{ and so on.}$$

$\therefore$ By Maclaurin's Theorem,

$$f(x) = f(0) + xf'(0) + \frac{x^2}{2!}f''(0) + \dots\dots\dots + \frac{x^n}{n!}f^n(0) + \dots.$$

$$\frac{e^x}{e^x+1} = \frac{1}{2} + x.\frac{1}{4} + \frac{x^2}{2!}.0 + \frac{x^3}{3!}.\left(-\frac{1}{8}\right) + \dots.$$

$$\frac{e^x}{e^x+1} = \frac{1}{2} + \frac{x}{4} - \frac{x^3}{48} + \dots.$$

Example 4. Expand $\cot^{-1} x$ in powers of x upto third degree term.

Solution. Let $f(x) = \cot^{-1} x$ $\qquad f(0) = \cot^{-1} 0 = \frac{\pi}{2}$

$$f'(x) = -\frac{1}{1+x^2} \qquad f'(0) = -1$$

$$f''(x) = \frac{2x}{(1+x^2)^2} \qquad f''(0) = 0$$

$$f'''(x) = \frac{2-8x^2}{(1+x^2)^3} \qquad f'''(0) = 2$$

$$f^{iv}(x) = \frac{32x^3-28x}{(1+x^2)^4} \qquad f^{iv}(0) = 0 \text{ and so on.}$$

$\therefore$ By Maclaurin's Theorem, we get

$$\cot^{-1} x = \frac{\pi}{2} + x.(-1) + \frac{x^2}{2!}.0 + \frac{x^3}{3!}.2 + \dots\dots\dots$$

$$\cot^{-1} x = \frac{\pi}{2} - x + \frac{x^3}{3} + \dots\dots\dots$$

Example 5. Expand $f(x) = 2x^3 + 7x^2 + x - 1$ in powers of $(x-2)$.

Solution. $f(x) = 2x^3 + 7x^2 + x - 1 \qquad f(2) = 2.2^3 + 7.2^2 + 2 - 1 = 45$

$$f'(x) = 6x^2 + 14x - 1 \qquad f'(2) = 6.2^2 + 14.2 + 1 = 53$$

$$f''(x) = 12x + 14 \qquad f''(2) = 12.2 + 14 = 38$$

$$f'''(x) = 12 \qquad f'''(2) = 12$$

$$f^{iv}(x) = 0 \qquad f^{iv}(2) = 0 \text{ and so on.}$$

$\therefore$ By Taylor's Theorem, we have

$$f(x) = f(a)+(x-a)f'(a)+\frac{(x-a)^2}{2!}f''(a)+\frac{(x-a)^3}{3!}f'''(a)+....$$

$$f(x) = 45+(x-2).53+\frac{(x-2)^2}{2!}.38+\frac{(x-2)^3}{3!}.12$$

$$f(x) = 45+53(x-2)+19(x-2)^2+2(x-2)^3.$$

Example 6. Expand $f(x)=a^x$ in powers of $(x - 1)$ upto the term containing x^3.

Solution.

$$f(x) = a^x \qquad f(1) = a$$

$$f'(x) = a^x(\log a) \qquad f'(1) = a(\log a)$$

$$f''(x) = a^x(\log a)^2 \qquad f''(1) = a(\log a)^2$$

$$f'''(x) = a^x(\log a)^3 \qquad f'''(1) = a(\log a)^3$$ and so on.

$\therefore$ By Taylor's Theorem, we have

$$a^x = a+(x-1).a(\log a)+\frac{(x-1)^2}{2!}.a(\log a)^2 + \frac{(x-1)^3}{3!}.a(\log a)^3+.....$$

Example 7. Expand $f(x)=\cos x$ about the point $x=\frac{\pi}{4}$.

Solution.

$$f(x)=\cos x \qquad f\left(\frac{\pi}{4}\right)=\frac{1}{\sqrt{2}}$$

$$f'(x)=-\sin x \qquad f'\left(\frac{\pi}{4}\right)=-\frac{1}{\sqrt{2}}$$

$$f''(x)=-\cos x \qquad f''\left(\frac{\pi}{4}\right)=-\frac{1}{\sqrt{2}}$$

$$f'''(x)=\sin x \qquad f'''\left(\frac{\pi}{4}\right)=\frac{1}{\sqrt{2}}$$ and so on.

$\therefore$ By Taylor's Theorem, we have

$$\cos x=\frac{1}{\sqrt{2}}+\left(x-\frac{\pi}{4}\right)\left(-\frac{1}{\sqrt{2}}\right)+\frac{\left(x-\frac{\pi}{4}\right)^2}{2!}.\left(-\frac{1}{\sqrt{2}}\right)+\frac{\left(x-\frac{\pi}{4}\right)^3}{3!}.\frac{1}{\sqrt{2}}+.........$$

$$\cos x=\frac{1}{\sqrt{2}}\left\{1-\left(x-\frac{\pi}{4}\right)-\frac{\left(x-\frac{\pi}{4}\right)^2}{2!}+\frac{\left(x-\frac{\pi}{4}\right)^3}{3!}+.........\right\}.$$

Example 8. Expand $1 + x^3$ in powers of $(x + 3)$.

Solution. In powers of $(x + 3)$ means about the point $x = -3$.

Let
$$f(x) = 1 + x^3 \qquad f(-3) = 1 - 27 = -26$$
$$f'(x) = 3x^2 \qquad f'(-3) = 27$$
$$f''(x) = 6x \qquad f''(-3) = -18$$
$$f'''(x) = 6 \qquad f'''(-3) = 6$$
$$f^{iv}(x) = 0 \qquad f^{iv}(-3) = 0 \text{ and so on.}$$

∴ By Taylor's Theorem, we have

$$1 + x^3 = (-26) + (x+3).27 + \frac{(x+3)^2}{2!}.(-18) + \frac{(x+3)^3}{3!}.6$$

$$1 + x^3 = -26 + 27(x+3) - 9(x+3)^2 + (x+3)^3.$$

EXERCISE 1.6

1. Expand the following functions by Maclaurin's Theorem :
   (*i*) a^x (*ii*) $\sec x$ (*iii*) $\log(1 - x)$ (*iv*) $\tan^{-1} x$.

2. Expand $\sin x$ in powers of $\left(x - \frac{\pi}{2}\right)$.

3. Prove that : $\log \cos x = -\frac{x^2}{2!} - \frac{2x^4}{4!} - \frac{16x^6}{6!} - \ldots\ldots$

4. Expand $e^{\sin x}$ in powers of x upto fourth degree term.

5. Expand $f(x) = e^x$ about the point $x = 1$.

6. Expand $e^x \sin x$ about the point $x = 0$.

7. Prove that : $\cos^2 x = 1 - x^2 + \frac{x^4}{3} + \ldots\ldots$

8. Expand $\tan^{-1}(1 + x)$ by using Maclaurin's Theorem.

9. Expand $4x^2 + 7x + 5$ in powers of $(x - 3)$.

10. Expand $\tan x$ in powers of $\left(x - \frac{\pi}{4}\right)$.

ANSWERS

1. (*i*) $1+x\log a+\dfrac{x^2}{2!}(\log a)^2+\dfrac{x^3}{3!}(\log a)^3+\ldots\ldots\ldots$

(*ii*) $1+\dfrac{x^2}{2!}+\dfrac{5x^4}{4!}+\ldots\ldots\ldots$ (*iii*) $x+\dfrac{x^2}{2}+\dfrac{x^3}{3}+\dfrac{x^4}{4}+\ldots\ldots\ldots$

(*iv*) $x-\dfrac{x^3}{3}+\dfrac{x^5}{5}-\dfrac{x^7}{7}+\ldots\ldots\ldots$

2. $1-\dfrac{\left(x-\dfrac{\pi}{2}\right)^2}{2!}-\dfrac{\left(x-\dfrac{\pi}{2}\right)^4}{4!}-\ldots\ldots\ldots$

4. $1+x+\dfrac{x^2}{2}-\dfrac{x^4}{8}+\ldots\ldots\ldots$

5. $e\left\{1+(x-1)+\dfrac{(x-1)^2}{2}+\dfrac{(x-1)^3}{3}+\ldots\ldots\ldots\right\}$

6. $x+x^2-\dfrac{2x^3}{3!}-\dfrac{2^2x^5}{5!}+\ldots\ldots\ldots$

8. $\dfrac{\pi}{4}+\dfrac{x}{2}-\dfrac{x^2}{4}+\dfrac{x^3}{12}+\ldots\ldots\ldots$

9. $62+31(x-3)+4(x-3)^2$

10. $1+2\left(x-\dfrac{\pi}{4}\right)+2\left(x-\dfrac{\pi}{4}\right)^2+\dfrac{8}{3}\left(x-\dfrac{\pi}{4}\right)^3+\ldots\ldots\ldots$

1.17 EXPANSION OF FUNCTION OF TWO VARIABLES

Taylor's Theorem (Taylor's Series) of Two variables: If $f(x, y)$ possesses finite and continuous partial derivatives upto n^{th} the order, then

$$f(a+h,b+k)=f(a,b)+\left(h\frac{\partial}{\partial x}+k\frac{\partial}{\partial y}\right)f(a,b)+\frac{1}{2!}\left(h\frac{\partial}{\partial x}+k\frac{\partial}{\partial y}\right)^2 f(a,b)$$

$$+\frac{1}{3!}\left(h\frac{\partial}{\partial x}+k\frac{\partial}{\partial y}\right)^3 f(a,b)+\ldots\ldots\ldots \quad \ldots (1)$$

OR

$$f(a+h,b+k)=f(a,b)+\left(h\frac{\partial}{\partial x}f(a,b)+k\frac{\partial}{\partial y}f(a,b)\right)$$

$$+\frac{1}{2!}\left(h^2\frac{\partial^2}{\partial x^2}f(a,b)+k^2\frac{\partial^2}{\partial y^2}f(a,b)+2hk\frac{\partial^2}{\partial x\partial y}f(a,b)\right)$$

$$+\frac{1}{3!}\left(h^3\frac{\partial^3}{\partial x^3}f(a,b)+k^3\frac{\partial^3}{\partial y^3}f(a,b)+3h^2k\frac{\partial^3}{\partial x^2\partial y}f(a,b)+3hk^2\frac{\partial^3}{\partial x\partial y^2}f(a,b)\right)+\ldots\ldots.$$

Corollary 1: Putting $h=x-a,\ \ k=y-b$ in (1), we get

$$f(x,y)=f(a,b)+\left[(x-a)\frac{\partial}{\partial x}+(y-b)\frac{\partial}{\partial y}\right]f(a,b)$$

$$+\frac{1}{2!}\left[(x-a)\frac{\partial}{\partial x}+(y-b)\frac{\partial}{\partial y}\right]^2 f(a,b)$$

$$+\frac{1}{3!}\left[(x-a)\frac{\partial}{\partial x}+(y-b)\frac{\partial}{\partial y}\right]^3 f(a,b)+\ldots\ldots\ldots$$

OR

$$f(x,y)=f(a,b)+\left[(x-a)f_x(a,b)+(y-b)f_y(a,b)\right]$$

$$+\frac{1}{2!}\left[(x-a)^2 f_{xx}(a,b)+2(x-a)(y-b)f_{xy}(a,b)+(y-b)^2 f_{yy}(a,b)\right]$$

$$+\frac{1}{3!}\Big[(x-a)^3 f_{xxx}(a,b)+(y-b)^3 f_{yyy}(a,b)+3(x-a)^2(y-b)f_{xxy}(a,b)$$

$$+3(x-a)(y-b)^2 f_{xyy}(a,b)\Big]+\ldots\ldots\ldots\ldots \qquad \ldots(2)$$

It is called ***Taylor's series*** of $f(x, y)$ about the point (a, b).

Corollary 2: Putting $a=0, b=0$ in eqn. (2), we get

$$f(x,y)=f(0,0)+\left[x\frac{\partial}{\partial x}+y\frac{\partial}{\partial y}\right]f(0,0)+\frac{1}{2!}\left[x\frac{\partial}{\partial x}+y\frac{\partial}{\partial y}\right]^2 f(0,0)$$

$$+\frac{1}{3!}\left[x\frac{\partial}{\partial x}+y\frac{\partial}{\partial y}\right]^3 f(0,0)+\ldots.$$

OR

$$f(x,y)=f(0,0)+\left[xf_x(0,0)+yf_y(0,0)\right]$$

$$+\frac{1}{2!}\left[x^2 f_{xx}(0,0)+2xyf_{xy}(0,0)+y^2 f_{yy}(0,0)\right]$$

$$+\frac{1}{3!}\left[x^3 f_{xxx}(0,0)+y^3 f_{yyy}(0,0)+3x^2 yf_{xxy}(0,0)+3xy^2 f_{xyy}(0,0)\right]+\dots\dots\dots \quad \dots(3)$$

It is called Taylor's series of $f(x, y)$ about the point (0, 0) and is called ***Maclaurin's Theorem*** for two variables.

ILLUSTRATIVE EXAMPLES

Example 1. Expand $e^x \cos y$ in powers of x and y upto third degree term.

Solution. Let

$f(x,y)=e^x\cos y$	$f(0,0)=1$
$f_x(x,y)=e^x\cos y$	$f_x(0,0)=1$
$f_y(x,y)=-e^x\sin y$	$f_y(0,0)=0$
$f_{xx}(x,y)=e^x\cos y$	$f_{xx}(0,0)=1$
$f_{yy}(x,y)=-e^x\cos y$	$f_{yy}(0,0)=-1$
$f_{xy}(x,y)=-e^x\sin y$	$f_{xy}(0,0)=0$
$f_{xxx}(x,y)=e^x\cos y$	$f_{xxx}(0,0)=1$
$f_{yyy}(x,y)=e^x\sin y$	$f_{yyy}(0,0)=0$
$f_{xxy}(x,y)=-e^x\sin y$	$f_{xxy}(0,0)=0$
$f_{xyy}(x,y)=-e^x\cos y$	$f_{xyy}(0,0)=-1$

$\therefore$ By Maclaurin's series, we have

$$f(x,y)=f(0,0)+\left[xf_x(0,0)+yf_y(0,0)\right]$$

$$+\frac{1}{2!}\left[x^2 f_{xx}(0,0)+2xyf_{xy}(0,0)+y^2 f_{yy}(0,0)\right]$$

$$+\frac{1}{3!}\left[x^3 f_{xxx}(0,0)+y^3 f_{yyy}(0,0)+3x^2 yf_{xxy}(0,0)+3xy^2 f_{xyy}(0,0)\right]+\dots\dots\dots$$

$$\Rightarrow\ e^x\cos y=1+[x.1+y.0]+\frac{1}{2!}\left[x^2.1+2xy.0+y^2(-1)\right]$$

$$+\frac{1}{3!}\left[x^3.1+y^3.0+3x^2y.0+3xy^2(-1)\right]+\dots\dots$$

$$=1+x+\frac{1}{2!}\left(x^2-y^2\right)+\frac{1}{3!}\left(x^3-3xy^2\right)+\dots\dots\dots$$

Example 2. Expand e^{x+y} about the point (0, 0) as far as the terms of third degree.

Solution. Let $f(x,y)=e^{x+y}$ $\quad f(0,0)=1$

$f_x(x,y)=e^{x+y}$ $\quad f_x(0,0)=1$

$f_y(x,y)=e^{x+y}$ $\quad f_y(0,0)=1$

Similarly, $f_{xx}=f_{xy}=f_{yy}=f_{xxx}=f_{yyy}=f_{xxy}=f_{xyy}=e^{x+y}$

and hence $f_{xx}(0,0)=f_{xy}(0,0)=f_{yy}(0,0)=f_{xxx}(0,0)=f_{yyy}(0,0)$

$$=f_{xxy}(0,0)=f_{xyy}(0,0)=1$$

∴ By Maclaurin's series, we have

$$f(x,y)=f(0,0)+\left[xf_x(0,0)+yf_y(0,0)\right]$$
$$+\frac{1}{2!}\left[x^2f_{xx}(0,0)+2xyf_{xy}(0,0)+y^2f_{yy}(0,0)\right]$$
$$+\frac{1}{3!}\left[x^3f_{xxx}(0,0)+y^3f_{yyy}(0,0)+3x^2yf_{xxy}(0,0)+3xy^2f_{xyy}(0,0)\right]+\dots\dots\dots$$

$$\Rightarrow e^{x+y}=1+[x.1+y.1]+\frac{1}{2!}\left[x^2.1+2xy.1+y^2.1\right]$$
$$+\frac{1}{3!}\left[x^3.1+y^3.1+3x^2y.1+3xy^2.1\right]+\dots\dots$$
$$=1+(x+y)+\frac{1}{2!}\left(x^2+2xy+y^2\right)+\frac{1}{3!}\left(x^3+y^3+3x^2y+3xy^2\right)+\dots\dots\dots$$

Example 3. Expand $\cos x \sin y$ in powers of x and y upto third degree term.

Solution. Let $f(x,y)=\cos x\sin y$ $\quad f(0,0)=0$

$f_x(x,y)=-\sin x\sin y$ $\quad f_x(0,0)=0$

$f_y(x,y)=\cos x\cos y$ $\quad f_y(0,0)=1$

$f_{xx}(x,y)=-\cos x\sin y$ $\quad f_{xx}(0,0)=0$

$f_{yy}(x,y)=-\cos x\sin y$ $\quad f_{yy}(0,0)=0$

$f_{xy}(x,y)=-\sin x\cos y$ $\quad f_{xy}(0,0)=0$

$f_{xxx}(x,y)=\sin x\sin y$ $\quad f_{xxx}(0,0)=0$

$f_{yyy}(x,y)=-\cos x\cos y$ $\quad f_{yyy}(0,0)=-1$

$f_{xxy}(x,y)=-\cos x\ \cos y$ $\quad f_{xxy}(0,0)=-1$

$f_{xyy}(x,y)=\sin x\sin y$ $\quad f_{xyy}(0,0)=0$

∴ By Maclaurin's series, we have

$$f(x,y)=f(0,0)+\left[xf_x(0,0)+yf_y(0,0)\right]$$
$$+\frac{1}{2!}\left[x^2f_{xx}(0,0)+2xyf_{xy}(0,0)+y^2f_{yy}(0,0)\right]$$
$$+\frac{1}{3!}\left[x^3f_{xxx}(0,0)+y^3f_{yyy}(0,0)+3x^2yf_{xxy}(0,0)+3xy^2f_{xyy}(0,0)\right]+\dots\dots\dots$$

Therefore,

$$\cos x \sin y = 0 + [x.0 + y.1] + \frac{1}{2!}\left[x^2.0 + 2xy.0 + y^2.0\right]$$

$$+\frac{1}{3!}\left[x^3.0 + y^3.(-1) + 3x^2y.(-1) + 3xy^2.0\right] + ...$$

$$= y - \frac{1}{3!}\left(y^3 + 3x^2y\right) + ...$$

Example 4. Expand $x^2 - y^2 + xy$ in powers of $(x - 1)$ and $(y + 2)$ upto second degree term.

Solution. Let

$$\begin{aligned}
f(x,y) &= x^2 - y^2 + xy & f(1,-2) &= -5 \\
f_x(x,y) &= 2x + y & f_x(1,-2) &= 0 \\
f_y(x,y) &= -2y + x & f_y(1,-2) &= 5 \\
f_{xx}(x,y) &= 2 & f_{xx}(1,-2) &= 2 \\
f_{xy}(x,y) &= 1 & f_{xy}(1,-2) &= 1 \\
f_{yy}(x,y) &= -2 & f_{yy}(1,-2) &= -2
\end{aligned}$$

$\therefore$ By Taylor's series, we have

$$f(x,y) = f(a,b) + \left[(x-a)f_x(a,b) + (y-b)f_y(a,b)\right]$$

$$+\frac{1}{2!}\left[(x-a)^2 f_{xx}(a,b) + 2(x-a)(y-b)f_{xy}(a,b)\right.$$

$$\left. +(y-b)^2 f_{yy}(a,b)\right] + ...$$

Therefore,

$$x^2 - y^2 + xy = -5 + \left[(x-1).0 + (y+2).5\right]$$

$$+\frac{1}{2!}\left[(x-1)^2.2 + 2(x-1)(y+2).1 + (y+2)^2(-2)\right] + ...$$

$$= -5 + 5(y+2) + (x-1)^2 + (x-1)(y+2) - (y+2)^2 + ...$$

Example 5. Expand $f(x,y) = \tan^{-1}\dfrac{y}{x}$ about the point (1, 1) upto the second degree terms. Hence, evaluate $f(1.1, 0.9)$.

Solution. We have

$$f(x,y) = \tan^{-1}\frac{y}{x} \qquad\qquad f(1,1) = \tan^{-1}1 = \frac{\pi}{4}$$

$$f_x(x,y)=\frac{1}{1+\frac{y^2}{x^2}}\left(-\frac{y}{x^2}\right)=\frac{-y}{x^2+y^2} \qquad f_x(1,1)=-\frac{1}{2}$$

$$f_y(x,y)=\frac{1}{1+\frac{y^2}{x^2}}\left(\frac{1}{x}\right)=\frac{x}{x^2+y^2} \qquad f_y(1,1)=\frac{1}{2}$$

$$f_{xx}(x,y)=\frac{2xy}{\left(x^2+y^2\right)^2} \qquad f_{xx}(1,1)=\frac{1}{2}$$

$$f_{yy}(x,y)=-\frac{2xy}{\left(x^2+y^2\right)^2} \qquad f_{yy}(1,1)=-\frac{1}{2}$$

$$f_{xy}(x,y)=\frac{y^2-x^2}{\left(x^2+y^2\right)^2} \qquad f_{xy}(1,1)=0$$

By Taylor's series, we have

$$f(x,y)=f(1,1)+\left[(x-1)f_x(1,1)+(y-1)f_y(1,1)\right]$$

$$+\frac{1}{2!}\left[(x-1)^2 f_{xx}(1,1)+2(x-1)(y-1)f_{xy}(1,1)\right.$$

$$\left.+(y-1)^2 f_{yy}(1,1)\right]+\dots\dots\dots\dots$$

$$\Rightarrow \quad \tan^{-1}\frac{y}{x}=\frac{\pi}{4}+\left[(x-1)\left(-\frac{1}{2}\right)+(y-1).\frac{1}{2}\right]$$

$$+\frac{1}{2!}\left[(x-1)^2.\frac{1}{2}+2(x-1)(y-1).0\right.$$

$$\left.+(y-1)^2\left(-\frac{1}{2}\right)\right]+\dots\dots\dots\dots$$

$$\Rightarrow \quad \tan^{-1}\frac{y}{x}=\frac{\pi}{4}-\frac{1}{2}(x-1)+\frac{1}{2}(y-1)+\frac{1}{4}(x-1)^2-\frac{1}{4}(y-1)^2+\dots\dots\dots \qquad \dots.(1)$$

Now putting $x=1.1$, $y=0.9$ so that $x-1=0.1$, $y-1=-0.1$ in (1), we have

$$f(1.1,0.9)=\frac{\pi}{4}-\frac{1}{2}(0.1)+\frac{1}{2}(-0.1)+\frac{1}{4}(0.1)^2-\frac{1}{4}(-0.1)^2+\dots\dots\dots$$

$$=0.6854 \text{ approximately.}$$

Example 6. Expand $e^x \log(1+y)$ in the neighbourhood of the point (0, 0) upto the first six terms.

Solution. Let

$$f(x,y)=e^x \log(1+y) \qquad f(0,0)=0$$

$$f_x(x,y)=e^x \log(1+y) \qquad f_x(0,0)=0$$

$$f_y(x,y)=\frac{e^x}{1+y} \qquad f_y(0,0)=1$$

$$f_{xx}(x,y)=e^x \log(1+y) \qquad f_{xx}(0,0)=0$$

$$f_{yy}(x,y)=-\frac{e^x}{(1+y)^2} \qquad f_{yy}(0,0)=-1$$

$$f_{xy}(x,y)=\frac{e^x}{1+y} \qquad f_{xy}(0,0)=1$$

By Maclaurin's series, we have

$$f(x,y)=f(0,0)+\left[xf_x(0,0)+yf_y(0,0)\right]$$

$$+\frac{1}{2!}\left[x^2 f_{xx}(0,0)+2xyf_{xy}(0,0)+y^2 f_{yy}(0,0)\right]+.....$$

$$\Rightarrow \quad e^x \log(1+y)=0+[x.0+y.1]+\frac{1}{2!}\left[x^2.0+2xy.1+y^2(-1)\right]+.....$$

$$=y+xy-\frac{y^2}{2} \text{ approximately.}$$

Example 7. Find Taylor's expansion of the function $f(x,y)=e^{xy}$ in powers of $(x-1)$ and $(y-1)$ upto third degree terms.

Solution.

$$f(x,y)=e^{xy} \qquad f(1,1)=e$$

$$f_x(x,y)=ye^{xy} \qquad f_x(1,1)=e$$

$$f_y(x,y)=xe^{xy} \qquad f_y(1,1)=e$$

$$f_{xx}(x,y)=y^2e^{xy} \qquad f_{xx}(1,1)=e$$

$$f_{yy}(x,y)=x^2e^{xy} \qquad f_{yy}(1,1)=e$$

$$f_{xy}(x,y)=e^{xy}+xye^{xy} \qquad f_{xy}(1,1)=2e$$

$$f_{xxx}(x,y)=y^3e^{xy} \qquad f_{xxx}(1,1)=e$$

$$f_{yyy}(x,y)=x^3e^{xy} \qquad f_{yyy}(1,1)=e$$

$$f_{xxy}(x,y)=2ye^{xy}+xy^2e^{xy} \qquad f_{xxy}(1,1)=3e$$

$$f_{xyy}(x,y)=2xe^{xy}+yx^2e^{xy} \qquad f_{xyy}(1,1)=3e$$

By Taylor's series, we have

$$f(x,y)=f(a,b)+\left[(x-a)\frac{\partial}{\partial x}+(y-b)\frac{\partial}{\partial y}\right]f(a,b)$$

$$+\frac{1}{2!}\left[(x-a)\frac{\partial}{\partial x}+(y-b)\frac{\partial}{\partial y}\right]^2 f(a,b)$$

$$+\frac{1}{3!}\left[(x-a)\frac{\partial}{\partial x}+(y-b)\frac{\partial}{\partial y}\right]^3 f(a,b)+\ldots\ldots\ldots$$

$$\Rightarrow e^{xy}=e+\left[(x-1)e+(y-1)e\right]$$

$$+\frac{1}{2!}\left[(x-1)^2 e+(y-1)^2 e+2(x-1)(y-1)2e\right]$$

$$+\frac{1}{3!}\left[(x-1)^3 e+(y-1)^3 e+3(x-1)^2(y-1)3e+3(x-1)(y-1)^2 3e\right]+\ldots.$$

$$=e\left[1+(x-1)+(y-1)+\frac{(x-1)^2}{2}+\frac{(y-1)^2}{2}+2(x-1)(y-1)\right.$$

$$\left.+\frac{(x-1)^3}{6}+\frac{(y-1)^3}{6}+\frac{3}{2}(x-1)^2(y-1)+\frac{3}{2}(x-1)(y-1)^2+\ldots\right]$$

Example 8. Expand sin xy in powers of $(x-1)$ and $\left(y-\frac{\pi}{2}\right)$ upto third degree terms.

Solution. Let $f(x,y)=\sin xy$ $\qquad f\left(1,\frac{\pi}{2}\right)=\sin\frac{\pi}{2}=1$

$$f_x(x,y)=y\cos xy \qquad f_x\left(1,\frac{\pi}{2}\right)=\frac{\pi}{2}\cos\frac{\pi}{2}=0$$

$$f_y(x,y)=x\cos xy \qquad f_y\left(1,\frac{\pi}{2}\right)=1.\cos\frac{\pi}{2}=0$$

$$f_{xx}(x,y)=-y^2\sin xy \qquad f_{xx}\left(1,\frac{\pi}{2}\right)=-\left(\frac{\pi}{2}\right)^2$$

$$f_{xy}(x,y)=\cos xy-xy\sin xy \qquad f_{xy}\left(1,\frac{\pi}{2}\right)=-\frac{\pi}{2}$$

$$f_{yy}(x,y)=-x^2\sin xy \qquad f_{yy}\left(1,\frac{\pi}{2}\right)=-1$$

By Taylor's series, we get

$$\sin xy = 1 + \left[(x-1).0 + \left(y - \frac{\pi}{2}\right).0\right]$$

$$+\frac{1}{2!}\left[-(x-1)^2\left(\frac{\pi}{2}\right)^2 + \left(y-\frac{\pi}{2}\right)^2(-1) - 2(x-1)\left(y-\frac{\pi}{2}\right)\frac{\pi}{2}\right] + \ldots.$$

$$= 1 - \frac{\pi^2}{8}(x-1)^2 - \frac{1}{2}\left(y-\frac{\pi}{2}\right)^2 - \frac{\pi}{2}(x-1)\left(y-\frac{\pi}{2}\right) + \ldots\ldots.$$

Example 9. Expand x^y about the point (1, 1) upto the term containing third degree.

Solution. Let

$$f(x,y) = x^y \qquad f(1,1) = 1$$

$$f_x(x,y) = yx^{y-1} \qquad f_x(1,1) = 1$$

$$f_y(x,y) = x^y \log x \qquad f_y(1,1) = 0$$

$$f_{xx}(x,y) = y(y-1)x^{y-2} \qquad f_{xx}(1,1) = 0$$

$$f_{yy}(x,y) = x^y(\log x\ )^2 \qquad f_{yy}(1,1) = 0$$

$$f_{xy}(x,y) = x^{y-1} + yx^{y-1}\log x \qquad f_{xy}(1,1) = 1$$

$$f_{xxx}(x,y) = y(y-1)(y-2)x^{y-3} \qquad f_{xxx}(1,1) = 0$$

$$f_{yyy}(x,y) = x^y(\log x\ )^3 \qquad f_{yyy}(1,1) = 0$$

$$f_{xxy}(x,y) = y(y-1)x^{y-2}\log x + (2y-1)x^{y-2} \qquad f_{xxy}(1,1) = 1$$

$$f_{xyy}(x,y) = yx^{y-1}(\log x)^2 + 2\log x.\frac{1}{x}x^y \qquad f_{xyy}(1,1) = 0$$

By Taylor's series, we get

$$x^y = 1 + [(x-1).1 + (y-1).0]$$

$$+\frac{1}{2!}\left[(x-1)^2.0 + (y-1)^2.0 + 2(x-1)(y-1).1\right] + \ldots.$$

$$+\frac{1}{3!}\left[(x-1)^3.0 + (y-1)^3.0 + 3(x-1)^2(y-1).1 + 3(x-1)(y-1)^2.0\right] + \ldots\ldots\ldots$$

$$\Rightarrow x^y = 1 + (x-1) + (x-1)(y-1) + \frac{1}{2}(x-1)^2(y-1) + \ldots\ldots.$$

EXERCISE 1.7

1. Expand $e^x \cos y$ in powers of x and y upto second degree terms.
2. Expand $\cos x \cos y$ about origin upto third degree terms.
3. Prove that : $e^y \log(1+x) = x + xy - \dfrac{x^2}{2}$ approximately
4. Expand $x^2y + \sin y + e^x$ in powers of $(x-1)$ and $(y-\pi)$ upto second degree terms.
5. Expand the function $f(x,y) = x^3 + 3x^2y^2 + 4xy^2 + y^3$ about the point (1, 1).
6. Expand $f(x,y) = \tan^{-1}(xy)$ about the point (1, –1) and hence compute $f(0.9,-1.2)$.
7. Expand $e^{2x}\cos 3y$ in powers of x and y upto quadratic terms.
8. Expand y^x by using Maclaurin's Theorem.
9. Expand $x^2 + y^2 + xy$ in powers of $(x-2)$ and $(y-3)$ upto the second degree terms.

ANSWERS

1. $1 + x + \dfrac{x^2}{2} - \dfrac{y^2}{2} + \ldots\ldots$

2. $1 - \dfrac{x^2}{2} - \dfrac{y^2}{2} + \dfrac{x^4}{24} + \dfrac{x^2y^2}{24} + \dfrac{y^4}{24} + \ldots\ldots\ldots$

4. $\pi + e + (2\pi + e)(x-1) + \dfrac{1}{2}(2\pi + e)(x-1)^2 + 2(x-1)(y-\pi) + \ldots\ldots.$

5. $9 + 13(x-1) + 14(y-1) + 6(x-1)^2 + 7(y-1)^2 + 14(x-1)(y-1) + (x-1)^3$
$+3(x-1)^2(y-1) + 4(x-1)(y-1)^2 + (y-1)^3$

6. $-\dfrac{\pi}{4} - \dfrac{1}{2}(x-1) + \dfrac{1}{2}(y+1) + \dfrac{1}{4}(x-1)^2 + \dfrac{1}{4}(y+1)^2 + \ldots\ldots\ldots; -0.8229.$

7. $1 + 2x + 2x^2 - \dfrac{9}{2}y^2 + \ldots\ldots.$

8. $1 + (y-1) + (x-1)(y-1) + \dfrac{1}{2}(y-1)^2(x-1) + \ldots\ldots.$

9. $19 + 7(x-2) + 8(y-3) + (x-2)^2 + (y-3)^2 + (x-2)(y-3) + \ldots\ldots\ldots.$

1.18 CURVE TRACING

It is a process of determining the approximate shape of the curve without plotting a large number of points.

Important Points

(*i*) **Singular point :** A point on a curve at which the curve exhibits unusual behaviour is called a singular point.

(*ii*) **Multiple point :** A point through which a curve passes more than one time.

(*iii*) **Double point :** If a curve passes two times through a point, then this point is called a double point.

(*a*) **Node :** A double point at which two real tangents (not coincident) can be drawn.

(*b*) **Cusp :** A double point is called cusp if the two tangents at it are coincident.

(*iv*) **Point of inflexion :** A point where the curve crosses the tangent is called a point of inflexion.

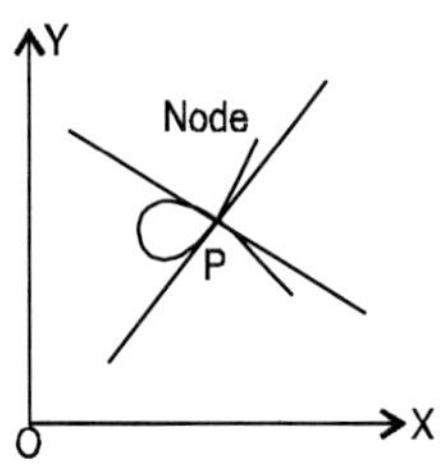

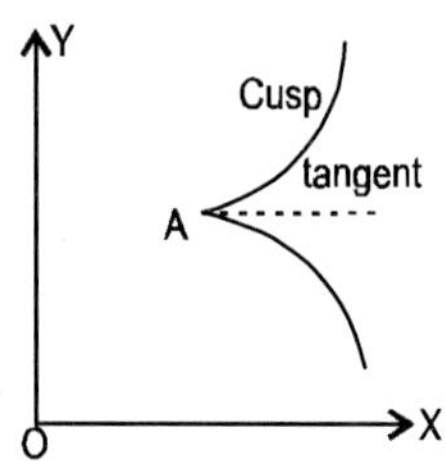

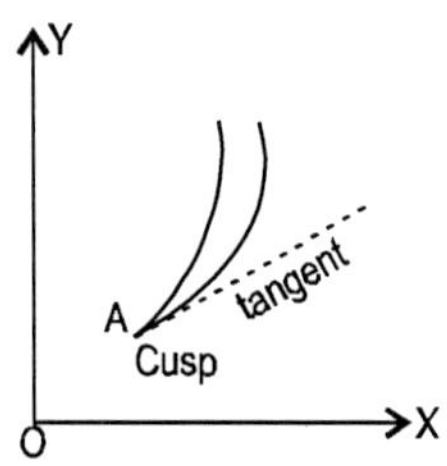

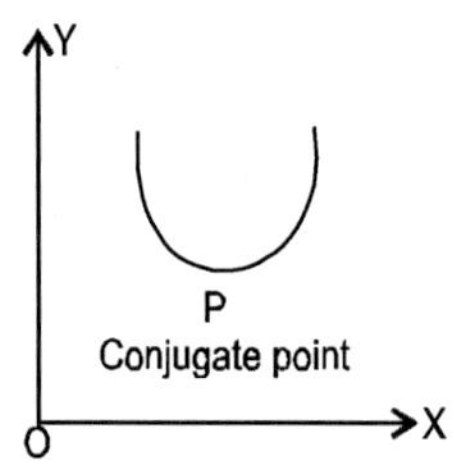

(*v*) **Conjugate point :** This is an isolated point. In its neighbour there is no real point of the curve.

At each double point of the curve $y = f(x)$, we get

$$D = \left(\frac{\partial^2 f}{\partial x \partial y}\right)^2 = \frac{\partial^2 f}{\partial x^2} \times \frac{\partial^2 f}{\partial y^2}$$

(*a*) If D is +ve, double point is a node or conjugate point.

(*b*) If D is zero, double point is a cusp or conjugate point.

(*c*) If D is – ve, double point is a conjugate point.

1.19 METHOD OF TRACING A CURVE HAVING CARTESIAN EQUATIONS

1. Symmetry

(*i*) The curve is symmetrical about the *x-axis*, if the equation of the curve remains unchanged when y is replaced by $-y$ *i.e.*, if the equation of the curve contains only even powers of y.

e.g., $y^2 = 4ax$ is symmetrical about x-axis where as $y^2 + 2xy = x^2$ is not symmetrical about x-axis.

(*ii*) The curve is symmetrical about the y-axis if the equation of the curve does not change when x is replaced by $-x$ *i.e.*, if the curve contains only even power of x.

e.g., $x^2 = 4ay$, but $y = x^3 + ax^2$ is not symmetrical about y-axis.

(*iii*) The curve is symmetrical about the line $y = x$ if the equation of the curve remains unchanged when x is changed to y any y is changed to x.

e.g., $x^3 + y^3 = 3axy$

(*iv*) The curve is symmetrical about the line $y = -x$ if the equation of the curve remains unchanged when x is changed to $-y$ and y is changed to $-x$.

e.g., $x^4 + y^4 = 4a^2xy$

(*v*) The curve is symmetrical in opposite quadrants if the equation of the curve remains unchanged when x is changed to $-x$ and y is changed to $-y$.

e.g., $y = x^3$.

2. Origin

(*a*) **Curve through origin :** The curve passes through origin if the equation does not contain constant term.

e.g., $y^2 = 4ax$ passes through the origin.

(*b*) **Tangent at the origin :** To know the nature of a multiple point it is necessary to find the tangent at that point.

The equation of the tangent at the origin can be obtained by equating to zero, the lowest degree term in the equation of the curve.

3. Point of intersection with the axes

Find intersections of the curve

(*i*) with the x-axis ($y = 0$);

(*ii*) with the y-axis ($x = 0$)

(*iii*) with the line $y = x$ particularly if the curve is symmetrical about it.

(*iv*) with the line $y = -x$ particularly if the curve is symmetrical about it.

(*v*) with the asymptotes (if necessary).

4. Region

Find regions in the four quadrants to which the curve is limited. This is done by

solving for y or for x separately and considering both positive and negative values of x (or y). Values of x (or y) which make y (or x) imaginary are to be rejected.
Note : If the value of y is imaginary for certain value of x then the curve does not exist for such values.

e.g.,

(1) $y^2 = 4x$

for negative value of x, y is imaginary so there is no curve in second and third quadrant.

(2) $a^2x^2 = y^3 (2a - y)$

(*i*) for $y > 2a$, x is imaginary *i.e.*, there is no curve beyond $y = 2a$.

(*ii*) for negative values of y, x is imaginary *i.e.*, there is no curve in third and fourth quadrant.

5. Asymptotes

Asymptotes are the tangents to the curve at infinity. Find all the asymptotes of the curve. The curve will not go beyond its asymptotes.

(*a*) Asymptotes parallel to x-axis is obtained by equating to zero, the coefficient of the highest power of x.

e.g., $yx^2 + 2x^2 + x + 1 = 0$

i.e., $(y + 2)\, x^2 + x + 1 = 0$

The coefficient of the highest power of x *i.e.*, x^2 is $y + 2 = 0$

$\therefore$ $y + 2 = 0$ is the asymptote parallel to x-axis.

(*b*) Asymptotes parallel to y-axis is obtained by equating to zero, the coefficient of the highest power of y.

e.g., $xy^2 + 4y^2 + x + 3 = 0$

i.e., $(x + 4)\, y^2 + x + 3 = 0$

The coefficient of the highest power of y *i.e.*, y^2 is $x + 4 = 0$

$\therefore$ $x + 4 = 0$ is the asymptote parallel to y-axis.

(*c*) oblique Asymptote $y = mx + c$

(*i*) Find $\phi_n(m)$ by putting $x = 1, y = m$ in the highest degree (n) terms of the equation. Similarly, find $\phi_{n-1}(m)$.

(*ii*) Solve $\phi_n(m) = 0$ for m.

(*iii*) Find c by the formula $c = -\dfrac{\phi_{n-1}(m)}{\phi_n^1(m)}$. If the roots of the m are equal,

then find c by $\dfrac{c^2}{2}\ \phi_n^{11}(m) + c\ \phi_{n-1}^1(m) + \phi_{n-2}(m) = 0$

(*iv*) Obtain the equation of asymptote by putting the value of m and c in $y = m\,x + c$.

Remark

The above steps can be remembered as: POSTER, where

P = point of Intersection, O = origin, S = symmetry, T = tangents., A = Asymptote, R = Region

ILLUSTRATIVE EXAMPLES

Example 1. Trace the curve $ay^2 = x^3$ (Semi-cubical parabola).

Solution. We have $ay^2 = x^3$ ...(1)

1. **Origin :** Since the equation (1) does not contain any constant term, therefore it passes through origin.
2. **Symmetry :** Since the given equation contains even powers of y, therefore it is symmetrical about x-axis.
3. **Tangent at the origin :** Equation of the tangent is obtained by equating to zero the lowest degree terms in the equation (1).

   $$ay^2 = 0$$

   i.e., $y^2 = 0$, $y = 0$ is double point.

   Thus the origin is a double point and it may be a cusp since there are two coincident tangents at the origin.
4. **Point of intersection :** The curve does not intersect the coordinate axes any where except the origin.
5. **Region :** Solving the equation of the curve for y, we get $y^2 = \dfrac{x^3}{a}$.

   When $x = 0$, $y^2 = 0$

   When $x > 0$, y^2 is + *ve i.e.*, y is real.

   Therefore, the curve exists in the region $x > 0$.

   As x increases, y^2 also increases and when $x \longrightarrow \infty$, $y^2 \longrightarrow \infty$.

   When $x < 0$, y^2 is –*ve i.e.*, y is imaginary. Therefore, the curve does not exist in the region $x < 0$.
6. **Asymptotes :** The curve has no asymptotes.

   Thus, the shape of curve is as shown in the figure,

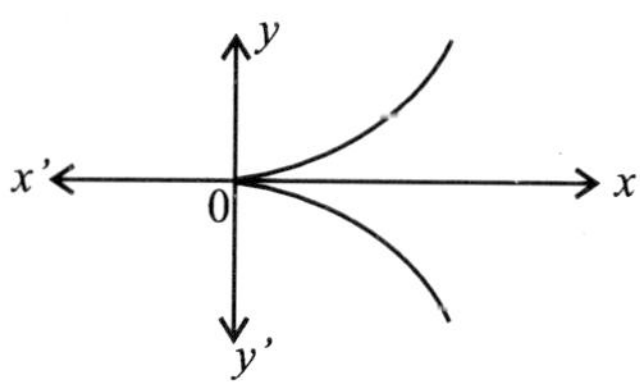

Example 2. Trace the curve $y^2 (2a - x) = x^3$. (cissoid)

Solution. The equation of the curve is

$y^2 (2a - x) = x^3$ ...(*i*)

1. **Origin:** Since the equation (*i*) do not contain any constant term, therefore it passes through origin.
2. **Symmetry:** Equation (*i*) contains only even powers of y, therefore, it is symmetrical about x-axis.

3. **Tangent at the origin:** Equation of the tangent is obtained by equating to zero the lowest degree term in the equation $2ay^2 - xy^2 = x^3$, we get
   $2ay^2 = 0 \Rightarrow y^2 = 0, y = 0$, is double point.
   As two tangents are coincident, therefore, origin is cusp.
4. **Points of intersections:** The curve meets x-axis and y-axis at the origin only.
5. **Region:** From equation (*i*), we have

$$y = x\sqrt{\frac{x}{2a-x}}$$

When $x < 0$, y is imaginary

$\therefore$ No portion of the curve lies to the left of the line $x = 0$ *i.e.*, y-axis.

When $0 < x < 2a$, y is real

When $x > 2a$, y is imaginary

$\therefore$ No portion of the curve lies to the right of the line x.

6. **Asymptotes:** Equating to zero, the coefficient of highest degree term in y (asymptote 11 to y-axis) *i.e.*, y^2, we get
   $2a - x = 0$
   $x = 2a$
   There is no other asymptote of the curve.
   Thus, the shape of curve is as shown in the figure.

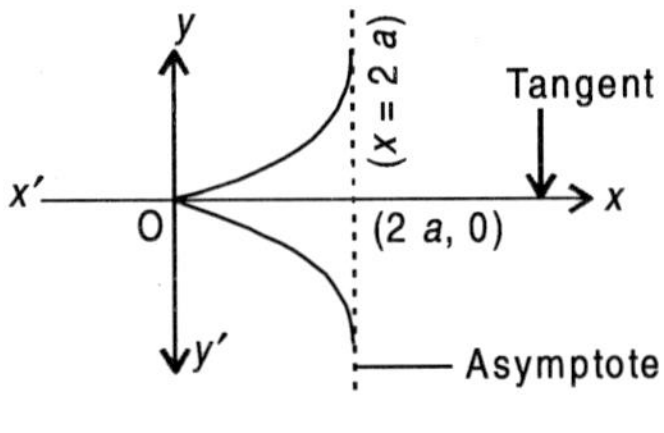

Example 3. Trace the curve $x^2y^2 = x^2 - a^2$.

Solution. We have $x^2y^2 = x^2 - a^2$ ... (*i*)

1. **Origin:** The curve does not pass through the origin because (0, 0) does not satisfy the given equation.
2. **Symmetry:** Since (*i*) contains even powers of x as well as even powers of y, therefore, the curve is symmetrical about both the axes. Also the curve is symmetrical in opposite quadrants.
3. **Point of intersection:**
   With x-axis: Putting $y = 0$ in (*i*) $x^2 = a^2$
   $\therefore x = \pm a$
   $\therefore$ The curve meets the x-axis in the points $(a, 0)$ and $(-a, 0)$.
   Also $x = a$ and $x = -a$ are the tangents at the points $(a, 0)$ and $(-a, 0)$ respectively.
   With y-axis: Putting $x = 0$ in (*i*), we get $0 = -a^2$, which is impossible.
   $\therefore$ The curve does not meet the y-axis.

4. **Asymptotes:** Asymptotes 11 to x-axis are given by $y^2 - 1 = 0$

   $\therefore$ $y = 1$ and $y = -1$ are the asymptotes.

   Asymptotes 11 to y-axis is $x = 0$.

   Oblique asymptotes $\phi_4(m) = m^2$

   $\phi_4(m) = 0$ gives $m = 0$ and this value of m gives asymptotes 11 to x-axis (which we have already obtained).

5. **Region:** From (*i*),

$$y = \sqrt{\frac{x^2 - a^2}{x}} \qquad \ldots (ii)$$

(This value of y corresponds to the part of the curve in Ist-two quadrants).

For y to be real, $x^2 - a^2$ must be +ve *i.e.*, $x^2 > a^2$

$\therefore x > a$ or $x < -a$

$\therefore$No portion of the curve lies between the lines

$x = a$ and $x = -a$ $\qquad \ldots (iii)$

$$\text{Again from } (i),\ x^2 (1 - y^2) = a^2 \Rightarrow x = \frac{a}{\sqrt{1 - y^2}}$$

$\therefore x$ to be real, $1 - y^2 > 0$ *i.e.*, $y^2 < 1$

$\therefore -1 < y < 1$.

$\therefore$ The curve lies entirely between the lines $y = -1$ and $y = 1$. Thus, the shape of the curve is as shown in the figure.

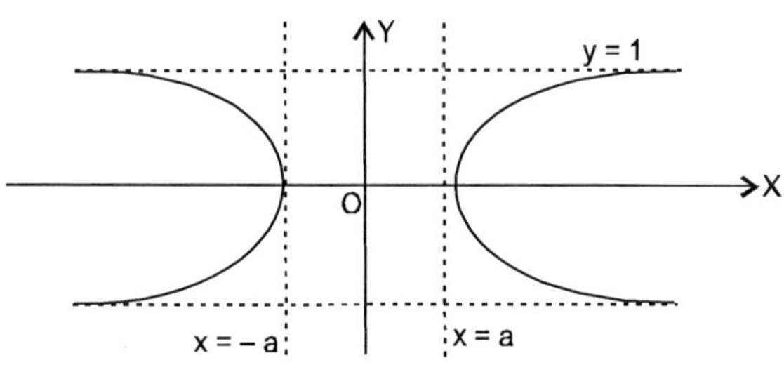

Example 4. Trace the curve $x^{2/3} + y^{2/3} = a^{2/3}$. (Astroid).

Solution. We have $x^{2/3} + y^{2/3} = a^{2/3}$ $\qquad \ldots (i)$

1. **Origin:** The curve does not pass through the origin.
2. **Symmetry:** The curve is symmetrical about both the axes. Also the curve is symmetrical about the lines $y = x$ and $y = -x$.
3. **Tangent at the origin:**

   From (*i*), when $y = 0$, $x = \pm a$.

   Hence, the tangents to the curve are 11 to x-axis at the points $(\pm a, 0)$

   From (*i*), when $x = 0$, $y = \pm a$.

   Hence, the tangents to the curve are 11 to y-axis at the points $(0, \pm a)$
4. **Point of intersection:** The curve meets x-axis at the points $(a, 0)$ and $(-a, 0)$. The curve meets y-axis at the points $(0, a)$ and $(0, -a)$.

5. **Asymptotes:** The curve has no asymptotes.
6. **Region:** From (*i*) $y^{2/3} = a^{2/3} - x^{2/3}$

   If $|x| > a$, $y^{2/3}$ is – ve

   $\therefore$ y^2 is – ve

   $\therefore$ y is imaginary

   Hence, the curve does not lie beyond the lines $y = \pm a$.

   Again, when $x = a$, $y = a$. As x increases from 0 to a, y decreases from a to 0.

   Thus, the shape of the curve is as shown in the figure.

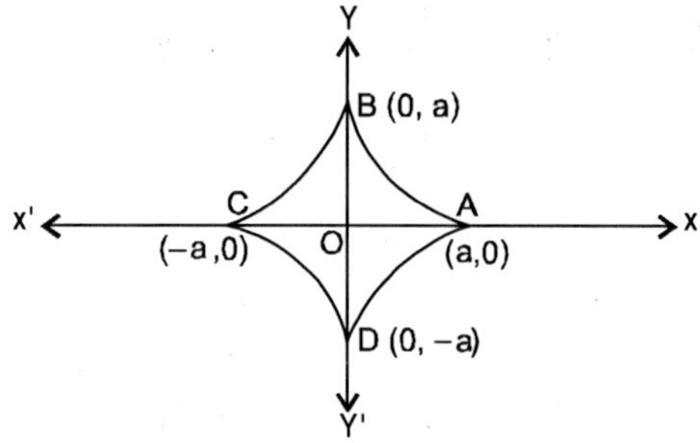

Example 5. Trace the curve $xy^2 = a^2 (a - x)$ [Witch of Ag nesi]

Solution. We have $xy^2 = a^2 (a - x)$ ...(*i*)

1. **Origin:** The curve does not pass through the origin.
2. **Symmetry:** Since (*i*) contains only even powers of *y*. So the curve is symmetrical about *x*-axis.
3. **Axes intersection:** The curve meets *x*-axis at the point A $(a, 0)$. The curve does not meet *y*-axis. Shifting the origin to A $(a, 0)$, equation (*i*) transforms to

   $$(X + a)Y^2 = a^2 [a - (X + a)]$$

   $$\Rightarrow \quad (X + a)Y^2 + a^2 X = 0$$

   Equating to zero, the lowest degree terms, the tangent at the new origin is $X = a$.

   Hence, at the new origin, *i.e.*, at A $(a, 0)$, the tangent is $\perp$ to *x*-axis.
4. **Asymptotes:** The only asymptotes is $x = 0$.
5. **Region:** From (*i*)

   $$y = a \sqrt{\frac{a - x}{x}}$$

   When $x < 0$ or when $x > a$, y is imaginary. Thus, no portion of the curve lies to the left of *y*-axis and to the right of $x = a$.

   When $x = a$. $y = 0$. As x decreases from a to 0, y increases from 0 to ∞.

   Thus, the shape of the curve is as shown in the figure.

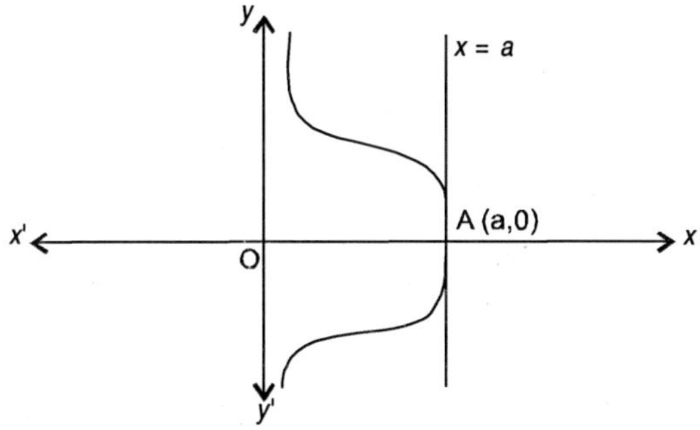

Example 6. Trace the curve $y^2 (a^2 + x^2) = x^2 (a^2 - x^2)$

Solution. $y^2 (a^2 + x^2) = x^2 (a^2 - x^2)$ ... (*i*)

1. **Origin:** The equation of the given curve does not contain constant term, therefore, the curve passes through the origin.
2. **Symmetry:** The equation contains even powers of x as well as y, so the curve is symmetrical about both the axes.
3. **Point of intersection:** Putting $y = 0$ in equation (*i*), we get

   $x^2 (a^2 - x^2) = 0 \Rightarrow x = \pm a, 0, 0$
4. **Tangent at the origin:** Equation of the tangent is obtained by equating to zero the lowest degree term.

   $a^2y^2 - a^2x^2 = 0$

   $\Rightarrow y = \pm x$

   There are two tangents $y = x$ and $y = -x$ at the origin.

   Since, there are two real and different tangents at the origin so origin is the node.
5. **Region:** For values of $x > a$ and $x < -a$, y^2 becomes negative, hence the entire curve remains between $x = -a$ and $x = a$.
6. **Asymptotes:** The given curve is of degree four and so it cannot have more than four asymptotes, real or imaginary.

   The asymptotes 11 to x-axis is given by $-x^4 = 0$

   $x = 0, 0, 0, 0$

   The asymptotes 11 to y-axis is given by

   $x^2 + a^2 = 0$

   $x = \pm ia$

   Thus, the shape of the curve is as shown in the figure.

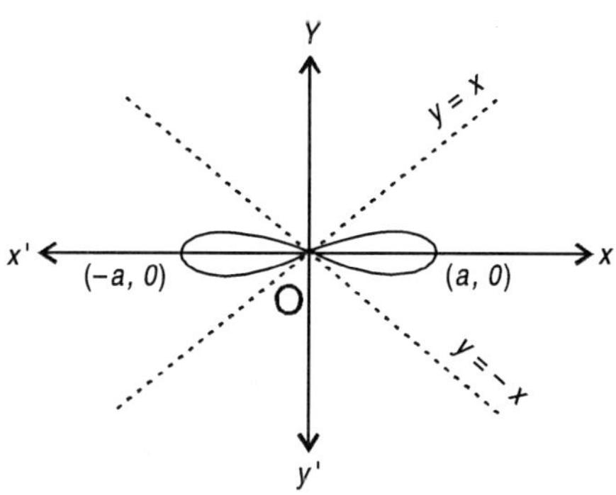

Example 7. Trace the curve $x^{1/2} + y^{1/2} = a^{1/2}$

Solution. We have $x^{1/2} + y^{1/2} = a^{1/2}$ ...(*i*)

1. **Origin:** The curve does not pass through origin.
2. **Symmetry:** The curve is symmetrical about the line $y = x$.
3. **Asymptotes:** No asymptotes parallel to coordinate axes.
4. **Point of intersection:** When $x = 0, y = a$

   When $y = 0, x = a$.

   Hence, the curve meets the coordinate axes at $(0, a)$ and $(a, 0)$
5. **Special points:**

$x:$	a	$a/4$	$a/2$	$3a/4$	0
$y:$	0	$a/4$	$\left(\frac{\sqrt{2}-1}{\sqrt{2}}\right)^2 a$	$\left(\frac{2-\sqrt{3}}{\sqrt{2}}\right)^2 a$	a

Thus, the shape of the curve is as shown in the figure.

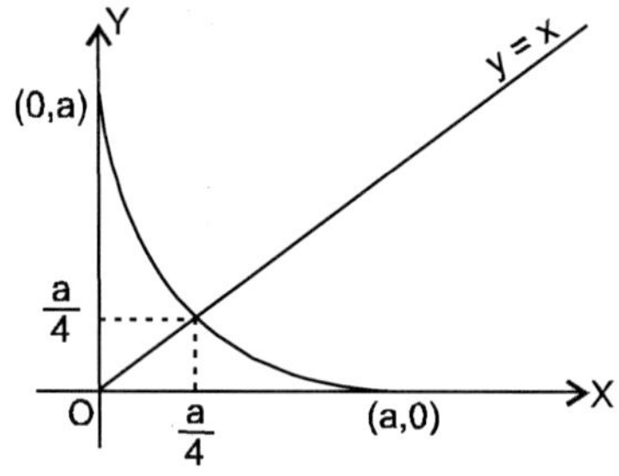

1.20. METHOD OF TRACING A CURVE HAVING PARAMETRIC EQUATIONS

Let $x = f(t)$ and $y = \phi(t)$, where t is the parameter in the equations of the curve.

Case I

If conveniently possible, the parameter is eliminated and the corresponding Cartesian equation is obtained and then the procedure explained earlier is followed.

Case II

If it is not conveniently possible to eliminate the parameter, then the following procedure is followed.

1. **Origin:** If on putting $x = 0$ (or $y = 0$) a real value of t can be found out, which makes $y = 0$ (or $x = 0$), then the curve passes through the origin.

   OR Put both x and y equal to zero and find values of t.

   If there is any common value of t, then the curve passes through the origin.
2. **Symmetry:**

   (*i*) If on changing t to $-t$ (or t to $\pi - t$); x remains unchanged and y changes to $-y$; then the curve is **symmetrical about x-axis.**

   e.g., the parabola $x = at^2$, $y = 2at$ is symmetrical about x-axis.

(*ii*) If on changing t to $-t$ (or t to $\pi - t$), x changes to $-x$ and y remains unchanged, then the curve is **symmetrical about y-axis.**

e.g., (1), the cycloid $x = a(t + \sin t)$, $y = a(1 - \cos t)$ is symmetrical about y-axis.

(2), the ellipse $x = a\cos t$, $y = b\sin t$ is symmetrical about y-axis. (Changing t to $\pi - t$).

(*iii*) If on changing t to $-t$, x changes to $-x$ and y to $-y$, the curve is symmetrical in opposite quadrants.

e.g., the rectangular hyperbola $x = ct$, $y = c/t$ is symmetrical in opposite quadrants.

3. **Asymptotes:** Find asymptotes if any.
4. **Point of intersection:** Find the points of intersection of the curve with coordinate axes.
5. **Region:** If easily possible, find the greatest and least values of x and y and therefore the lines parallel to the axes between which the curve lies or does not lie.

e.g., If the curve is $x = a\cos\theta$, $y = b\sin\theta$; then the greatest value of x is a and the least value is $-a$. ($\because$ greatest and least values of $\cos\theta$ are 1 and -1)

$\therefore$ The curve lies entirely between the lines $x = \pm a$.

Similarly, the curve lies entirely between the lines $y = \pm b$.

ILLUSTRATIVE EXAMPLES

Example 1. Trace the cycloid $x = a(\theta + \sin\theta)$, $y = a(1 + \cos\theta)$.

Solution. The parametric equations of the cycloid are

$$x = a(\theta + \sin\theta),\ y = a(1 + \cos\theta) \qquad \ldots(i)$$

Let us trace this curve firstly for values of θ in the interval $[-\pi, \pi]$

1. **Origin:** Putting $y = 0$, $a(1 + \cos\theta) = 0$

$\therefore$ $\cos\theta = -1$ or $\theta = \pi$

For $\theta = \pi$, $x = a(\theta + \sin\theta) = a\pi \neq 0$.

$\therefore$ The curve does not pass through the origin.

2. **Symmetry:** on changing θ to $-\theta$ in (*i*), x changes to $-x$ and y remains unchanged.

$\therefore$ The curve is symmetrical about y-axis.

3. **Asymptotes:** The curve has no asymptotes.
4. **Points of intersections:**

With x-axis putting $y = 0$ in (*i*), we get $\theta = \pi$ which gives $x = a\pi$.

$\therefore$ Intersection with x-axis is $(a\pi, 0)$.

With y-axis. Putting $x = 0$ in (*i*), we have $\theta + \sin\theta = 0$ which is satisfied by only $\theta = 0$ and for $\theta = 0$, $y = a(1 + \cos\theta) = 2a$.

$\therefore$ Intersection with y-axis is $(0, 2a)$.

5. **Region:** We know that $-1 \le \cos\theta \le 1$

$$1-1 \le (1+\cos\theta) \le 1+1$$

$$0 \le (1+\cos\theta) \le 2$$

$\therefore$ $0 \le a\,(1+\cos\theta) \le 2a$

or $0 \le y \le 2a$

$\therefore$ Curve lies entirely between the lines $y = 0$ and $y = 2a$.

6. **Tangent** from (*i*)

$$\frac{dx}{d\theta} = a\,(1+\cos\theta), \quad \frac{dy}{d\theta} = -a\sin\theta.$$

$$\therefore \quad \frac{dy}{dx} = \frac{dy/d\theta}{dx/d\theta} - \frac{a\sin\theta}{a(1+\cos\theta)} = -\tan\frac{\theta}{2} \qquad \ldots(ii)$$

Corresponding values of x, y and $\frac{dy}{dx}$ for different values of θ are given below :

	$\theta = -\pi$	$\frac{-\pi}{2}$	0	$\frac{\pi}{2}$	π
From (*i*)	$x = -a\pi$	$-a\left(\frac{\pi}{2}+1\right)$	0	$a\left(\frac{\pi}{2}+1\right)$	$a\pi$
	$y = 0$	a	$2a$	a	0
From (*ii*)	$\frac{dy}{dx} = \infty$	1	0	-1	$-\infty$

$\therefore$Points on the curve are $[-a\pi, 0]$, $\left[-a\left(\frac{\pi}{2}+1\right), a\right]$, $[0, 2a]$,

$\left[a\left(\frac{\pi}{2}+1\right), a\right]$, $[a\pi, 0]$.

Tangents at the points $[-a\pi, 0]$ and $[a\pi, 0]$ are parallel to y-axis

$\left(\because \frac{dy}{dx} = \infty \text{ or } -\infty\right)$

Tangent at the point $(0, 2a)$ is || to x-axis $\left(\because \frac{dy}{dx} = 0\right)$

Hence, the shape of the curve is as shown in the figure.

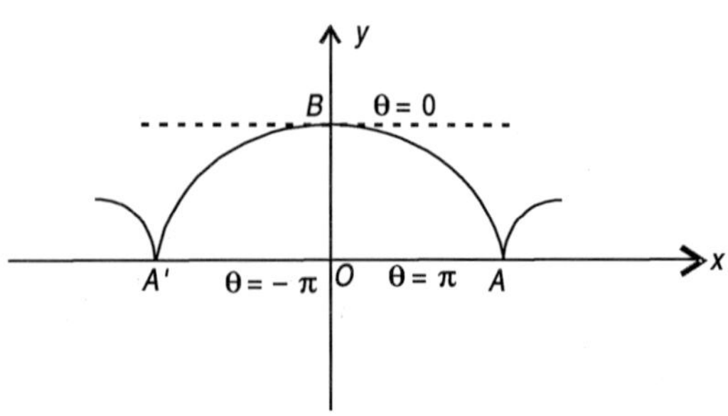

Some Important Curves

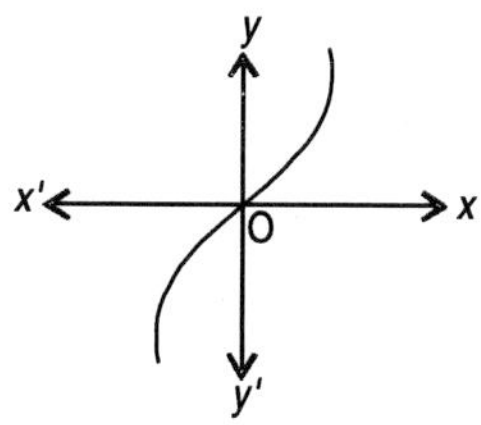

Cubical parabola
$a^2y = x^3$

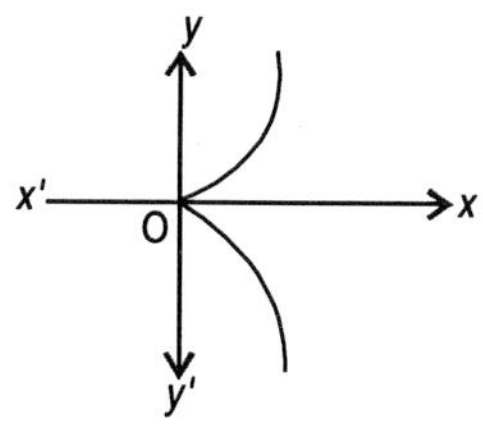

Semi-cubical parabola
$ay^2 = x^3$

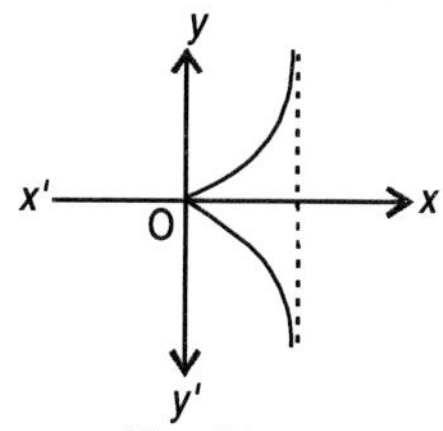

Cissoid
$y^2 (2a - x) = x^3$

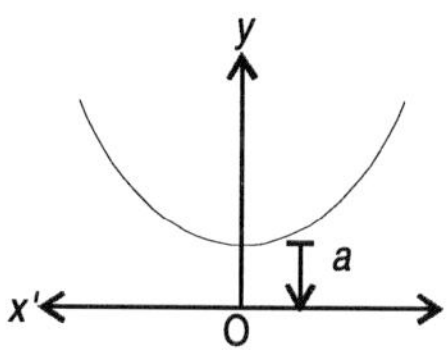

Catenary
$y = a \cos h \dfrac{x}{a}$

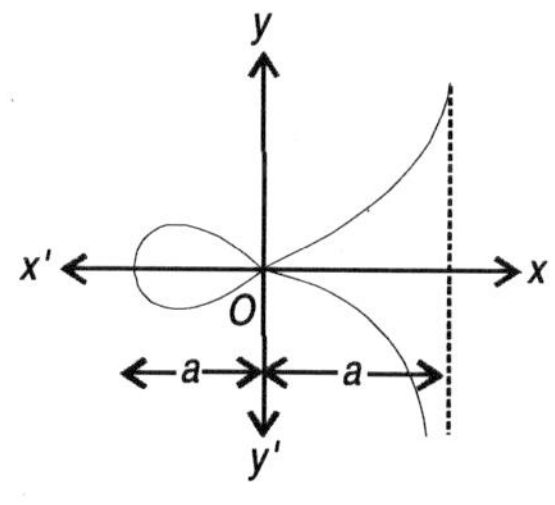

Strophoid
$y^2(a-x) = x^2 (a + x)$

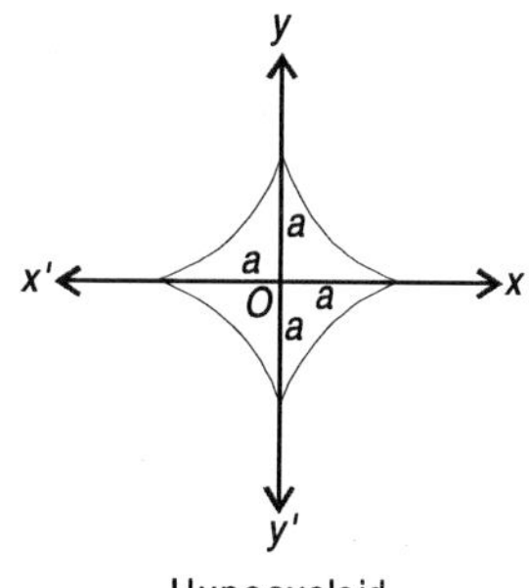

Hypocycloid
$x^{2/3} + y^{2/3} = a^{2/3}$
$x = a \cos^3\theta,\ y = \sin^3\theta$

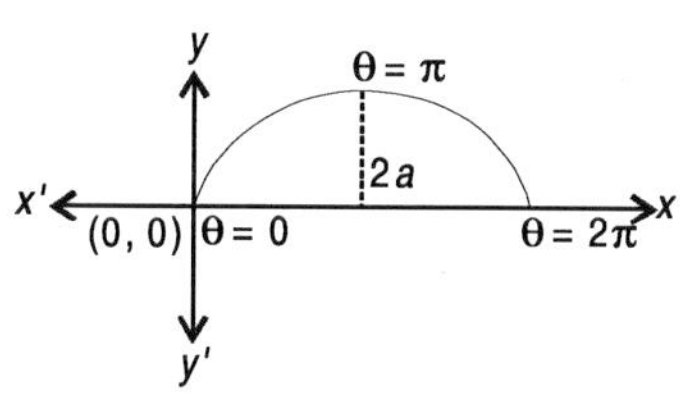

Cycloid
$x = a (\theta - \sin \theta)$
$y = a (1 - \cos \theta)$

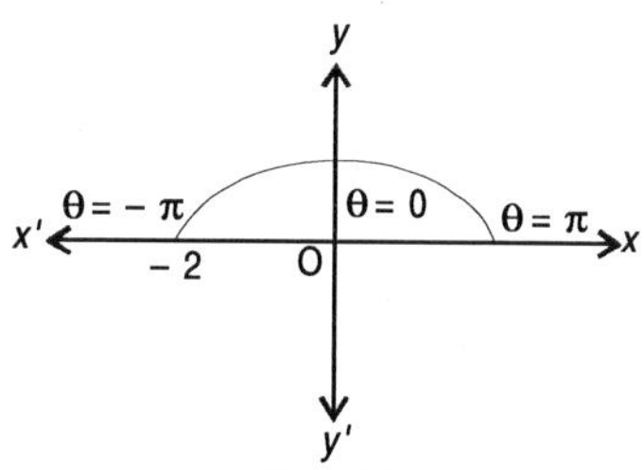

Cycloid
$x = a (\theta + \sin \theta)$
$y = a (1 + \cos \theta)$

EXERCISE 1.8

Trace the following curves:

1. $ay^2 = x^2 (a - x)$

2. $y^2 = x^3$

3. $a^4y^2 = a^2x^4 - x^6$

4. $a^2y^2 = x^2 (a^2 - x^2)$

5. $a^2x^2 = y^3 (2a - y)$

6. $\left(\dfrac{x}{a}\right)^{2/3} + \left(\dfrac{y}{b}\right)^{2/3} = 1$

7. $x = a(\theta + \sin \theta)$
$y = a(1 - \cos \theta)$

ANSWERS

1.

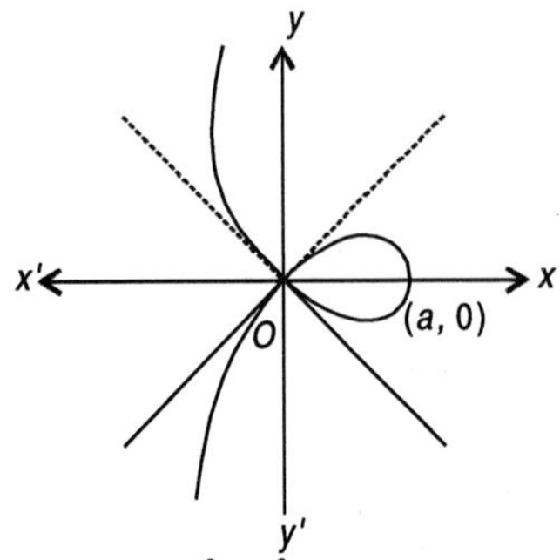

$ay^2 = x^2(a - x)$

2.

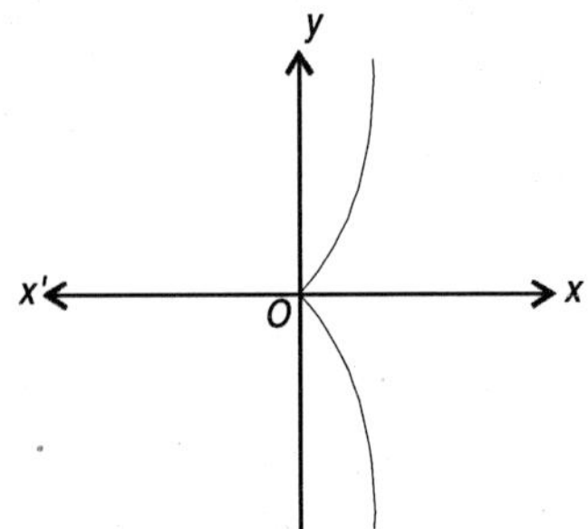

$y^2 = x^3$

Semi-cubical parabola

3.

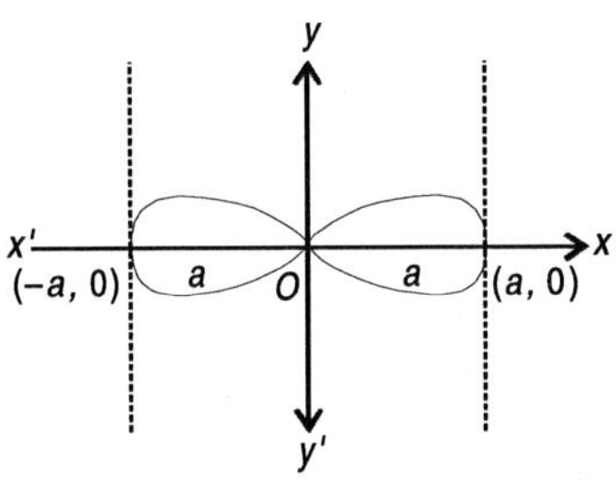

4.

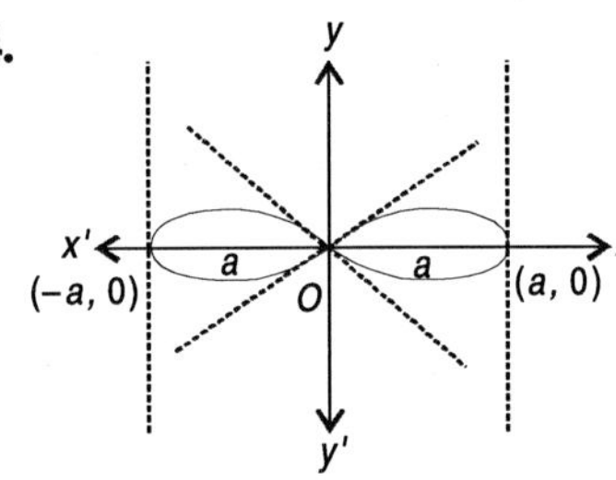

5.

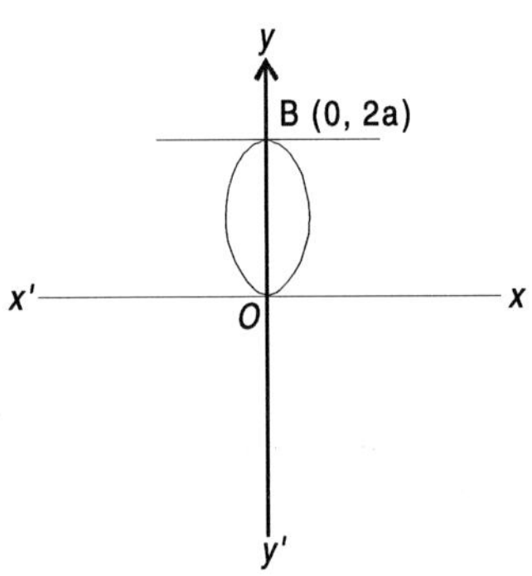

6.

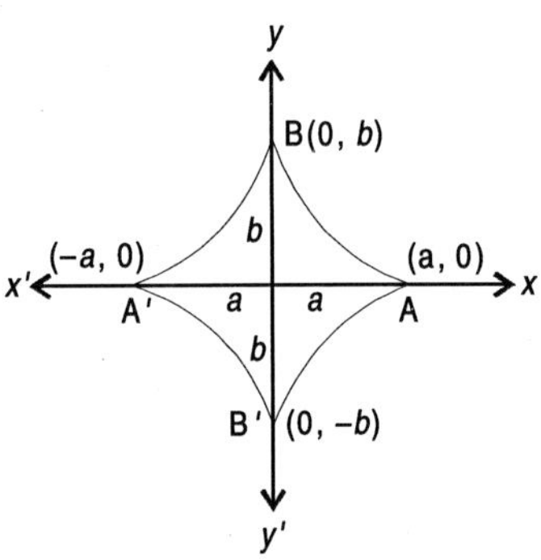

7.

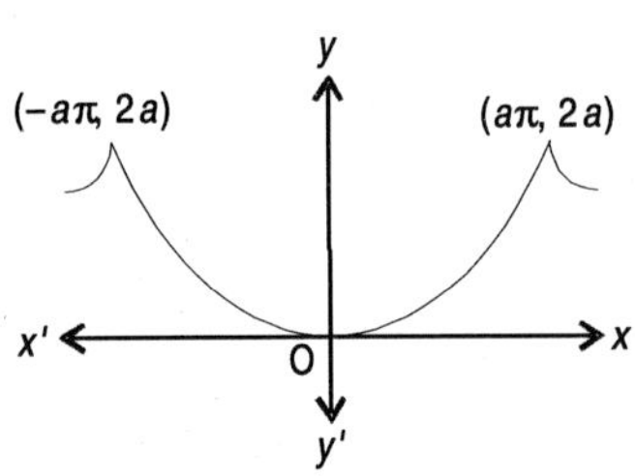

1.21 POLAR COORDINATE

If we have horizontal line OX called the initial line and another line called the revolving line, makes an angle θ with the initial line, then the polar coordinates of a point P are (r, θ).

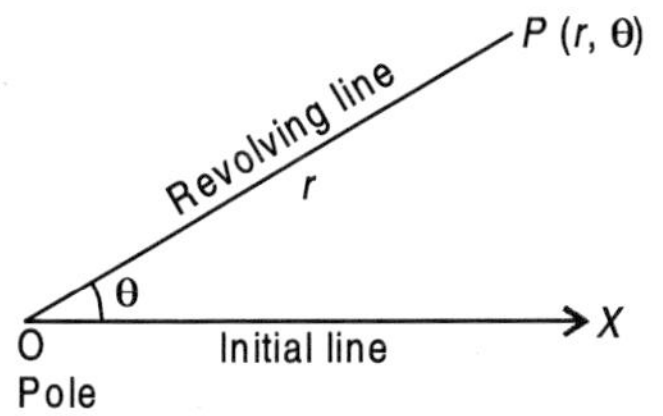

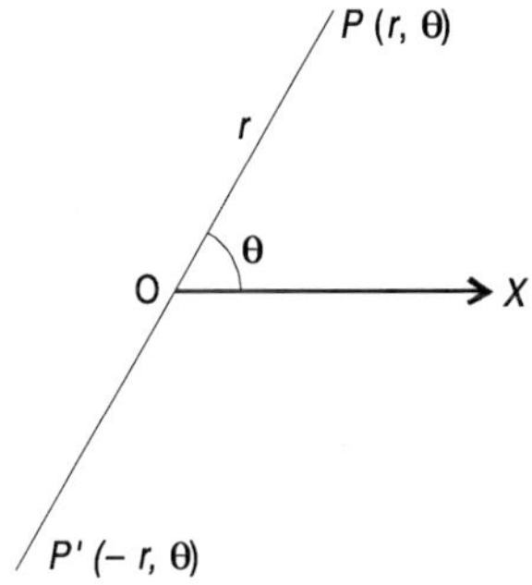

The point O is called the pole and the angle θ is called the vectorial angle of the point P and the length r is called its radius vector.
Polar coordinates of the point P are (r, θ) and polar coordinates of P′ are $(-r, \theta)$ or $(r_1\ \theta + \pi)$.
Polar coordinates of the point P are (r, θ) and Polar coordinate of P′ are $(-r, \theta)$ or $(r, \theta + \pi)$

1.22 METHOD OF TRACING A CURVE HAVING POLAR EQUATIONS

1. **Symmetry:**
   (*i*) The curve is symmetrical about the initial line $\theta = 0$ (*i.e.*, x-axis) if the equation remains unchanged on changing θ to $-\theta$.

   e.g., $r = a\,(1 + \cos\theta)$, $r = a\cos 3\theta$ are symmetrical about the initial line.

   (*ii*) The curve is symmetrical about the pole (origin) if the equation remains unchanged on changing r to $-r$ or θ to $\pi + \theta$. [*i.e.*, the curve is symmetrical about the pole (origin) if the powers of r are even.]

   e.g., $r^2 = a^2 \cos 2\theta$.

   (*iii*) If on changing θ to $-\theta$ and r to $-r$ simultaneously, the equation remains unchanged, the curve is symmetrical about a line passing through pole and ⊥ to the initial line

   e.g., $r^2 = a^2 \cos 2\,\theta$.

   If on changing θ by $\pi - \theta$, the equations remains unchanged the above symmetry exists.

   e.g., $r = a \sin\theta$.

2. **Pole or origin:**
   (*i*) Find whether the curve passes through the pole or not. This can be done by putting $r = 0$ in the equation and if on doing so, we get some real value (or values) of θ, then the curve passes through the pole.

   If it is not possible to find real value of θ for which $r = 0$, the curve does not pass through the pole.

(*ii*) Find the tangents at the pole. Putting $r = 0$, the real values of θ give the tangents at the pole.

e.g., consider the curve $r = a(1 - \cos\theta)$

Putting $r = 0$, we get $\cos\theta = 1 \Rightarrow \theta = 0$.

Hence the curve passes through the pole and the line $\theta = 0$. *i.e.*, the initial line is the tangent at the pole.

Again consider $r = a\,(3 - \sin\theta)$.

Putting $r = 0$. we get $\sin\theta = 3$ which does not give any real value of θ. $[\because |\sin\theta| \le 1]$

Hence, the curve does not pass through the pole.

(*iii*) Find the points where the curve cuts the initial line and the line $\theta = \frac{\pi}{2}$.

3. **Asymptotes:** If $\theta \longrightarrow \theta_1$ (some finite value) when $r \longrightarrow \infty$, then there is an asymptote. We find all the asymptotes of the curve.
4. **Points of Intersection:** Find some points on the curve for convenient values of θ.
5. **Region:**

(*i*) Solve the given equation for r or θ (if possible). Find the regions in which the curve does not lie. If for a certain value of θ, r^2 is negative, the curve does not lie there.

(*ii*) Find the greatest or least values of r, so the curve lies within or without a certain circle. If the greatest and least numerical values of r be a and b respectively, the curve lies entirely within the circle $r = a$ and entirely outside the circle $r = b$.

(*iii*) Trace the variations of r when θ varies in the intervals $(0, \infty)$ and $(-\infty, 0)$ making the values of θ for which $r = 0$ or attains maximum and minimum values. Plot the points so obtained.

Value of ϕ. Find $\tan\phi = r\,\frac{d\theta}{dr}$ and ϕ. (ϕ is the angle between the tangent and radius vector). Find the points where ϕ is 0 or $\frac{\pi}{2}$.

Again, r increases as θ increases if $\frac{dr}{d\theta}$ is positive and r deceases as θ increases if $\frac{dr}{d\theta}$ is negative.

ILLUSTRATIVE EXAMPLES

Example 1. Trace the curve $r = a\,(1 - \cos\theta)$. [Cardoid]

Solution. We have $r = a\,(1 - \cos\theta)$ ...(*i*)

1. **Symmetry:** The curve is symmetrical about the initial line because the equation of curve remains unchanged when θ is changed to $-\theta$.

2. Pole or origin:

(*i*) When $\theta = 0$, $r = 0$, hence the curve passes through the pole and the tangent at the pole is the line $\theta = 0$, *i.e.*, the initial line.

(*ii*) The curve meets the initial line $\theta = 0$ at (0, 0) and the lines $\theta = \frac{\pi}{2}$ and π in the points $\left(a, \frac{\pi}{2}\right)$ and $(2a, \pi)$ respectively.

3. Asymptotes: Since for any finite value of θ, r does not tend to infinity. Therefore, the curve has no asymptote.

4. Region:

(*i*) $\because |\cos\theta| \leq 1$

$\therefore$ from (*i*), $r \leq 2a$

$\therefore$ The curve lies entirely within the circle $r = 2a$.

(*ii*) When θ increases from 0 to π, r remains +ve and increases from 0 to $2a$ when θ increases from π to 2π, r is +ve and decreases from $2a$ to 0.

5. Value of ϕ:

From (*i*), $\frac{dr}{d\theta} = a\sin\theta.$

$$\therefore \tan\phi = r\frac{d\theta}{dr} = \frac{r}{\frac{dr}{d\theta}} = \frac{a(1-\cos\theta)}{a\sin\theta} = \tan\theta/2$$

$\therefore \phi = \theta/2$

Now $\phi = 0$ when $\theta = 0$ and $\phi = \frac{\pi}{2}$ when $\theta = \pi$.

$\therefore$ At $(2\,a, \pi)$ the tangent is $\perp$ to the line $\theta = \pi$.

6. Special points:

The corresponding values of θ and r are given below :

θ :	0	$\pi/3$	$\pi/2$	$2\pi/3$	π
r :	0	$a/2$	a	$3a/2$	$2a$

[We need not trace the curve for values of θ from π to 2π as the curve is symmetrical about the initial line]

Thus, the shape of the curve is as shown in the figure.

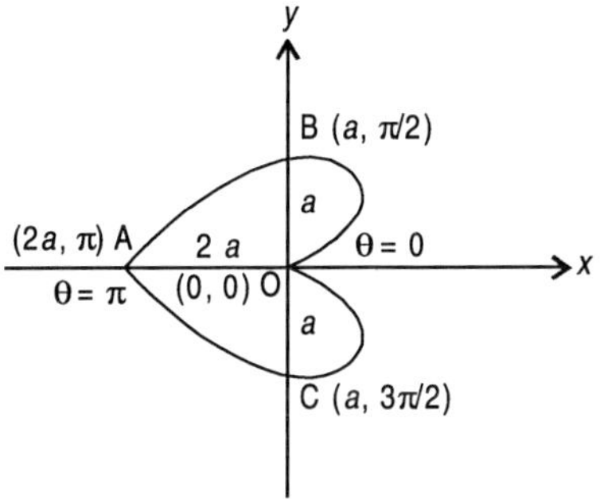

Example 2. Trace the curve $r^2 = a^2 \cos 2\,\theta$. [Bernoulli's Lemniscate]

Solution. We have $r^2 = a^2 \cos 2\,\theta$ ...(*i*)

1. **Symmetry:** The curve is symmetrical about the initial line the line $\theta = \frac{\pi}{2}$ and pole.
2. **Pole or origin:** Put $r = 0$ in (*i*), we get

   $$\cos 2\theta = 0 = \cos\left(\pm\frac{\pi}{2}\right)$$

   $$\Rightarrow 2\,\theta = \pm\frac{\pi}{2}$$

   $$\Rightarrow \theta = \pm\frac{\pi}{4}.\ \text{which are real.}$$

   Hence, the curve passes through the pole and the tangents at the pole are

   $$\theta = \pm\frac{\pi}{4}$$

   When $\theta = 0$, $r = \pm a$

   Hence, the curve meets the initial at $(\pm a, 0)$
3. **Asymptotes:** The curve has no asymptote.
4. **Value of ϕ:**

   $$\phi = \frac{\pi}{2} + \theta$$

   When $\theta = 0$, $\phi = \frac{\pi}{2}$

   For $\theta = 0$, $r = \pm\, a$

   Hence, at the points $(a, 0)$ and $(-\,a, 0)$ the tangents are $\perp$ to the initial line.
5. **Region and special points:**

   From (*i*), $r = a\sqrt{\cos 2\theta}$

   [Taking +ve root as the curve is symmetrical about the pole]

   $$\Rightarrow \frac{dr}{d\theta} = \frac{-a\sin 2\theta}{\sqrt{\cos 2\theta}}$$

   For $0 < \theta < \frac{\pi}{4}$, $\frac{dr}{d\theta}$ is – ve $\Rightarrow r$ decreases in this range.

   For $\frac{3\pi}{4} < \theta < \pi$, $\frac{dr}{d\theta}$ is +ve $\Rightarrow r$ increases in this range.

   When $\theta = 0$, $r = a$

As θ increases from 0 to $\frac{\pi}{4}$, r decreases from a to 0.

For $\frac{\pi}{4} < \theta < \frac{3\pi}{4}$, r is imaginary. Hence no portion of the curve lies between the lines $\theta = \frac{\pi}{4}$ and $\theta = \frac{3\pi}{4}$.

Again, as θ increases from $\frac{3\pi}{4}$ to π, r is + ve and increases from 0 to a. Thus we can trace the part of the curve about the initial line. The part of the curve below the initial line can be traced by symmetry.

Hence, shape of the curve is as shown in the figure.

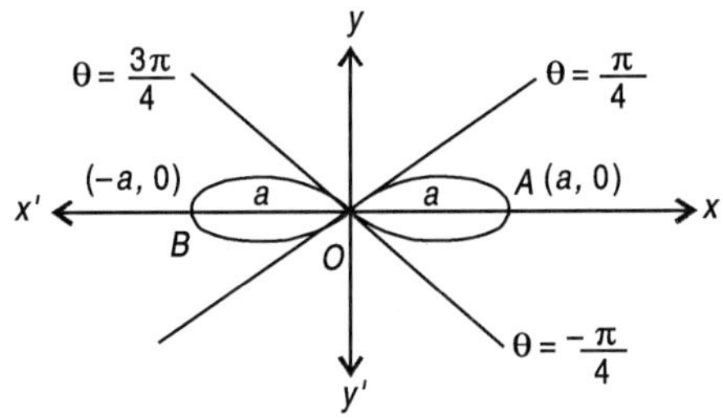

EXERCISE 1.9

Trace the following curves :

1. $r = a\,(1 + \cos\theta)$
2. $r = a\,(1 - \sin\theta)$
3. $r = 2\,a\cos\theta$
4. $2\,a = r\,(1 + \cos\theta)$
5. $r = a\cos 2\theta$.
6. $r = a\,(\cos\theta + \sec\theta)$

ANSWERS

1.

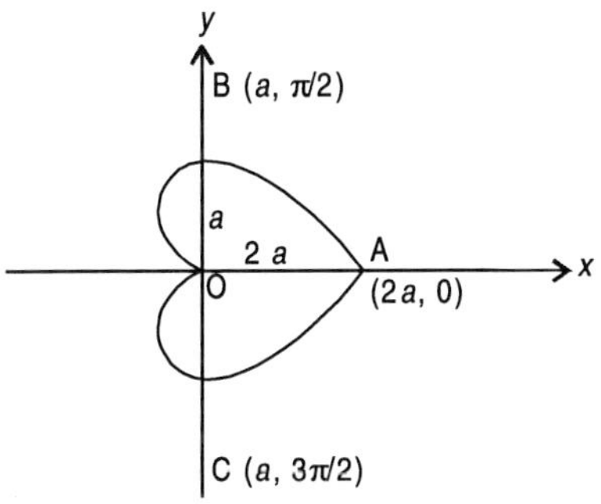

2.

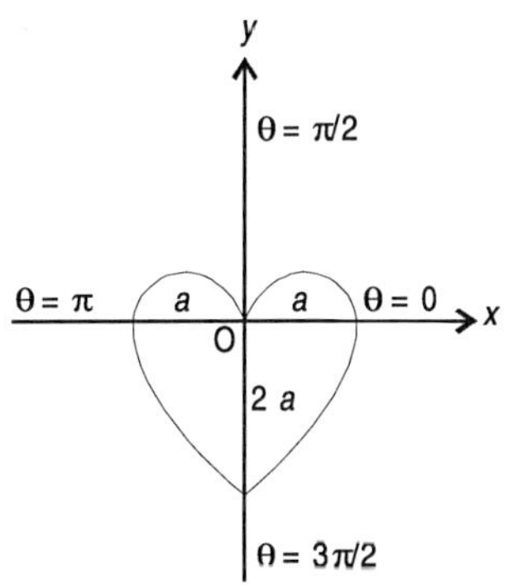

3.

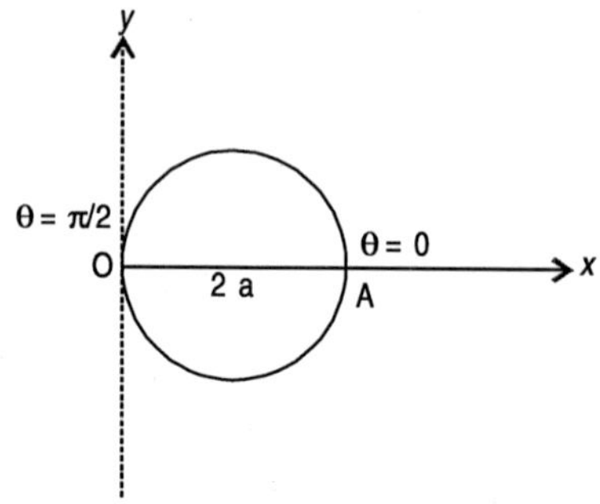
y
θ = π/2
O
2 a
θ = 0
A
x

4.

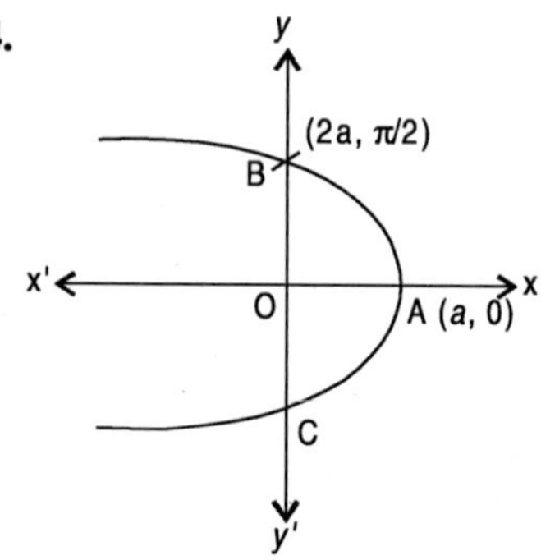
y
(2a, π/2)
B
x'
O
A (a, 0)
x
C
y'

5.

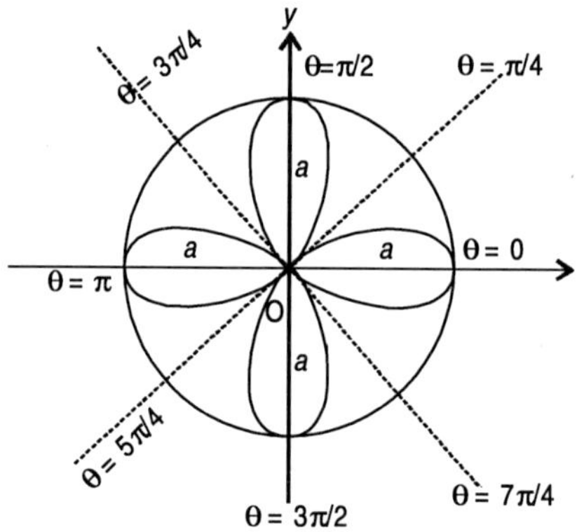
y
θ = 3π/4
θ=π/2
θ = π/4
a
a
a
θ = 0
θ = π
O
a
θ = 5π/4
θ = 7π/4
θ = 3π/2

6.

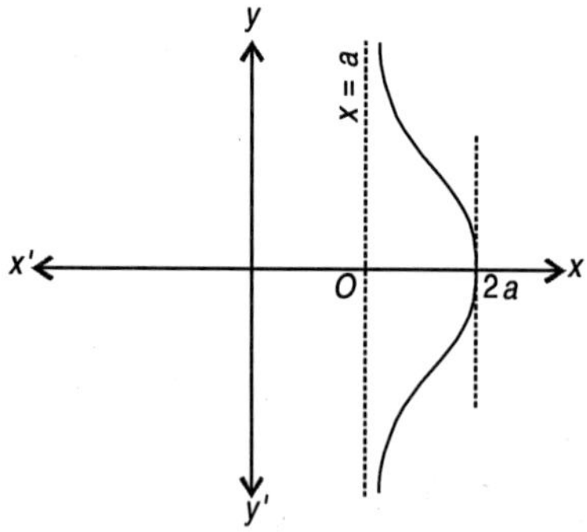
y
x = a
x'
O
2a
x
y'

UNIT – II

Ordinary Differential Equations

2.1 DIFFERENTIAL EQUATION

A differential equation is an equation containing the dependent and independent variables and different derivatives of the dependent variables.

2.2 ORDER AND DEGREE OF A DIFFERENTIAL EQUATION

The order of a differential Equation is the order of the highest derivative (or differential coefficient) occurring in the equation.

The degree of a differential equation is the degree of the highest derivative which occurs in it, after the differential equation has been rationalized *i.e.*, made free from radicals and fractions so far as derivatives are concerned.

2.3 ORDINARY DIFFERENTIAL EQUATION

A differential equation is called ordinary, if it has only one independent variable.

2.4 LINEAR AND NON-LINEAR DIFFERENTIAL EQUATIONS

A differential equation is said to be linear if the dependent variable and all its derivatives occur in the first degree, otherwise it is non-linear.

2.5 SOLUTIONS OF DIFFERENTIAL EQUATION

A function is called a *solution (or the primitive)* of a differential equation if, when substituted into the equation, it reduces the equation to an identity, and the process of finding all the solutions is called *integrating (or solving)* the differential equation.

General Solution or Complete Primitive : A solution of a differential equation , containing independent arbitrary constants equal in number to the order of the differential equation is called its *general solution or complete primitive.*

Particular Solution : A solution obtained by giving particular values to the arbitrary constants in the general solution is called a *particular solution or particular integral.*

Singular Solution : A solution which cannot be obtained by giving a particular value to the arbitrary constant in the general solution is called a *singular solution.*

Arbitrary Constants : The solution of a differential equation may contain as many arbitrary constants as is the order of the differential equation *i.e.*, the solution of an n th order differential equation may contain n arbitrary constants .

2.6 FORMATION OF DIFFERENTIAL EQUATION

Differential equations are formed by eliminating the arbitrary constants.

ILLUSTRATIVE EXAMPLES

Example 1. Find the differential equation of the family of curves $y = e^x(A\cos x + B\sin x)$, for different values of A and B.

Solution. We have the relation

$$y = e^x(A\cos x + B\sin x) \qquad \ldots(1)$$

Differentiating (1) w.r.t.x, we have

$$\frac{dy}{dx} = e^x(A\cos x + B\sin x) + e^x(-A\sin x + B\cos x)$$

$$= y + e^x(-A\sin x + B\cos x) \qquad \ldots(2)$$

Differentiating again w.r.t.x, we have

$$= \frac{dy}{dx} + \left(\frac{dy}{dx} - y\right) - y \qquad \text{[using (1) and (2)]}$$

or $$\frac{d^2y}{dx^2} - 2\frac{dy}{dx} + 2y = 0$$

which is the required differential equation.

Example 2. By eliminating the constants a and b obtain the differential equation of which $xy = ae^x + be^{-x} + x^2$ is a solution.

Solution. We have $xy = ae^x + be^{-x} + x^2$ ...(1)

Differentiating (1), we have

$$y + x\frac{dy}{dx} = ae^x - be^{-x} + 2x \qquad \ldots(2)$$

Differentiating (2), we have

$$\frac{dy}{dx} + \frac{dy}{dx} + x\frac{d^2y}{dx^2} = ae^x + be^{-x} + 2$$

$$= (xy - x^2) + 2, \quad \text{from (1)}$$

Thus we have

$$x\frac{d^2y}{dx^2} + 2\frac{dy}{dx} - xy + x^2 - 2 = 0.$$

which is the required differential equation.

Example 3. Find the differential equation of the family of curves $y = Ae^x + \frac{B}{e^x}$, for different values of A and B.

Solution. We have $y = Ae^x + Be^{-x}$ ...(1)

Differentiating (1), we have

$$\frac{dy}{dx} = Ae^x - Be^{-x} \quad ...(2)$$

Differentiating (2), we have

$$\frac{d^2y}{dx^2} = Ae^x + Be^{-x} \quad ...(3)$$

Thus, from (1) and (3) ,we have

$$\frac{d^2y}{dx^2} = y$$

which is the required differential equation.

2.7 DIFFERENTIAL EQUATION OF THE FIRST DEGREE AND FIRST ORDER

A differential equation of the form $\frac{dy}{dx} = f(x, y)$ or $Mdx + Ndy = 0$, where M, N are functions of x and y, is called a differential equation of the first order and first degree.

2.8 SOLUTION OF DIFFERENTIAL EQUATION OF THE FIRST DEGREE AND FIRST ORDER

As we cannot integrate every function so it is not possible to solve every differential equation . Even the differential equation of first order and first degree cannot be solved in every case.They can be solved, however, if they belong to one of the standard forms discussed below .

2.9 VARIABLES SEPARABLE FORM

If it is possible to write a differential equation of the first order and first degree in the form

$$f_1(x)dx = f_2(y)dy$$

we say that the variables are separable . Such equations can be solved immediately by integration and the solution is given by

$$\int f_1(x)dx = \int f_2(y)dy + c$$

where c is an arbitrary constant of integration.

Steps for Solution

1. Separate the variables as $f_1(x)dx = f_2(y)dy$
2. Integrate both sides as $\int f_1(x)dx = \int f_2(y)dy$
3. Add an arbitrary constant to any of the sides.

ILLUSTRATIVE EXAMPLES

Example 1. Solve : $\frac{dy}{dx} = \frac{1+y^2}{1+x^2}$

Solution. We have $\frac{dy}{dx} = \frac{1+y^2}{1+x^2}$ ...(1)

Separating the variables, we get

$$\frac{dy}{1+y^2} = \frac{dx}{1+x^2}$$

Integrating both sides , we get

$$\tan^{-1} y = \tan^{-1} x + \tan^{-1} c$$

where $\tan^{-1} c$ is an arbitrary constant of integration.

$$\Rightarrow \quad \tan^{-1} y - \tan^{-1} x = \tan^{-1} c$$

$$\Rightarrow \quad \tan^{-1}\left(\frac{y-x}{1+xy}\right) = \tan^{-1} c$$

$$\Rightarrow \quad \frac{y-x}{1+xy} = c$$

$$\Rightarrow \quad y-x = c(1+xy)$$

which is the required solution.

Example 2. Solve : $\frac{dy}{dx} = e^{2x+3y}$

Solution. We have $\frac{dy}{dx} = e^{2x+3y}$

Separating the variables , we get

$$\frac{dy}{e^{3y}} = e^{2x} dx$$

$$\Rightarrow \quad e^{-3y} dy = e^{2x} dx$$

Integrating both sides, we get

$$\frac{e^{-3y}}{-3} = \frac{e^{2x}}{2} + c,$$

where c is an arbitrary constant of integration.

$$\Rightarrow \quad 2e^{-3y} = -3e^{2x} - 3c$$

$$\Rightarrow \quad 3e^{2x} + 2e^{-3y} = 3c$$

$$\Rightarrow \quad 3e^{2x} + 2e^{-3y} = c$$, which is the required solution.

Example 3. Solve : $\sec^2 x \tan y dx + \sec^2 y \tan x dy = 0$

Solution. Separating the variables, we get

$$\frac{\sec^2 x}{\tan x} dx + \frac{\sec^2 y}{\tan y} dy = 0$$

Integrating both sides , we get

$$\log \tan x + \log \tan y = \log c,$$

where log c is an arbitrary constant of integration.

$$\Rightarrow \quad \log \tan x \tan y = \log c$$

$$\Rightarrow \quad \tan x \tan y = c$$

which is the required solution.

Example 4. Solve : $(1+x) y dx + (1-y) x dy = 0$

Solution. Separating the variables, we get

$$\frac{1+x}{x} dx + \frac{1-y}{y} dy = 0$$

$$\Rightarrow \quad \left(\frac{1}{x}+1\right) dx + \left(\frac{1}{y}-1\right) dy = 0$$

Integrating both sides, we get

$$(\log x + x) + (\log y - y) = \log c,$$

where log c is an arbitrary constant of integration.

$$\Rightarrow \quad \log x + \log y - \log c = y - x$$

$$\Rightarrow \quad \log \frac{xy}{c} = y - x$$

$$\Rightarrow \quad \frac{xy}{c} = e^{y-x}$$

$$\Rightarrow \quad xy = ce^{y-x}$$

which is the required solution.

Example 5. Solve : $\tan y \dfrac{dy}{dx} = \sin(x+y) + \sin(x-y)$

Solution. We have $\quad \tan y \dfrac{dy}{dx} = \sin(x+y) + \sin(x-y)$

$$\Rightarrow \quad \tan y \frac{dy}{dx} = 2 \sin x \cos y$$

[using $\sin C + \sin D = 2 \sin\left(\frac{C+D}{2}\right) \cos\left(\frac{C-D}{2}\right)$]

Separating the variables, we get

$$\frac{\tan y dy}{\cos y} = 2\sin x dx$$

$$\Rightarrow \qquad \sec y \tan y dy = 2\sin x dx$$

Integrating both sides, we get

$$\sec y = -2\cos x + c,$$

where c is an arbitrary constant of integration.

$$\Rightarrow \qquad 2\cos x + \sec y = c$$

which is the required solution.

EXERCISE 2.1

Solve the following differential equations :

1. $\frac{dy}{dx} = e^{x-y} + x^2 e^{-y}$
2. $ydx - xdy = 0$
3. $\frac{dy}{dx} = e^{ax-by}$
4. $y - x\frac{dy}{dx} = a\left(y^2 + \frac{dy}{dx}\right)$
5. $(x+y)(dx-dy) = dx+dy$
6. $(x+y+1)^2 \frac{dy}{dx} = 1$
7. $y\sqrt{1-x^2}\,dy + x\sqrt{1-y^2}\,dx = 0$
8. $xy\frac{dy}{dx} = 1 = x + y + xy$
9. $(1-x)dy + (1-y)dx = 0$
10. $x\cos^2 y dx = y\cos^2 x dy$

ANSWERS

1. $e^y = e^x + \frac{x^3}{3} + c$
2. $x = cy$
3. $be^{ax} - ae^{by} = c$
4. $y = c(1-ay)(x+a)$
5. $x + y = ce^{x-y}$
6. $y = \tan^{-1}(x+y+1) + c - 1$ [*Hint.* *put* $x+y+1 = t$]
7. $\sqrt{1-x^2} + \sqrt{1-y^2} = c$
8. $y = x + \log\{x(1+y)\} + c$
9. $(1-x)(1-y) = c$
10. $y\tan y + \log\cos y = x\tan x + \log\cos x + c$

2.10 HOMOGENEOUS FUNCTION

A function in which every term is of the same degree is known as homogeneous function of that degree.For example,

Let $f(x,y) = a_0x^n + a_1x^{n-1}y + a_2x^{n-2}y^2 + ... + a_{n-1}xy^{n-1} + a_ny^n$

be a homogeneous function of x and y of degree n.

Now,

$$f(x,y) = x^n[a_0 + a_1\left(\frac{y}{x}\right) + a_2\left(\frac{y}{x}\right)^2 + ... + a_{n-1}\left(\frac{y}{x}\right)^{n-1} + a_n\left(\frac{y}{x}\right)^n] = x^nF\left(\frac{y}{x}\right)$$

Thus, every homogeneous function of x and y of degree n can be expressed in the form $x^nF\left(\frac{y}{x}\right)$.

2.11 TEST OF HOMOGENEITY OF A FUNCTION OF *x* AND *y*

Let $f(x,y)$ be any function of x and y. Then $f(x,y)$ is said to be homogeneous function of x and y of degree n, if

$$f(tx,ty) = t^nf(x,y)$$

2.12 HOMOGENEOUS DIFFERENTIAL EQUATIONS

A differential equation of the form

$$\frac{dy}{dx} = \frac{f_1(x,y)}{f_2(x,y)}$$

where f_1 and f_2 are homogeneous functions of x and y of the same degree, is called a *Homogeneous Differential Equation.*

For example : $\dfrac{dy}{dx} = \dfrac{-y(x+y)}{x^2}$

Steps for Solution

1. Put $y = vx$ so that $\dfrac{dy}{dx} = v + x\dfrac{dv}{dx}$
2. Separate the variables and solve the given equation in v and x.
3. In the last, put $v = \dfrac{y}{x}$ and simplify.

ILLUSTRATIVE EXAMPLES

Example 1. Solve : $x + y\dfrac{dy}{dx} = 2y$

Solution. We have $x + y\dfrac{dy}{dx} = 2y$

$$\Rightarrow \qquad \frac{dy}{dx} = \frac{2y-x}{y} \qquad ...(1)$$

which is a homogeneous differential equation .

Put $y = vx$, so that,

$$\frac{dy}{dx} = v + x\frac{dv}{dx}$$

Now (1) becomes

$$v + x\frac{dv}{dx} = \frac{2vx - x}{vx} = \frac{2v-1}{v}$$

$$\Rightarrow \qquad x\frac{dv}{dx} = \frac{2v-1}{v} - v = \frac{2v-1-v^2}{v}$$

$$= -\left(\frac{v^2+1-2v}{v}\right) = -\frac{(v-1)^2}{v}$$

Separating the variables, we get

$$\frac{v}{(v-1)^2}dv = -\frac{dx}{x}$$

$$\Rightarrow \qquad \frac{(v-1)+1}{(v-1)^2}dv = -\frac{dx}{x}$$

$$\Rightarrow \qquad \left(\frac{1}{v-1} + \frac{1}{(v-1)^2}\right)dv = -\frac{dx}{x}$$

Integrating we have,

$$\log(v-1) - \frac{1}{v-1} = -\log x + c, \text{ } c \text{ is an arbitrary constant}$$

$$\Rightarrow \qquad \log(v-1)x = \frac{1}{v-1} + c$$

Putting the value of v, we get

$$\log\left(\frac{y}{x} - 1\right)x = \frac{1}{\frac{y}{x} - 1} + c$$

$$\Rightarrow \qquad \log(y-x) = \frac{x}{y-x} + c, \text{ this gives the complete solution.}$$

Example 2. Solve: $y - x\frac{dy}{dx} = y\frac{dy}{dx} + x$

Solution. We have $y - x\frac{dy}{dx} = y\frac{dy}{dx} + x$

$$\Rightarrow \qquad \frac{dy}{dx}(y+x) = y - x$$

$$\Rightarrow \qquad \frac{dy}{dx} = \frac{y-x}{y+x} \qquad \ldots(1)$$

which is a homogeneous differential equation.

Put $y = vx$, so that,

$$\frac{dy}{dx} = v + x\frac{dv}{dx}$$

Now (1) becomes

$$v + x\frac{dv}{dx} = \frac{vx-x}{vx+x} = \frac{v-1}{v+1}$$

$$\Rightarrow \qquad x\frac{dv}{dx} = \frac{v-1}{v+1} - v = \frac{v-1-v^2-v}{v+1} = -\frac{v^2+1}{v+1}$$

Separating the variables, we get

$$\frac{v+1}{v^2+1}dv = -\frac{dx}{x}$$

$$\Rightarrow \qquad \frac{v}{v^2+1}dv + \frac{1}{v^2+1}dv = -\frac{dx}{x}$$

Integrating we have,

$$\frac{1}{2}\int\frac{2v}{v^2+1}dv + \int\frac{dv}{v^2+1} = -\int\frac{dx}{x} + c, \ c \text{ is an arbitrary constant}$$

$$\Rightarrow \qquad \frac{1}{2}\log(v^2+1) + \tan^{-1}v = -\log x + c$$

Putting the value of *v*, we get

$$\frac{1}{2}\log\left(\frac{y^2}{x^2}+1\right) + \tan^{-1}\left(\frac{y}{x}\right) = -\log x + c$$

$$\Rightarrow \qquad \frac{1}{2}\log(y^2+x^2) - \frac{1}{2}\log x^2 + \tan^{-1}\left(\frac{y}{x}\right) + \log x = c$$

$$\Rightarrow \qquad \frac{1}{2}\log(y^2+x^2) - \log x + \tan^{-1}\left(\frac{y}{x}\right) + \log x = c$$

$$\Rightarrow \qquad \frac{1}{2}\log(y^2+x^2) + \tan^{-1}\left(\frac{y}{x}\right) = c,$$

this gives the complete solution.

Example 3. Solve : $x^2 dy + y(x+y)dx = 0$

Solution. We have $x^2 dy + y(x+y)dx = 0$

$$\Rightarrow \qquad \frac{dy}{dx} = -\frac{y(x+y)}{x^2} \qquad \ldots(1)$$

which is a homogeneous differential equation.

Put $y = vx$, so that,

$$\frac{dy}{dx} = v + x\frac{dv}{dx}$$

Now (1) becomes

$$v + x\frac{dv}{dx} = -\frac{vx(x+vx)}{x^2} = -v(1+v)$$

$$\Rightarrow \quad x\frac{dv}{dx} = -v(1+v) - v = -(v^2+2v) = -v(v+2)$$

Separating the variables, we get

$$\frac{-1}{v(v+2)}dv = \frac{dx}{x}$$

$$\Rightarrow \quad \frac{1}{2}\left(\frac{1}{v} - \frac{1}{v+2}\right)dv = -\frac{dx}{x}$$

Integrating we have,

$$\frac{1}{2}\int\frac{1}{v}dv - \frac{1}{2}\int\frac{dv}{v+2} = -\int\frac{dx}{x} + \log c \text{, } c \text{ is an arbitrary constant}$$

$$\Rightarrow \quad \frac{1}{2}\log v - \frac{1}{2}\log(v+2) = -\log x + \log c$$

Putting the value of v, we get

$$\frac{1}{2}\log\left(\frac{y}{x}\right) - \frac{1}{2}\log\left(\frac{y}{x}+2\right) = -\log x + \log c$$

$$\Rightarrow \quad \frac{1}{2}\log y - \frac{1}{2}\log x - \frac{1}{2}\log(y+2x) + \frac{1}{2}\log x = -\log x + \log c$$

$$\Rightarrow \quad \frac{1}{2}\log y + \log x - \frac{1}{2}\log(y+2x) = \log c$$

$$\Rightarrow \quad \log\sqrt{y} + \log x - \log\sqrt{(y+2x)} = \log c$$

$$\Rightarrow \quad \log\frac{x\sqrt{y}}{\sqrt{(y+2x)}} = \log c$$

$$\Rightarrow \quad \frac{x\sqrt{y}}{\sqrt{(y+2x)}} = c$$

$$\Rightarrow \quad \frac{x^2 y}{(y+2x)} = c^2$$

$$\Rightarrow \quad x^2 y = c^2(y+2x),$$

this gives the complete solution.

EXERCISE 2.2

Solve the following differential equations :

1. $x(x-y)dy+y^2dx=0$

2. $y^2dx+x(x+y)dy=0$

3. $\frac{dy}{dx}=\frac{y^3+3x^2y}{x^3+3xy^2}$

4. $(x^2-y^2)dx+2xydy=0$

5. $x^2\frac{dy}{dx}=\frac{y(x+y)}{2}$

6. $x(x-y)dy=y(x+y)dx$

ANSWERS

1. $y=ce^{\frac{y}{x}}$

2. $xy^2=c(2y+x)$

3. $xy=c^2(x^2-y^2)^2$

4. $x^2+y^2=cx$

5. $(y-x)^2=cxy^2$

6. $\frac{x}{y}+\log(xy)=c$

2.13 EQUATION REDUCIBLE TO HOMOGENEOUS FORM

A differential equation of the form

$$\frac{dy}{dx}=\frac{ax+by+c}{Ax+By+C} \quad \text{... (1)}$$

is called a differential equation reducible to homogeneous form .

While solving such problems , two cases arise:

Case I: When $\frac{a}{A}\neq\frac{b}{B}$

In this case , equation (1) can be reduced to homogeneous form by introducing two new variables X and Y, related to x and y, by the equations

$$x=X+h$$

$$\text{and} \quad y=Y+k \quad \text{... (2)}$$

where h and k are the constants which are yet to be found.

From (2), $dx=dX$ and $dy=dY$

$$\therefore \quad \frac{dy}{dx}=\frac{dY}{dX}$$

Hence equation (1) becomes

$$\frac{dY}{dX}=\frac{a(X+h)+b(Y+k)+c}{A(X+h)+B(Y+k)+C}$$

$$=\frac{(aX+bY)+(ah+bk+c)}{(AX+BY)+(Ah+Bk+C)} \quad \text{... (3)}$$

Now choose h and k such that

$$ah + bk + c = 0 \qquad \dots(4)$$

and

$$Ah + Bk + C = 0 \qquad \dots(5)$$

Then (3) becomes

$$\frac{dY}{dX} = \frac{aX + bY}{AX + BY} \qquad \dots(6)$$

which is clearly a homogeneous differential equation in X and Y.

Solving (4) and (5) for h and k by cross multiplication method, we get

$$\Rightarrow \quad \frac{h}{bC - Bc} = \frac{k}{Ac - aC} = \frac{1}{aB - Ab}$$

$$\Rightarrow \quad \left.\begin{aligned} h &= \frac{bC - Bc}{aB - Ab} \\ \text{and} \quad k &= \frac{Ac - aC}{aB - Ab} \end{aligned}\right\} \qquad \dots(7)$$

To solve (6), put $Y = vX$ so that $\dfrac{dY}{dX} = v + X\dfrac{dv}{dX}$

We solve equation (6) in v and X. then put $\dfrac{Y}{X}$ for v.

In the last, we shall put x-h and y-k for X and Y respectively (where h and k have the values given by (7)) to get the required solution of the given differential equation.

Case II: When $\dfrac{a}{A} = \dfrac{b}{B}$ *i.e.,* $aB - Ab = 0$

$\therefore$ From above equation (7), h and k both become infinite.

Hence the method followed in case I fails

So let $\dfrac{a}{A} = \dfrac{b}{B} = \dfrac{1}{m}$

$\therefore \quad A = ma, \; B = mb$

$\therefore$ The given equation assumes the form

$$\frac{dy}{dx} = \frac{ax + by + c}{m(ax + by) + c} \qquad \dots(8)$$

To solve such equations, we put $ax + by = t$ so that

$$\frac{dy}{dx} = \frac{1}{b}\left(\frac{dt}{dx} - a\right)$$

We solve equation (8) by separating the variables. In the last we put $ax + by$ for t.

ILLUSTRATIVE EXAMPLES

Example 1. Solve : $(2x+y-3)\frac{dy}{dx}=x+2y-3$

Solution. We have $\frac{dy}{dx}=\frac{x+2y-3}{2x+y-3}$ ... (1)

Here, $\frac{1}{2}\neq\frac{2}{1}$ so case I follows.

Put $x=X+h$ and $y=Y+k$

So that $\frac{dy}{dx}=\frac{dY}{dX}$

Hence equation (1) becomes

$$\frac{dY}{dX}=\frac{X+h+2(Y+k)-3}{2(X+h)+Y+k-3}=\frac{X+2Y+(h+2k-3)}{2X+Y+(2h+k-3)} \quad \text{... (2)}$$

Choose h and k such that

$$h+2k-3=0 \quad \text{... (3)}$$

and $$2h+k-3=0 \quad \text{... (4)}$$

Solving (3) and (4), we get

$$\Rightarrow \quad \frac{h}{-6+3}=\frac{k}{-6+3}=\frac{1}{1-4}$$

$$\Rightarrow \quad h=1, \quad k=1$$

∴ Equation (2) reduces to

$$\frac{dY}{dX}=\frac{X+2Y}{2X+Y} \quad \text{... (5)}$$

which is a homogeneous differential equation in X and Y.

Put $Y=vX$, so that,

$$\frac{dY}{dX}=v+X\frac{dv}{dX}$$

Now (5) becomes

$$v+X\frac{dv}{dx}=\frac{X+2vX}{2X+vX}=\frac{1+2v}{2+v}$$

$$\Rightarrow \quad X\frac{dv}{dX}=\frac{1+2v}{2+v}-v=\frac{1+2v-2v-v^2}{2+v}=\frac{1-v^2}{2+v}$$

Separating the variables, we get

$$\frac{2+v}{1-v^2}dv=\frac{dX}{X}$$

$$\Rightarrow \qquad \frac{dX}{X} = \frac{2+v}{(1-v)(1+v)}$$

$$\Rightarrow \qquad \frac{1}{2}\left(\frac{3}{1-v}+\frac{1}{1+v}\right)dv = \frac{dX}{X} \qquad \text{(By partial fractions)}$$

Integrating , we have

$$\frac{1}{2}[-3\log(1-v)+\log(1+v)] = \log X + \log c,$$

where $\log c$ is a constant of integration.

$$\Rightarrow -\log(1-v)^3 + \log(1+v) = 2(\log X + \log c)$$

$$\Rightarrow \qquad \log\left\{\frac{1+v}{(1-v)^3}\right\} = 2\log cX = \log c^2X^2$$

$$\Rightarrow \qquad \frac{1+v}{(1-v)^3} = c^2X^2$$

Putting $v = \dfrac{Y}{X}$, we have

$$\frac{1+\dfrac{Y}{X}}{\left(1-\dfrac{Y}{X}\right)^3} = c^2X^2$$

$$\Rightarrow \qquad \frac{X+Y}{(X-Y)^3} = c^2$$

$$\Rightarrow \qquad X+Y = c^2(X-Y)^3$$

Putting the value of X and Y, we get

$$(x-1+y-1) = c^2\{(x-1)-(y-1)\}^3$$

$$\Rightarrow \qquad x+y-2 = c^2(x-y)^3$$

which is the required solution.

Example 2. Solve : $(2x+y+1)dx+(4x+2y-1)dy=0$

Solution. We have $\dfrac{dy}{dx} = -\dfrac{2x+y+1}{4x+2y-1} = -\dfrac{2x+y+1}{2(2x+y)-1}$ ... (1)

Here, $\dfrac{2}{4}=\dfrac{1}{2}$ so case II follows

Put $\qquad 2x+y = t$

$$\therefore \qquad 2+\frac{dy}{dx} = \frac{dt}{dx}$$

$$\Rightarrow \qquad \frac{dy}{dx} = \frac{dt}{dx} - 2$$

Now putting in equation (1), we have

$$\frac{dt}{dx} - 2 = -\frac{t+1}{2t-1}$$

$$\Rightarrow \qquad \frac{dt}{dx} = -\frac{t+1}{2t-1} + 2$$

$$\Rightarrow \qquad \frac{dt}{dx} = \frac{3(t-1)}{2t-1}$$

Separating the variables , we have

$$\frac{2t-1}{t-1} dt = 3dx$$

$$\Rightarrow \qquad \left(2 + \frac{1}{t-1}\right) dt = 3dx$$

Integrating both sides, we have

$$2t + \log(t-1) = 3x + c \text{, where } c \text{ is a constant of integration.}$$

$$\Rightarrow \quad 2(2x+y) + \log(2x+y-1) = 3x + c \text{ (putting the value of } t\text{)}$$

$$\Rightarrow \qquad \log(2x+y-1) = -(x+2y) + c$$

$$\Rightarrow \qquad \log(2x+y-1)(x+2y) = c$$

which is the required solution.

EXERCISE 2.3

Solve the following differential equations :

1. $(6x+2y-10)dy - (2x+9y-20)dx = 0$

2. $(2x+3y-5)\dfrac{dy}{dx} + 3x + 2y - 5 = 0$

3. $\dfrac{dy}{dx} = \dfrac{x-y-2}{2x-2y-3}$

4. $\left(\dfrac{4x+6y+5}{3y+2x+4}\right)\dfrac{dy}{dx} = 1$

ANSWERS

1. $x+2y-5 = c(2x-y)^2$

2. $\dfrac{3}{2}(x^2+y^2) + 2xy - 5(x+y) = c$

3. $\log(x-y-1) = x - 2y + c$

4. $42y - 21x - 9\log(14x+21y+22) = c$

2.14 LINEAR DIFFERENTIAL EQUATIONS

A differential equation in which dependent variable and its derivative occur only in the first degree and are not multiplied together, is called a linear differential equation.

The general form of these types of equation is $\frac{dy}{dx} + Py = Q$ ... (1)

where P and Q are functions of x only or constants.

Steps for Solution

1. First make the coefficient of $\frac{dy}{dx}$, unity.
2. Find *Integrating Factor (I.F.)* $= e^{\int P dx}$
3. Required solution is $y(I.F.) = \int Q(I.F.)dx + c$

Note : Sometimes a linear differential equation takes the form $\frac{dx}{dy} + Px = Q$ where P and Q are functions of y only or constants. In this form x is considered as dependent variable and y independent variable.

The integrating factor in this case is $e^{\int P dy}$ and the solution is

$$x(I.F.) = \int Q(I.F.)dy + c$$

ILLUSTRATIVE EXAMPLES

Example 1. Solve : $\frac{dy}{dx} + \frac{y}{x} = x^3 - 3$.

Solution. Comparing with $\frac{dy}{dx} + Py = Q$, we get

$$P = \frac{1}{x}, \quad Q = x^3 - 3$$

Integrating Factor (I.F.) $= e^{\int P dx} = e^{\int \frac{1}{x} dx} = e^{\log x} = x$

Hence solution is given by

$$y(I.F.) = \int Q(I.F.)dx + c$$

$$\Rightarrow \quad yx = \int (x^3 - 3)x dx + c$$

$$= \int (x^4 - 3x)dx + c$$

$$= \frac{x^5}{5} - 3\frac{x^2}{2} + c$$

$$\Rightarrow \quad xy = \frac{x^5}{5} - \frac{3x^2}{2} + c$$

Example 2. Solve : $\sec x \dfrac{dy}{dx} = y + \sin x.$

Solution. Dividing by sec x on both sides, we get

$$\frac{dy}{dx} - y\cos x = \sin x \cos x$$

Comparing with $\dfrac{dy}{dx} + Py = Q$, we get

$$P = -\cos x, \quad Q = \sin x \cos x$$

Integrating Factor (I.F.) = $e^{\int P dx} = e^{-\int \cos x dx} = e^{-\sin x}$

Hence solution is given by

$$y(I.F.) = \int Q(I.F.)dx + c$$

$$\Rightarrow \qquad ye^{-\sin x} = \int \sin x \cos x . e^{-\sin x} dx + c$$

$$= \int te^{-t} dt + c \text{ on putting } \sin x = t, \ \cos x dx = dt$$

$$= -te^{-t} + \int e^{-t} + c$$

$$= -te^{-t} - e^{-t} + c$$

$$\Rightarrow \qquad ye^{-\sin x} = -\sin x e^{-\sin x} - e^{-\sin x} + c$$

$$\Rightarrow \qquad y = -\sin x - 1 + ce^{\sin x}$$

$$\Rightarrow \qquad y = ce^{\sin x} - (1 + \sin x)$$

Example 3. Solve : $x\dfrac{dy}{dx} + 2y = x^2 \log x$

Solution. Dividing by x on both sides, we get

$$\frac{dy}{dx} + \frac{2}{x}y = x\log x$$

Comparing with $\dfrac{dy}{dx} + Py = Q$, we get

$$P = \frac{2}{x}, \quad Q = x\log x$$

Integrating Factor (I.F.) = $e^{\int P dx} = e^{\int \frac{2}{x} dx} = e^{2\log x} = e^{\log x^2} = x^2$

Hence solution is given by

$$y(I.F.) = \int Q(I.F.)dx + c$$

$$\Rightarrow \qquad yx^2 = \int x \log x . x^2 dx + c$$

$$= \int x^3 \log x dx + c$$

$$= \log x . \frac{x^4}{4} - \int \frac{1}{x} . \frac{x^4}{4} dx$$

$$= \frac{x^4}{4} \log x - \int \frac{x^3}{4} dx = \frac{x^4}{4} \log x - \frac{x^4}{16} + c$$

$\Rightarrow$ $$yx^2 = \frac{x^4}{4} \log x - \frac{x^4}{16} + c$$

Example 4. Solve : $y \log y dx = (\log y - x) dy$

Solution. We have $$\frac{dx}{dy} = \frac{\log y - x}{y \log y}$$

$\Rightarrow$ $$\frac{dx}{dy} + \frac{x}{y \log y} = \frac{1}{y}$$

Comparing with $\frac{dx}{dy} + Px = Q$, we get

$$P = \frac{1}{y \log y}, \quad Q = \frac{1}{y}$$

Integrating Factor (I.F.) = $e^{\int P dy} = e^{\int \frac{1}{y \log y} dy} = e^{\log(\log y)} = \log y$

Hence solution is given by

$$x(I.F.) = \int Q(I.F.) dy + c$$

$\Rightarrow$ $$x \log y = \int \frac{1}{y} \log y dy + c$$

$\Rightarrow$ $$x \log y = \frac{1}{2} (\log y)^2 + c$$

Example 5. Solve : $\frac{dy}{dx} + 2xy = 2e^{-x^2}$

Solution. Comparing with $\frac{dy}{dx} + Py = Q$, we get

$$P = 2x, \quad Q = 2e^{-x^2}$$

Integrating Factor (I.F.) = $e^{\int P dx} = e^{\int 2x dx} = e^{x^2}$

Hence solution is given by

$$y(I.F.) = \int Q(I.F.) dx + c$$

$$\Rightarrow \quad ye^{x^2} = \int 2e^{-x^2}.e^{x^2}\,dx + c$$

$$= 2\int 1.dx + c$$

$$\Rightarrow \quad ye^{x^2} = 2x + c$$

$$\Rightarrow \quad y = (2x + c)e^{-x^2}$$

Example 6. Solve : $\cos^2 \dfrac{dy}{dx} + y = \tan x$

Solution. Dividing by $\cos^2 x$ on both sides, we get

$$\frac{dy}{dx} + y\sec^2 x = \sec^2 x \tan x$$

Comparing with $\dfrac{dy}{dx} + Py = Q$, we get

$$P = \sec^2 x, \quad Q = \sec^2 x \tan x$$

Integrating Factor (I.F.) = $e^{\int P dx} = e^{\int \sec^2 x dx} = e^{\tan x}$

Hence solution is given by

$$y(I.F.) = \int Q(I.F.)dx + c$$

$$\Rightarrow \quad ye^{\tan x} = \int \sec^2 x \tan x.e^{\tan x}\,dx + c$$

$$= \int te^t.dt + c \text{ on putting } \tan x = t, \quad \sec^2 x dx = dt$$

$$= te^t - e^t + c$$

$$\Rightarrow \quad ye^{\tan x} = \tan x.e^{\tan x} - e^{\tan x} + c$$

$$\Rightarrow \quad y = \tan x - 1 + ce^{-\tan x}$$

EXERCISE 2.4

Solve the following differential equations :

1. $\dfrac{dy}{dx} + 2y\tan x = \sin x$
2. $\dfrac{dy}{dx} + 3y = \tan x$
3. $\dfrac{dy}{dx} + y\cot x = \cos x$
4. $x\log x\dfrac{dy}{dx} + y = 2\log x$
5. $\dfrac{dy}{dx} + \dfrac{3x^2}{1+x^3}y = \dfrac{\sin^2 x}{1+x^3}$
6. $(x+1)\dfrac{dy}{dx} - y = e^x(x+1)^2$

7. $\frac{dy}{dx} + y\cos x = \frac{1}{2}\sin 2x$

8. $(1+x^2)\frac{dy}{dx} + y = e^{\tan^{-1} x}$

9. $(2y - 3x)dx + xdy = 0$

10. $\sin x \frac{dy}{dx} - y\cos x = 2e^x \sin^3 x$

ANSWERS

1. $y\sec^2 x = \sec x + c$

2. $y = \frac{1}{5}e^{2x} + ce^{-3x}$

3. $y\sin x = \frac{\sin^2 x}{2} + c$

4. $y\log x = (\log x)^2 + c$

5. $y(1+x^3) = \frac{x}{2} - \frac{\sin 2x}{4} + c$

6. $y = (e^x + c)(1+x)$

7. $y = \sin x - 1 + ce^{-\sin x}$

8. $2ye^{\tan^{-1} x} = e^{2\tan^{-1} x} + c$

9. $yx^2 = x^3 + c$

10. $y\text{cosec}\, x = e^x(\sin x - \cos x) + c$

2.15 EQUATIONS REDUCIBLE TO THE LINEAR FORM (OR BERNOULLI'S EQUATION)

Equation of the form

$$\frac{dy}{dx} + Py = Qy^n \qquad \text{...(1)}$$

where P and Q are constants or functions of x only, can be reduced to the linear form and are known as Bernoulli's equation.

Method of Solving

Divide the equation (1) by y^n on both the sides and we get,

$$y^{-n}\frac{dy}{dx} + Py^{1-n} = Q \qquad \text{... (2)}$$

Put $y^{1-n} = t$, so that $(1-n)y^{-n}\frac{dy}{dx} = \frac{dt}{dx}$

$$\Rightarrow \qquad y^{-n}\frac{dy}{dx} = \frac{1}{1-n}\frac{dt}{dx}$$

$\therefore$ (2) becomes
$$\frac{1}{1-n}\frac{dt}{dx} + Pt = Q$$

or
$$\frac{dt}{dx} + P(1-n)t = Q(1-n)$$

which is a linear differential equation and can be solved as discussed earlier.

ILLUSTRATIVE EXAMPLES

Example 1. Solve : $\dfrac{dy}{dx}=\dfrac{y}{2x}+\dfrac{y^2}{2x^2}$

Solution. We have $\dfrac{dy}{dx}-\dfrac{y}{2x}=\dfrac{y^2}{2x^2}$

Dividing by y^2 on both sides ,we have

$$y^{-2}\frac{dy}{dx}-\frac{y^{-1}}{2x}=\frac{1}{2x^2} \qquad \text{... (1)}$$

Put $y^{-1}=t$ so that

$$-y^{-2}\frac{dy}{dx}=\frac{dt}{dx}$$

$$\Rightarrow \qquad y^{-2}\frac{dy}{dx}=-\frac{dt}{dx}$$

Now eqn. (1) becomes

$$-\frac{dt}{dx}-\frac{t}{2x}=\frac{1}{2x^2}$$

$$\Rightarrow \qquad \frac{dt}{dx}+\frac{t}{2x}=-\frac{1}{2x^2}$$

which is linear.

Here $\qquad P=\dfrac{1}{2x}, \quad Q=-\dfrac{1}{2x^2}$

Integrating Factor (I.F.) = $e^{\int Pdx}=e^{\int \frac{1}{2x}dx}=e^{\frac{1}{2}\log x}=e^{\log\sqrt{x}}=\sqrt{x}$

Hence, solution is given by

$$t(I.F.)=\int Q(I.F.)dx+c$$

$$\Rightarrow \qquad t\sqrt{x}=-\int\frac{1}{2x^3}.\sqrt{x}dx+c$$

$$=-\frac{1}{2}\int x^{-\frac{5}{2}}dx+c=-\frac{1}{2}\frac{x^{-\frac{3}{2}}}{-\frac{3}{2}}+c$$

$$\Rightarrow \qquad \frac{1}{y}\sqrt{x}=\frac{1}{3}x^{-\frac{3}{2}}+c$$

$$\Rightarrow \qquad 3\sqrt{x}=y\left(x^{-\frac{5}{2}}+c\right)$$

Example 2. Solve : $\frac{dy}{dx}+y\tan x = y^3\cos x$

Solution. Dividing by y^3 on both sides ,we have

$$y^{-3}\frac{dy}{dx}+y^{-2}\tan x = \cos x \qquad \text{...(1)}$$

Put $y^{-2}=t$ so that

$$-2y^{-3}\frac{dy}{dx} = \frac{dt}{dx}$$

$$\Rightarrow \qquad y^{-3}\frac{dy}{dx} = -\frac{1}{2}\frac{dt}{dx}$$

Now eqn. (1) becomes

$$-\frac{1}{2}\frac{dt}{dx}+t\tan x = \cos x$$

$$\Rightarrow \qquad \frac{dt}{dx}-2t\tan x = -2\cos x$$

which is linear.

Here $\qquad P = -2\tan x, \quad Q = -2\cos x$

Integrating Factor (I.F.) = $e^{\int Pdx} = e^{-2\int \tan x dx} = e^{-2\log \sec x} = \frac{1}{\sec^2 x}$

Hence, solution is given by

$$t(I.F.) = \int Q(I.F.)dx + c$$

$$\Rightarrow \qquad t\frac{1}{\sec^2 x} = -2\int \cos x.\frac{1}{\sec^2 x}dx + c$$

$$= -2\int \cos^3 x dx + c = -2\int \cos x.(1-\sin^2 x)dx + c$$

$$= -2\int (1-t^2)dt + c \text{ where } \sin x = t$$

$$= -2\left(t-\frac{t^3}{3}\right)+c$$

$$\Rightarrow \qquad \frac{1}{y^2\sec^2 x} = -2\left(\sin x-\frac{\sin^3 x}{3}\right)+c$$

$$\Rightarrow \qquad \cos^2 x = -2y^2\sin x\left(1-\frac{\sin^2 x}{3}\right)+cy^2$$

Example 3. Solve : $\sec^2 y\dfrac{dy}{dx}+\tan x\tan y = e^x\cos x$

Solution. Put $\tan y = t$ so that $\sec^3 y\dfrac{dy}{dx}=\dfrac{dt}{dx}$

The given equation becomes $\dfrac{dt}{dx}+t\tan x = e^x\cos x$

which is linear.

Here $\quad P = \tan x,\quad Q = e^x\cos x$

Integrating Factor (I.F.) = $e^{\int Pdx} = e^{\int \tan x dx} = e^{\log\sec x} = \sec x$

Hence, solution is given by

$$t(I.F.) = \int Q(I.F.)dx + c$$

$$\Rightarrow \quad t\sec x = \int e^x\cos x.\sec x dx + c$$

$$= \int e^x dx + c = e^x + c$$

$$\Rightarrow \quad \tan y.\sec x = e^x + c$$

Example 4. Solve : $\dfrac{dy}{dx} = x^3y^3 - xy$

Solution. We have $\dfrac{dy}{dx}+xy = x^3y^3$

Dividing by y^{-3} on both sides ,we have

$$y^{-3}\frac{dy}{dx}+xy^{-2} = x^3 \qquad \ldots (1)$$

Put $y^{-2} = t$ so that $-2y^{-3}\dfrac{dy}{dx} = \dfrac{dt}{dx}$

$$\Rightarrow \quad y^{-3}\frac{dy}{dx} = -\frac{1}{2}\frac{dt}{dx}$$

Now eqn. (1) becomes

$$-\frac{1}{2}\frac{dt}{dx}+tx = x^3$$

$$\Rightarrow \quad \frac{dt}{dx}-2tx = -2x^3$$

which is linear.

Here $\quad P = -2x,\quad Q = -2x^3$

Integrating Factor (I.F.) = $e^{\int Pdx} = e^{-2\int xdx} = e^{-x^2}$

Hence, solution is given by

$$t(I.F.) = \int Q(I.F.)dx + c$$

$$\Rightarrow \quad te^{-x^2} = 2\int x^3.e^{-x^2}dx + c$$

$$= -\int ze^z dz + c \text{ where } -x^2 = z \quad \Rightarrow \quad -2xdx = dz$$

$$= -(ze^z - e^z) + c$$

$$\Rightarrow \quad \frac{1}{y^2}.e^{-x^2} = -\left(-x^2.e^{-x^2} - e^{-x^2}\right) + c$$

$$\Rightarrow \quad y^{-2} = -(-x^2 - 1) + ce^{x^2}$$

$$\Rightarrow \quad y^{-2} = x^2 + 1 + ce^{x^2}$$

EXERCISE 2.5

Solve the following differential equations :

1. $\frac{dy}{dx} + x\sin 2y = x^3 \cos^2 y$

2. $x\frac{dy}{dx} + y = y^2 \log x$

3. $\frac{dy}{dx} - x^2 y = y^2 e^{-\frac{1}{2}x^3}$

4. $x\frac{dy}{dx} + y = xy^3$

5. $(1 - x^2)\frac{dy}{dx} + xy = xy^2$

ANSWERS

1. $\tan y = \frac{(x^2 - 1)}{2} + ce^{-x^2}$

2. $y(\log ex + cx) = 1$

3. $y(c - x) = e^{\frac{1}{3}x^3}$

4. $xy^2(2 + cx) = 1$

5. $cy = (1 - y)\sqrt{1 - x^2}$

2.16 EXACT DIFFERENTIAL EQUATIONS

A differential equation is said to be exact if it can be derived from its primitive (solution) directly by differentiation , without any further multiplication, reduction, elimination etc.

A differential equation of the type

$$Mdx + Ndy = 0 \text{ is said to be exact}$$

if

$$\frac{\partial M}{\partial y} = \frac{\partial N}{\partial x}$$

Steps for Solving An Exact Differential Equation

Step 1: Integrate M w.r.t. x keeping y constant.

Step 2: Integrate w.r.t. y only those terms of N which does not contain x.

Step 3: Solution is given by, Result of 1 + result of 2 = constant.

or

$$\int_{y\ constant} Mdx + \int (those\ terms\ of\ N\ not\ containing\ x)dy = constant$$

ILLUSTRATIVE EXAMPLES

Example 1. Solve : $(x^2 + y^2)dx + 2xydy = 0$

Solution. Comparing with $Mdx + Ndy = 0$, we have

$$M = x^2 + y^2, \quad N = 2xy$$

$$\therefore \quad \frac{\partial M}{\partial y} = 2y, \quad \frac{\partial N}{\partial x} = 2y$$

$$\because \quad \frac{\partial M}{\partial y} = \frac{\partial N}{\partial x} \Rightarrow \text{ the given equation is exact.}$$

Now, the solution is given by

$$\int_{y\ constant} Mdx + \int (those\ terms\ of\ N\ not\ containing\ x)dy = constant$$

$$\Rightarrow \quad \int (x^2 + y^2)dx = c$$

$$\Rightarrow \quad \frac{x^3}{3} + xy^2 = c$$

Example 2. Solve : $(e^y + 1)\cos x dx + e^y \sin x dy = 0$

Solution. Comparing with $Mdx + Ndy = 0$, we have

$$M = (e^y + 1)\cos x,$$

$$N = e^y \sin x$$

$$\therefore \quad \frac{\partial M}{\partial y} = e^y \cos x,$$

$$\frac{\partial N}{\partial x} = e^y \cos x$$

$$\because \quad \frac{\partial M}{\partial y} = \frac{\partial N}{\partial x} \Rightarrow \text{ the given equation is exact.}$$

Now, the solution is given by

$$\int_{y\ constant} Mdx + \int (those\ terms\ of\ N\ not\ containing\ x)dy = constant$$

$$\Rightarrow \quad \int (e^y + 1)\cos x dx = c$$

$$\Rightarrow \quad (e^y + 1)\sin x = c$$

Example 3. Solve : $(y\sec^2 x + \sec x \tan x)dx + (\tan x + 2y)dy = 0$

Solution. Comparing with $Mdx + Ndy = 0$, we have

$$M = y\sec^2 x + \sec x \tan x,$$

$$N = \tan x + 2y$$

$$\therefore \quad \frac{\partial M}{\partial y} = \sec^2 x,$$

$$\frac{\partial N}{\partial x} = \sec^2 x$$

$$\because \quad \frac{\partial M}{\partial y} = \frac{\partial N}{\partial x} \Rightarrow \text{the given equation is exact.}$$

Now, the solution is given by

$$\int_{y\ constant} Mdx + \int (those\ terms\ of\ N\ not\ containing\ x)dy = constant$$

$$\Rightarrow \quad \int (y\sec^2 x + \sec x \tan x)dx + \int 2y dy = c$$

$$\Rightarrow \quad y\tan x + \sec x + 2.\frac{y^2}{2} = c$$

$$\Rightarrow \quad y\tan x + \sec x + y^2 = c.$$

Example 4. Solve : $\left(1+3e^{\frac{x}{y}}\right)dx + 3e^{\frac{x}{y}}\left(1-\frac{x}{y}\right)dy = 0$

Solution. Comparing with $Mdx + Ndy = 0$, we have

$$M = 1 + 3e^{\frac{x}{y}},$$

$$N = 3e^{\frac{x}{y}}\left(1-\frac{x}{y}\right)$$

$$\therefore \quad \frac{\partial M}{\partial y} = 3e^{\frac{x}{y}}.\left(-\frac{x}{y^2}\right) \text{ and}$$

$$\frac{\partial N}{\partial x} = 3e^{\frac{x}{y}}\left(-\frac{1}{y}\right) + 3e^{\frac{x}{y}}\left(\frac{1}{y}\right)\left(1-\frac{x}{y}\right) = 3e^{\frac{x}{y}}\left(-\frac{x}{y^2}\right)$$

$$\because \quad \frac{\partial M}{\partial y} = \frac{\partial N}{\partial x} \Rightarrow \text{the given equation is exact.}$$

Now, the solution is given by

$$\int_{y\ constant} Mdx + \int (those\ terms\ of\ N\ not\ containing\ x)dy = constant$$

$$\Rightarrow \quad \int\left(1+3e^{\frac{x}{y}}\right)dx = c$$

$$\Rightarrow \quad x+3e^{\frac{x}{y}}.y = c$$

$$\Rightarrow \quad x+3ye^{\frac{x}{y}} = c$$

Example 5. Solve: $xdx + ydy = \dfrac{a^2(xdy - ydx)}{x^2+y^2}$

Solution. We have, $\left(x+\dfrac{a^2y}{x^2+y^2}\right)dx+\left(y-\dfrac{a^2x}{x^2+y^2}\right)dy = 0$

Comparing with $Mdx + Ndy = 0$, we have

$$M = x+\frac{a^2y}{x^2+y^2},$$

$$N = y-\frac{a^2x}{x^2+y^2}$$

$$\therefore \quad \frac{\partial M}{\partial y} = \frac{a^2(x^2-y^2)}{(x^2+y^2)^2},$$

$$\frac{\partial N}{\partial x} = \frac{a^2(x^2-y^2)}{(x^2+y^2)^2}$$

$$\because \quad \frac{\partial M}{\partial y} = \frac{\partial N}{\partial x} \quad \Rightarrow \quad \text{the given equation is exact.}$$

Now, the solution is given by

$$\int_{y\ constant} Mdx + \int(\textit{those terms of } N \textit{ not containing } x)dy = \textit{constant}$$

$$\Rightarrow \quad \int\left(x+\frac{a^2y}{x^2+y^2}\right)dx+\int ydy = c$$

$$\Rightarrow \quad \frac{x^2}{2}+a^2y.\frac{1}{y}\tan^{-1}\left(\frac{x}{y}\right)+\frac{y^2}{2} = c$$

$$\Rightarrow \quad \frac{x^2+y^2}{2}+a^2\tan^{-1}\left(\frac{x}{y}\right) = c$$

Example 6. Solve : $(ax+hy+g)dx+(hx+by+f)dy=0$ *(I.U. 2007, 08)*

Solution. Comparing with $Mdx + Ndy = 0$, we have

$$M = ax + hy + g$$

and $$N = hx + by + f$$

$$\therefore \quad \frac{\partial M}{\partial y} = h \text{ and } \frac{\partial N}{\partial x} = h$$

$$\because \quad \frac{\partial M}{\partial y} = \frac{\partial N}{\partial x} \Rightarrow \text{the given equation is exact.}$$

Now, the solution is

$$\int (ax+hy+g)dx + \int (by+f)dy = 0$$

$$\Rightarrow \quad \frac{ax^2}{2} + hxy + gx + \frac{by^2}{2} + fy = c$$

$$\Rightarrow \quad ax^2 + 2hxy + by^2 + 2gx + 2fy = c$$

EXERCISE 2.6

Solve the following differential equations :

1. $(x+y-10)dx+(x-y-2)dy=0$
2. $(x^4+y^4-2xy^2)dx=(2x^2y+\sin y-4xy^3)dy=0$
3. $(x^2-ay)dx-(ax-y^2)dy=0$
4. $(y^2e^{xy^2}+4x^3)dx+(2xye^{xy^2}-3y^2)dy=0$
5. $(e^y+2)\sin x dx - e^y \cos x dy = 0$
6. $y\sin 2x dx-(1+y^2+\cos^2 x)dy=0$
7. $(2xy+e^y)dx+(x^2+xe^y)dy=0$
8. $(5x^4-2xy^3+3x^2y^2)dx+(2x^3y-3x^2y^2-5y^4)dy=0$
9. $(1+4xy+2y^2)dx+(1+4xy+2x^2)dy=0$

ANSWERS

1. $\dfrac{x^2-y^2}{2}+xy-10x-2y=c$
2. $\dfrac{x^5}{5}-x^2y^2+xy^4+\cos y=c$
3. $x^3+y^3-3axy=c$
4. $x^4-y^3+e^{xy^2}=c$
5. $(e^y+2)\cos x=c$
6. $-\dfrac{1}{2}y\cos 2x-\dfrac{3}{2}y-\dfrac{1}{3}y^3=c$
7. $x(xy+e^y)=c$
8. $x^5+x^3y^2-x^2y^3-y^5=c$
9. $x+y+2x^2y+2xy^2$

2.17 EQUATIONS REDUCIBLE TO THE EXACT DIFFERENTIAL EQUATIONS

Sometimes a differential equation is not exact but can be made exact by multiplying by a function, known as *Integrating Factor (I.F.)*.

There are certain rules for finding the I.F as follows :

Rule I: If M is of the form $M = yf_1(xy)$ and N is of the form $N = xf_2(xy)$ and $Mx - Ny \neq 0$

then, $$\text{I.F.} = \frac{1}{Mx - Ny}$$

ILLUSTRATIVE EXAMPLES

Example 1. Solve : $y(1+xy)dx + x(1+xy+x^2y^2)dy = 0$

Solution. We have, $y(1+xy)dx + x(1+xy+x^2y^2)dy = 0$ ... (1)

Comparing with $Mdx + Ndy = 0$, we have

$$M = y(1+xy),$$

$$N = x(1+xy+x^2y^2)$$

$$\therefore \quad \frac{\partial M}{\partial y} = 1+2xy,$$

$$\frac{\partial N}{\partial x} = 1+2xy+3x^2y^2$$

$$\because \quad \frac{\partial M}{\partial y} \neq \frac{\partial N}{\partial x}$$

$\Rightarrow$ equation (1) is not exact.

But $M = yf_1(xy)$ and $N = xf_2(xy)$

and $Mx - Ny = xy + x^2y^2 - xy - x^2y^2 - x^3y^3 = -x^3y^3 \neq 0$

$$\therefore \quad \text{I.F.} = \frac{1}{Mx - Ny} = -\frac{1}{x^3y^3}$$

Multiplying (1) by $-\dfrac{1}{x^3y^3}$, we get

$$\left(-\frac{1}{x^3y^2} - \frac{1}{x^2y}\right)dx + \left(-\frac{1}{x^2y^3} - \frac{1}{xy^2} - \frac{1}{y}\right)dy = 0$$

Here, $$M = -\frac{1}{x^3y^2} - \frac{1}{x^2y},$$

$$N = -\frac{1}{x^2y^3} - \frac{1}{xy^2} - \frac{1}{y}$$

$$\therefore \quad \frac{\partial M}{\partial y} = \frac{2}{x^3y^3} + \frac{1}{x^2y^2},$$

$$\frac{\partial N}{\partial x} = \frac{2}{x^3y^3} + \frac{1}{x^2y^2}$$

$$\because \quad \frac{\partial M}{\partial y} = \frac{\partial N}{\partial x}$$

$\Rightarrow$ equation (1) is exact.

Now, the solution is given by

$$\int_{y\ constant} Mdx + \int (those\ terms\ of\ N\ not\ containing\ x)dy = constant$$

$$\Rightarrow \quad \int\left(\frac{-1}{x^3y^2} - \frac{1}{x^2y}\right)dx + \int \frac{-1}{y}dy = c$$

$$\Rightarrow \quad \frac{1}{2x^2y^2} + \frac{1}{xy} - \log y = c$$

Rule II: If the equation $Mdx + Ndy = 0$ is a homogeneous equation and $Mx + Ny \neq 0$

then, $\quad$ I.F. $= \dfrac{1}{Mx + Ny}$

Example 2. Solve : $(x^2y - 2xy^2)dx - (x^3 - 3x^2y)dy = 0$

Solution. We have, $(x^2y - 2xy^2)dx - (x^3 - 3x^2y)dy = 0$ ... (1)

Comparing with $Mdx + Ndy = 0$, we have

$$M = x^2y - 2xy^2,$$

$$N = -x^3 + 3x^2y$$

$$\therefore \quad \frac{\partial M}{\partial y} = x^2 - 4xy,$$

$$\frac{\partial N}{\partial x} = -3x^2 + 6xy$$

$$\because \quad \frac{\partial M}{\partial y} \neq \frac{\partial N}{\partial x} \quad \Rightarrow \text{ equation (1) is not exact.}$$

But M and N are homogeneous function of x and y

and $$Mx + Ny = x^3y - 2x^2y^2 - x^3y + 3x^2y^2 = x^2y^2 \neq 0$$

$$\therefore \qquad \text{I.F.} = \frac{1}{Mx+Ny} = \frac{1}{x^2y^2}$$

Multiplying equation (1) by $\frac{1}{x^2y^2}$, we get

$$\left(\frac{1}{y} - \frac{2}{x}\right)dx + \left(-\frac{x}{y^2} + \frac{3}{y}\right)dy = 0$$

Here, $$M = \frac{1}{y} - \frac{2}{x},$$

$$N = -\frac{x}{y^2} + \frac{3}{y}$$

$$\therefore \qquad \frac{\partial M}{\partial y} = \frac{-1}{y^2},$$

$$\frac{\partial N}{\partial x} = \frac{-1}{y^2}$$

$$\because \qquad \frac{\partial M}{\partial y} = \frac{\partial N}{\partial x} \Rightarrow \text{ equation (1) is exact.}$$

Now, the solution is given by

$$\int_{y\ constant} Mdx + \int (\textit{those terms of N not containing x})dy = \textit{constant}$$

$$\Rightarrow \qquad \int \left(\frac{1}{y} - \frac{2}{x}\right)dx + \int \frac{3}{y}dy = c$$

$$\Rightarrow \qquad \frac{x}{y} - 2\log x + 3\log y = c$$

Rule III: If $\dfrac{\left(\dfrac{\partial M}{\partial y} - \dfrac{\partial N}{\partial x}\right)}{N}$ is a function x of alone, say $f(x)$, then

$$\text{I.F.} = e^{\int f(x)dx}$$

Example 4. Solve : $(2x\log x - xy)dy + 2ydx = 0$

Solution. We have, $(2x\log x - xy)dy + 2ydx = 0$ ... (1)

Comparing with $Mdx + Ndy = 0$, we have

$$M = 2y,$$

$$N = 2x\log x - xy$$

$$\therefore \quad \frac{\partial M}{\partial y} = 2,$$

$$\frac{\partial N}{\partial x} = 2 + 2\log x - y$$

$$\because \quad \frac{\partial M}{\partial y} \neq \frac{\partial N}{\partial x} \Rightarrow \text{ equation (1) is not exact.}$$

Now,

$$\frac{\left(\frac{\partial M}{\partial y} - \frac{\partial N}{\partial x}\right)}{N} = \frac{(2-2-2\log x + y)}{2x\log x - xy} = -\frac{1}{x} = f(x)$$

$$\text{I.F.} = e^{\int f(x)dx} = e^{\int \frac{-1}{x}dx} = e^{-\log x} = e^{\log\frac{1}{x}} = \frac{1}{x}$$

Multiplying eqn. (1) by $\frac{1}{x}$, we get

$$\left(\frac{2y}{x}\right)dx + (2\log x - y)dy = 0$$

Here,

$$M = \frac{2y}{x},$$

$$N = 2\log x - y$$

$$\therefore \quad \frac{\partial M}{\partial y} = \frac{2}{x},$$

$$\frac{\partial N}{\partial x} = \frac{2}{x}$$

$$\because \quad \frac{\partial M}{\partial y} = \frac{\partial N}{\partial x} \Rightarrow \text{ equation (1) is exact.}$$

Now, the solution is given by

$$\int_{y\ constant} Mdx + \int (those\ terms\ of\ N\ not\ containing\ x)dy = constant$$

$$\Rightarrow \quad \int \left(\frac{2y}{x}\right)dx + \int (-y)dy = c$$

$$\Rightarrow \quad 2y\log x - \frac{y^2}{2} = c$$

Rule IV: If $\dfrac{\left(\dfrac{\partial N}{\partial x}-\dfrac{\partial M}{\partial y}\right)}{M}$ is a function of y alone, say $f(y)$, then

$$\text{I.F.} = e^{\int f(y)dy}$$

Example 4. Solve : $(x^2y^2 + ye^x)dx - e^x dy = 0$

Solution. We have, $(x^2y^2 + ye^x)dx - e^x dy = 0$ ... (1)

Comparing with $Mdx + Ndy = 0$, we have

$$M = x^2y^2 + ye^x,$$

$$N = -e^x$$

$$\therefore \quad \frac{\partial M}{\partial y} = 2x^2y + e^x,$$

$$\frac{\partial N}{\partial x} = -e^x$$

$$\because \quad \frac{\partial M}{\partial y} \neq \frac{\partial N}{\partial x}$$

$\Rightarrow$ equation (1) is not exact.

Now,
$$\frac{\left(\dfrac{\partial N}{\partial x}-\dfrac{\partial M}{\partial y}\right)}{M} = \frac{(-e^x - 2x^2y - e^x)}{x^2y^2 + ye^x} = \frac{-2}{y} = f(y)$$

$$\therefore \quad \text{I.F.} = e^{\int f(y)dy} = e^{\int \frac{-2}{y}dy} = e^{-2\log y} = e^{\log\frac{1}{y^2}} = \frac{1}{y^2}$$

Multiplying equation (1) by $\dfrac{1}{y^2}$, we get

$$\left(x^2 + \frac{e^x}{y}\right)dx - \left(\frac{e^x}{y^2}\right)dy = 0$$

Here,
$$M = x^2 + \frac{e^x}{y},$$

$$N = -\frac{e^x}{y^2}$$

$$\therefore \quad \frac{\partial M}{\partial y} = \frac{-e^x}{y^2},$$

$$\frac{\partial N}{\partial x} = \frac{-e^x}{y^2}$$

$$\because \quad \frac{\partial M}{\partial y} = \frac{\partial N}{\partial x}$$

$\Rightarrow$ equation (1) is exact.

Now, the solution is given by

$$\int_{y\ constant} Mdx + \int (those\ terms\ of\ N\ not\ containing\ x)dy = constant$$

$$\Rightarrow \quad \int \left(x^2 + \frac{e^x}{y} \right) dx = c$$

$$\Rightarrow \quad \frac{x^3}{3} + \frac{e^x}{y} = c.$$

Rule V: If equation is of the form $x^m y^n (aydx + bxdy) + x^p y^q (rydx + sxdy) = 0$, then, I.F. $= x^h y^k$, where a, b, m, n, p, q, r, s are all constants.

Example 5. Solve : $x(3ydx + 2xdy) + 8y^4 (ydx + 3xdy) = 0$

Solution. We have, $x(3ydx + 2xdy) + 8y^4 (ydx + 3xdy) = 0$

or $$(3xy + 8y^5)dx + (2x^2 + 24xy^4)dy = 0 \quad \text{... (1)}$$

Comparing with $Mdx + Ndy = 0$, we have

$$M = 3xy + 8y^5,$$

$$N = 2x^2 + 24xy^4$$

$$\therefore \quad \frac{\partial M}{\partial y} = 3x + 40y^4,$$

$$\frac{\partial N}{\partial x} = 4x + 24y^4$$

$$\because \quad \frac{\partial M}{\partial y} \neq \frac{\partial N}{\partial x} \Rightarrow \text{equation (1) is not exact.}$$

$$\therefore \quad \text{I.F.} = x^h y^k$$

Multiplying equation (1) by $x^h y^k$, we get

$$(3x^{h+1}y^{k+1} + 8x^h y^{k+5})dx + (2x^{h+2}y^k + 24x^{h+1}y^{k+4})dy = 0 \quad \text{... (2)}$$

which is exact.

Here, $M = 3x^{h+1}y^{k+1} + 8x^h y^{k+5}$,

$N = 2x^{h+2}y^k + 24x^{h+1}y^{k+4}$

$$\therefore \quad \frac{\partial M}{\partial y} = \frac{\partial N}{\partial x}$$

$$\Rightarrow 3(k+1)x^{h+1}y^k + 8(k+5)x^h y^{k+4} = 3(h+2)x^{h+1}y^k + 24(h+1)x^h y^{k+4}$$

$$\Rightarrow \quad 3(k+1) = 3(h+2),$$

$$8(k+5) = 24(h+1)$$

$$\Rightarrow \quad k - h = 1,$$

$$k - 2h = -2$$

$$\Rightarrow \quad h = 3, \ k = 4$$

Putting these values in equation (2), we get

$(3x^4y^5 + 8x^3y^9)dx + (2x^5y^4 + 24x^4y^8)dy = 0$, which is exact.

Hence, the solution is given by

$$\int_{y\ constant} Mdx + \int(those\ terms\ of\ N\ not\ containing\ x)dy = constant$$

$$\Rightarrow \quad \int(3x^4y^5 + 8x^3y^9)dx = c$$

$$\Rightarrow \quad \frac{3x^5y^4}{5} + 2x^4y^9 = c.$$

EXERCISE 2.7

Solve the following differential equations :

1. $(x^2y^2 + xy + 1)ydx + (x^2y^2 - xy + 1)xdy = 0$ [By Rule I]
2. $(y^4 + 2y)dx + (xy^3 + 2y^4 - 4x)dy = 0$ [By Rule IV]
3. $x(3ydx + 2xdy) + 8y^4(ydx + 3xdy) = 0$ [By Rule V]
4. $(x^2y - 2xy^2)dx - (x^3 - 3x^2y)dy = 0$ [By Rule II]
5. $\left(y + \frac{y^3}{3} + \frac{x^2}{2}\right)dx + \frac{1}{4}(1 + y^2)xdy = 0$ [By Rule III]
6. $xy^2dy = (x^3 + y^3)dx$ [By Rule II]

7. $(xy^3+y)dx+2(x^2y^2+x+y^4)dy=0$ [By Rule IV]

8. $(xy\sin xy+\cos xy)ydx+(xy\sin xy-\cos xy)xdy=0$ [By Rule I]

9. $(2x^2y^2+y)dx-(x^3y-3x)dy=0$ [By Rule V]

10. $(x-y^2)dx+2xydy=0$ [By Rule III]

ANSWERS

1. $x^2y^2+xy\log\left(\dfrac{cx}{y}\right)=1$

2. $x\left(y+\dfrac{2}{y^2}\right)+y^2=c$

3. $x^3y^2+4x^2y^6=c$

4. $\dfrac{x}{y}-2\log x+3\log y=c$

5. $\dfrac{x^4y}{4}+\dfrac{x^4y^3}{12}+\dfrac{x^6}{12}=c$

6. $\log x-\dfrac{y^3}{3x^3}=c$

7. $3x^2y^4+6xy^2+2y^6=c$

8. $x\sec xy=cy$

9. $\dfrac{7}{5}x^{10/7}y^{-5/7}-\dfrac{7}{4}x^{-4/7}y^{-12/7}=c$

10. $y^2=cx-x\log x$

2.18 LINEAR DIFFERENTIAL EQUATIONS WITH CONSTANT COEFFICIENTS

The equation of the form

$$\frac{d^ny}{dx^n}+a_1\frac{d^{n-1}y}{dx^{n-1}}+a_2\frac{d^{n-2}y}{dx^{n-2}}+\ldots\ldots\ldots\ldots+a_ny=Q \qquad \ldots (1)$$

Where $a_1, a_2, \ldots\ldots, a_n$ are all constants and Q is a function of x alone; is called a linear differential equation of nth order with constant coefficients.

Note: The coefficient of $\dfrac{d^ny}{dx^n}$ should always be unity.

The Differential Operator D: In $\dfrac{dy}{dx}, \dfrac{d}{dx}$ is called *operator or differential operator.*

We write $\dfrac{d}{dx}\equiv D, \dfrac{d^2}{dx^2}\equiv D^2, \ldots\ldots\ldots\ldots, \dfrac{d^n}{dx^n}\equiv D^n$

In symbolic form equation (1) can be written as

$$D^ny+a_1D^{n-1}y+a_2D^{n-2}y+\ldots\ldots\ldots+a_{n-1}Dy+a_ny=Q$$

or

$$(D^n+a_1D^{n-1}+a_2D^{n-2}+\ldots\ldots\ldots+a_{n-1}D+a_n)y=Q$$

or $\qquad f(D) = Q$

where $f(D) \equiv D^n + a_1 D^{n-1} + a_2 D^{n-2} + ... + a_{n-1} D + a_n$

i.e., $f(D)$ is a polynomial in D.

Note: The polynomial $f(D)$ can be factorised by ordinary rules of algebra.

2.19 COMPLETE SOLUTION

The complete solution of these types of equations is made of two parts , one is *complementary function (C.F.)* and the other *particular integral (P.I.)*.

$\therefore$ The complete solution of equations (1), is y = **C.F.+P.I.**

Where C.F. is complete solution of the equation $f(D) = 0$ and P.I. is the particular solution of the equation $f(D) = Q$.

2.20 COMPLEMENTARY FUNCTION (C.F.)

Complementary function is actually the solution of the equation $f(D) = Q$ when Q is replaced by zero.for finding C.F. we first find auxiliary equation.

Note 1: C.F. and P.I. are not connected to each other. If $Q = 0$, P.I. is zero.

Note 2: The solution of the differential equation $f(D) = Q$ should always be written as y = C.F. + P.I. no matter P.I. is zero or not.

Auxiliary Equation : The equation obtained by equating to zero the symbolic coefficient of y is called the *auxiliary equation (A.E.)*. In this equation D is replaced by m.

Steps for Finding Auxiliary Equation

1. In $f(D) = 0$ replace D by m.
2. Replace D^2 by m^2 and so on D^n by m^n
3. Now we get an algebraic equation in m of degree n, called auxiliary equation.

Note: While finding out the auxiliary equation, we assume Q as zero in the equation $f(D) = Q$.

2.21 METHODS FOR FINDING THE COMPLEMENTARY FUNCTION

Consider the equation

$$(D^n + a_1 D^{n-1} + a_2 D^{n-2} + ... + a_n)y = 0$$

where all the a_i's are constant.

Its auxiliary equation is $\qquad m^n + a_1 m^{n-1} + a_2 m^{n-2} + ... + a_n = 0$

It is an algebraic equation in m of degree n. So, on solving, it will give n values of m.

Let $m = m_1, m_2, m_3, ..., m_n$ be the roots of the auxiliary equation. The C.F.now depends upon the nature of the roots of the A.E. Hence the following cases arise.

Case I: When the roots of the A.E. are real and distinct :

In this case ,

$$\text{C.F.} = c_1e^{m_1x} + c_2e^{m_2x} + ... + c_ne^{m_nx}$$

Case II: When the roots of the A.E. are equal :

(*a*) When two roots are equal, *i.e.*, $m_1 = m_2$,then

$$\text{C.F.} = (c_1 + c_2x)e^{m_1x} + c_3e^{m_3x} + ... + c_ne^{m_nx}$$

(*b*) When three roots are equal, *i.e.*, $m_1 = m_2 = m_3$, then

$$\text{C.F.} = (c_1 + c_2x + c_3x^2)e^{m_1x} + c_4e^{m_4x} + ... + c_ne^{m_nx}$$

Case III: When roots of the auxiliary equation are imaginary :

Let $\quad m_1 = \alpha + i\beta$ and $m_2 = \alpha - i\beta$

are two imaginary roots and the other roots are real and distinct, then

$$\text{C.F.} = e^{\alpha x}(c_1 \cos\beta x + c_2 \sin\beta x) + c_3e^{m_3x} + ... + c_ne^{m_nx}$$

Case IV: When roots of the auxiliary equation are repeated imaginary :

Let $\quad m_1 = m_2 = \alpha + i\beta$ and $m_3 = m_4 = \alpha - i\beta$, then

$$\text{C.F.} = e^{\alpha x}[(c_1 + c_2x)\cos\beta x + (c_3 + c_4x)\sin\beta x] + c_5e^{m_5x} + ... + c_ne^{m_nx}$$

ILLUSTRATIVE EXAMPLES

Example 1. Solve : $\dfrac{d^2y}{dx^2} - 7\dfrac{dy}{dx} + 12y = 0$

Solution. The auxiliary equation is

$$m^2 - 7m + 12 = 0$$

$$\Rightarrow \quad (m-3)(m-4) = 0$$

$$\Rightarrow \quad m = 3, 4$$

The roots are real and distinct

$\therefore$ Complementary function, (C.F.) = $c_1e^{3x} + c_2e^{4x}$

And particular integral, (P.I.) = 0

Hence the complete solution is

$$y = \text{C.F.} + \text{P.I.} = c_1e^{3x} + c_2e^{4x}$$

where c_1 and c_2 are arbitrary constants of integration.

Example 2. Solve : $\dfrac{d^3y}{dx^3} - 3\dfrac{d^2y}{dx^2} + 4y = 0$

Solution. The auxiliary equation is

$$m^3 - 3m^2 + 4 = 0$$

$\Rightarrow \qquad (m+1)(m-2)^2 = 0$

$\Rightarrow \qquad m = 1, 2, 2$

Complementary function,

$$(\text{C.F.}) = c_1e^{-x} + (c_2 + c_3x)e^{2x}$$

And particular integral,

$$(\text{P.I.}) = 0$$

Hence, the complete solution is

$$y = \text{C.F.} + \text{P.I.} = c_1e^{-x} + (c_2 + c_3x)e^{2x}$$

where c_1, c_2 and c_3 are arbitrary constants of integration.

Example 3. Solve : $\frac{d^3y}{dx^3} - 3\frac{d^2y}{dx^2} + 3\frac{dy}{dx} - y = 0$

Solution. The auxiliary equation is

$m^3 - 3m^2 + 3m - 1 = 0$

$\Rightarrow \qquad (m-1)^3 = 0$

$\Rightarrow \qquad m = -1, 1, 1$

$\therefore \qquad \text{C.F.} = (c_1 + c_2x + c_3x^2)e^x$

$\text{P.I.} = 0$

Hence, the complete solution is

$$y = \text{C.F.} + \text{P.I.}$$

$$= (c_1 + c_2x + c_3x^2)e^x$$

where c_1, c_2 and c_3 are arbitrary constants of integration.

Example 4. Solve : $(D^3 - 2D^2 + 4D - 8)y = 0$

Solution. The auxiliary equation is

$m^3 - 2m^2 + 4m - 8 = 0$

$\rightarrow \qquad (m-2)(m^2+4) - 0$

$\Rightarrow \qquad m = 2, \pm 2i$

$\therefore \qquad \text{C.F.} = c_1e^{2x} + c_2\cos 2x + c_3\sin 2x$

$\text{P.I.} = 0$

Hence, the complete solution is

$$y = \text{C.F.} + \text{P.I.} = c_1e^{2x} + c_2\cos 2x + c_3\sin 2x$$

where c_1, c_2 and c_3 are arbitrary constants of integration.

Example 5. Solve: $(D^2+1)^3 y = 0$

Solution. The auxiliary equation is

$$(m^2+1)^3 = 0$$

$$\Rightarrow \quad m = \pm i, \ \pm i, \ \pm i$$

$$\therefore \quad \text{C.F.} = (c_1 + c_2 x + c_3 x^2)\cos x + (c_4 + c_5 x + c_6 x^2)\sin x$$

$$\text{P.I.} = 0$$

Hence, the complete solution is

$$y = \text{C.F.+P.I.} = (c_1 + c_2 x + c_3 x^2)\cos x + (c_4 + c_5 x + c_6 x^2)\sin x$$

where c_1, c_2, c_3, c_4, c_5 and c_6 are arbitrary constants of integration.

Example 6. Solve : $(D^2+D+1)^2(D-2)y = 0$

Solution. The auxiliary equation is

$$(m^2+m+1)^2(m-2) = 0$$

$$\Rightarrow \quad (m^2+m+1)^2 = 0 \text{ and } m-2=0$$

$$\Rightarrow \quad m = \frac{-1 \pm \sqrt{1-4}}{2} \text{ and } m = 2$$

$$\Rightarrow \quad m = \frac{-1 \pm i\sqrt{3}}{2}, \ \frac{-1 \pm i\sqrt{3}}{2}, \ 2$$

$$i.e., \quad m = \frac{-1}{2} \pm i\frac{\sqrt{3}}{2}, \ \frac{-1}{2} \pm i\frac{\sqrt{3}}{2}, \ 2$$

$$\therefore \quad \text{C.F.} = e^{\frac{-x}{2}}\left\{(c_1 + c_2 x)\cos\frac{\sqrt{3}}{2}x + (c_4 + c_5 x)\sin\frac{\sqrt{3}}{2}x\right\} + c_5 e^{2x}$$

$$\text{P.I.} = 0$$

Hence, the complete solution is

$$y = \text{C.F.+P.I.} = e^{\frac{-x}{2}}\left\{(c_1 + c_2 x)\cos\frac{\sqrt{3}}{2}x + (c_4 + c_5 x)\sin\frac{\sqrt{3}}{2}x\right\} + c_5 e^{2x}$$

where c_1, c_2, c_3, c_4 and c_5 are arbitrary constants of integration.

Example 7. Solve : $\dfrac{d^4 y}{dx^4} + 4y = 0$

Solution. The auxiliary equation is

$$m^4 + 4 = 0$$

$$\Rightarrow \quad m^4 + 4 - 4m^2 + 4m^2 = 0$$

$$\Rightarrow \qquad (m^2+2)^2 - 4m^2 = 0$$

$$\Rightarrow \qquad (m^2+2)^2 - (2m)^2 = 0$$

$$\Rightarrow \quad (m^2+2-2m)(m^2+2+2m) = 0$$

$$\Rightarrow \qquad m^2 - 2m + 2 = 0, \quad m^2 + 2m + 2 = 0$$

$$\Rightarrow \qquad m = \frac{2 \pm \sqrt{4-8}}{2}; \quad m = \frac{-2 \pm \sqrt{4-8}}{2}$$

$$\Rightarrow \qquad m = 1 \pm i, -1 \pm i$$

$$\therefore \qquad \text{C.F.} = e^{-x}\left[c_1 \cos x + c_2 \sin x\right] + e^{x}\left[c_3 \cos x + c_4 \sin x\right]$$

$$\text{P.I.} = 0$$

Hence, the complete solution is

$$y = \text{C.F.+P.I.} = e^{-x}\left[c_1 \cos x + c_2 \sin x\right] + e^{x}\left[c_3 \cos x + c_4 \sin x\right]$$

where c_1, c_2, c_3 and c_4 are arbitrary constants of integration.

Example 8. Solve : $\dfrac{d^4y}{dx^4} - 4\dfrac{d^3y}{dx^3} + 8\dfrac{d^2y}{dx^2} - 8\dfrac{dy}{dx} + 4y = 0$

Solution. The auxiliary equation is

$$\Rightarrow \quad m^4 - 4m^3 + 8m^2 - 8m + 4 = 0$$

$$\Rightarrow \qquad (m^2 - 2m + 2)^2 = 0$$

$$\Rightarrow \qquad m = \frac{2 \pm \sqrt{4-8}}{2} \quad \text{(twice)}$$

$$\Rightarrow \qquad m = 1 \pm i, \; 1 \pm i$$

$$\therefore \qquad \text{C.F.} = e^{x}\left[(c_1 + c_2 x)\cos x + (c_3 + c_4 x)\sin x\right]$$

$$\text{P.I.} = 0$$

Hence, the complete solution is

$$y = \text{C.F. + P.I.} = e^{x}\left[(c_1 + c_2 x)\cos x + (c_3 + c_4 x)\sin x\right]$$

where c_1, c_2, c_3 and c_4 are arbitrary constants of integration.

EXERCISE 2.8

1. $\dfrac{d^3y}{dx^3} - 7\dfrac{dy}{dx} - 6y = 0$

2. $(D^4 - D^3 - 9D^2 - 11D - 4)y = 0$

3. $\dfrac{d^4y}{dx^4} + 13\dfrac{d^2y}{dx^2} + 36y = 0$

4. $(D^4 - n^4)y = 0$

5. $\frac{d^3y}{dx^3} - 4\frac{d^2y}{dx^2} + 4\frac{dy}{dx} = 0$

6. $\frac{d^4y}{dx^4} + 8\frac{d^2y}{dx^2} + 16y = 0$

7. $(D^3 + 2D^2 + D)y = 0$

8. $(D^2 + 1)^3 (D^2 + D + 1)^2 y = 0$

ANSWERS

1. $y = c_1 e^{-x} + c_2 e^{-2x} + c_3 e^{3x}$
2. $y = e^{-x}(c_1 + c_2 x + c_3 x^2) + c_4 e^{4x}$
3. $y = c_1 \cos 3x + c_2 \sin 3x + c_3 \cos 2x + c_4 \sin 2x$
4. $y = c_1 e^{nx} + c_2 e^{-nx} + c_3 \cos nx + c_4 \sin nx$
5. $y = c_1 + (c_2 + c_3 x)e^{2x}$
6. $y = (c_1 + c_2 x)\cos 2x + (c_3 + c_4 x)\sin 2x$
7. $y = c_1 + (c_2 + c_3 x)e^{-x}$
8. $y = (c_1 + c_2 x + c_3 x^2)\cos x + (c_4 + c_5 x + c_6 x^2)\sin x$

$$+e^{\frac{-x}{2}}\left\{(c_7 + c_8 x) cox \frac{\sqrt{3}}{2}x + (c_9 + c_{10} x)\sin\frac{\sqrt{3}}{2}x\right\}$$

2.22 RULES TO FIND THE PARTICULAR INTEGRAL (P.I.)

The P.I. of the differential equation $f(D)y = Q$ is given by

$$\text{P.I.} = \frac{1}{f(D)}Q$$

Case I : When $Q = e^{ax}$.

We know that
$$D(e^{ax}) = ae^{ax}$$
$$D^2(e^{ax}) = a^2 e^{ax}$$
$$D^3(e^{ax}) = a^3 e^{ax}$$
$$\dots\dots\dots\dots$$
$$\dots\dots\dots\dots$$
$$D^n(e^{ax}) = a^n e^{ax}$$

$\therefore \quad (D^n + a_1 D^{n-1} + a_2 D^{n-2} + \dots\dots\dots + a_n)e^{ax}$

$$= (a^n + a_1 a^{n-1} + a_2 a^{n-2} + \dots\dots\dots + a_n)e^{ax}$$

$\Rightarrow \quad f(D)e^{ax} = f(a)e^{ax}$

Operating on both sides by $\frac{1}{f(D)}$, we have

$$\frac{1}{f(D)} f(D)e^{ax} = \frac{1}{f(D)} f(a)e^{ax}$$

$$\Rightarrow \qquad e^{ax} = f(a)\frac{1}{f(D)}e^{ax}$$

$$\Rightarrow \qquad \frac{1}{f(a)}e^{ax} = \frac{1}{f(D)}e^{ax}$$

Hence $\qquad \text{P.I.} = \frac{1}{f(D)}e^{ax} = \frac{1}{f(a)}e^{ax}$, provided $f(a) \neq 0$

If $f(a) = 0$, the above method fails.

then $\text{P.I.} = \frac{1}{f(D)}e^{ax} = x.\frac{1}{f'(a)}e^{ax}$, provided $f'(a) \neq 0$

If $f'(a) = 0$, then $\text{P.I.} = \frac{1}{f(D)}e^{ax} = x^2.\frac{1}{f''(a)}e^{ax}$, provided $f''(a) \neq 0$ and so on.

ILLUSTRATIVE EXAMPLES

Example 1. Solve : $\frac{d^3y}{dx^3} - \frac{dy}{dx} + 6y = e^{4x}$

Solution. The auxiliary equation is

$$m^3 - m + 6 = 0$$

$$\Rightarrow \quad (m+2)(m^2 - 2m + 3) = 0$$

$$\Rightarrow \quad m = -2, \ m = \frac{2 \pm \sqrt{4-12}}{2}$$

$$\Rightarrow \quad m = -2, \ 1 \pm i\sqrt{2}$$

$$\therefore \quad \text{C.F.} = c_1e^{-2x} + e^x(c_2 \cos\sqrt{2}x + c_3 \sin\sqrt{2}x)$$

And particular integral,

$$\text{P.I.} = \frac{1}{D^3 - D + 6}e^{4x}$$

$$= \frac{1}{4^3 - 4 + 6}e^{4x} = \frac{1}{66}e^{4x}$$

Hence, the complete solution is

$$y = \text{C.F.} + \text{P.I}$$

i.e.,
$$y = c_1e^{-2x} + e^x(c_2\cos\sqrt{2}x + c_3\sin\sqrt{2}x) + \frac{1}{66}e^{4x}$$

where c_1, c_2 and c_3 are arbitrary constants of integration.

Example 2. Solve : $(D^3+1)y = (e^x+1)^2$

Solution. The auxiliary equation is

$$m^3 + 1 = 0$$

$$\Rightarrow \quad (m+1)(m^2-m+1) = 0$$

$$\Rightarrow \quad m = -1,\ \frac{1}{2} \pm i\frac{\sqrt{3}}{2}$$

$$\therefore \quad \text{C.F.} = c_1e^{-x} + e^{\frac{x}{2}}\left(c_2\cos\frac{\sqrt{3}}{2}x + c_3\sin\frac{\sqrt{3}}{2}x\right)$$

And
$$\text{P.I.} = \frac{1}{D^3+1}(e^x+1)^2 = \frac{1}{D^3+1}(e^{2x}+2e^x+1)$$

$$= \frac{1}{D^3+1}e^{2x} + 2\frac{1}{D^3+1}e^x + \frac{1}{D^3+1}e^{0x}$$

$$= \frac{1}{2^3+1}e^{2x} + \frac{1}{1^3+1}e^x + \frac{1}{0^3+1}e^{0x}$$

$$= \frac{1}{9}e^{2x} + \frac{1}{2}e^x + 1$$

Hence, the complete solution is

$$y = \text{C.F.} + \text{P.I}$$

i.e.,
$$y = c_1e^{-x} + e^{\frac{x}{2}}\left(c_2\cos\frac{\sqrt{3}}{2}x + c_3\sin\frac{\sqrt{3}}{2}x\right) + \frac{1}{9}e^{2x} + \frac{1}{2}e^x + 1$$

where c_1, c_2 and c_3 are arbitrary constants of integration.

Example 3. Solve : $(D^2 - D - 6)y = e^x\cosh 2x$

Solution. The auxiliary equation is

$$m^2 - m - 6 = 0$$

$$\Rightarrow \quad (m-3)(m+2) = 0$$

$$\Rightarrow \quad m = 3, -2$$

$$\therefore \quad \text{C.F.} = c_1e^{3x} + c_2e^{-2x}$$

And $\quad \text{P.I.} = \dfrac{1}{D^2 - D - 6} e^x \cosh 2x$

$$= \frac{1}{D^2 - D - 6} e^x \left(\frac{e^{2x} + e^{-2x}}{2} \right)$$

$$= \frac{1}{D^2 - D - 6} \left(\frac{e^{3x} + e^{-x}}{2} \right)$$

$$= \frac{1}{2} \left[\frac{1}{D^2 - D - 6} e^{3x} \right] + \frac{1}{2} \left[\frac{1}{D^2 - D - 6} e^{-x} \right]$$

$$= \frac{1}{2} \left[x \frac{1}{2D - 1} e^{3x} \right] + \frac{1}{2} \left[\frac{1}{(-1)^2 - (-1) - 6} e^{-x} \right]$$

(case of failure in first term)

$$= \frac{1}{2} \left[x \frac{1}{2.3 - 1} e^{3x} \right] + \frac{1}{2} \left[-\frac{1}{4} e^{-x} \right] = \frac{1}{2} \left[\frac{x}{5} e^{3x} \right] - \frac{1}{8} e^{-x}$$

$\therefore \quad \text{P.I.} = \dfrac{x}{10} e^{3x} - \dfrac{1}{8} e^{-x}$

Hence, the complete solution is

$$y = \text{C.F.} + \text{P.I.}$$

i.e., $\quad y = c_1 e^{3x} + c_2 e^{-2x} + \dfrac{x}{10} e^{3x} - \dfrac{1}{8} e^{-x}$

where c_1 and c_2 are arbitrary constants of integration.

Example 4. Solve : $(D^3 + 3D^2 + 3D + 1)y = e^{-x}$

Solution. The auxiliary equation is

$$m^3 + 3m^2 + 3m + 1 = 0$$

$\Rightarrow \quad (m+1)^3 = 0$

$\Rightarrow \quad m = -1, \ -1, \ -1$

$\therefore \quad \text{C.F.} = (c_1 + c_2 x + c_3 x^2) e^{-x}$

And $\quad \text{P.I.} = \dfrac{1}{D^3 + 3D^2 + 3D + 1} e^{-x}$

$$= \frac{1}{D^3 + 3D^2 + 3D + 1} e^{-x}$$ (case of failure)

$$= x \frac{1}{3D^2 + 6D + 3} e^{-x}$$ (Again case of failure)

$$= x^2 \frac{1}{6D+6} e^{-x} \qquad \text{(Again case of failure)}$$

$$= x^3 \frac{1}{6} e^{-x}$$

$\therefore$ $$\text{P.I.} = x^3 \frac{1}{6} e^{-x}$$

Hence, the complete solution is

$$y = \text{C.F.} + \text{P.I}$$

i.e., $$y = (c_1 + c_2 x + c_3 x^2)e^{-x} + \frac{x^3}{6} e^{-x}$$

where c_1, c_2 and c_3 are arbitrary constants of integration.

Example 5. Solve : $(D^3 - 3D^2 + 2D)y = 4 + 60e^{5x}$

Solution. The auxiliary equation is

$$m^3 - 3m^2 + 2m = 0$$

$\Rightarrow$ $$m = 0, 1, 2$$

$\therefore$ $$\text{C.F.} = c_1 + c_2 e^x + c_3 e^{2x}$$

And $$\text{P.I.} = \frac{1}{D^3 - 3D^2 + 2D}(4 + 60e^{5x})$$

$$= 4.\frac{1}{D^3 - 3D^2 + 2D} e^{0x} + \frac{60}{D^3 - 3D^2 + 2D} e^{5x}$$

$$= 4.\frac{x}{3D^2 - 6D + 2} e^{0x} + \frac{60}{5^3 - 3.5^2 + 2.5} e^{5x}$$

(case of failure)

$$= 4.\frac{x}{3.0^2 - 6.0 + 2} e^{0x} + \frac{60}{60} e^{5x}$$

$\therefore$ $$\text{P.I.} = 2xe^{0x} + \frac{1}{40} e^{5x} = 2x + e^{5x}$$

Hence, the complete solution is

$$y = \text{C.F.} + \text{P.I}$$

i.e., $$y = c_1 + c_2 e^x + c_3 e^{2x} + 2x + e^{5x}$$

where c_1, c_2 and c_3 are arbitrary constants of integration.

Example 6. Solve : $(D^2 + 4D + 4)y = e^{-2x} + e^{2x}$

Solution. The auxiliary equation is

$$m^2 + 4m + 4 = 0$$

$\Rightarrow$ $$(m+2)^2 = 0$$

$$\Rightarrow \qquad m = -2, -2$$

$$\therefore \qquad \text{C.F.} = (c_1 + c_2 x)e^{-2x}$$

And

$$\text{P.I.} = \frac{1}{D^2+4D+4}e^{-2x} + \frac{1}{D^2+4D+4}e^{2x}$$

$$= \frac{x}{2D+4}e^{-2x} + \frac{1}{2^2+4.2+4}e^{2x}$$

(case of failure in first term)

$$= \frac{x^2}{2}e^{-2x} + \frac{1}{16}e^{2x} \quad \text{(again case of failure in first term)}$$

$$\therefore \qquad \text{P.I.} = \frac{x^2}{2}e^{-2x} + \frac{1}{16}e^{2x}$$

Hence, the complete solution is

$$y = \text{C.F.} + \text{P.I}$$

i.e.,

$$y = (c_1 + c_2 x)e^{-2x} + \frac{x^2}{2}e^{-2x} + \frac{1}{16}e^{2x}$$

where c_1 and c_2 and are arbitrary constants of integration.

EXERCISE 2.9

1. $\dfrac{d^2y}{dx^2} - 3\dfrac{dy}{dx} + 2y = e^{3x}$

2. $(D-2)(D+1)^2 y = e^{2x} + e^x$

3. $(D^3 - 5D^2 + 8D - 4)y = (e^x + 1)^2$

4. $(D^2 - a^2)y = e^{ax} - e^{-ax}$

5. $(2D+1)^2 y = 4e^{-\frac{x}{2}}$

6. $(D^2 + 2D + 2)y = \sinh x$

7. $\dfrac{d^3y}{dx^3} - 3\dfrac{d^2y}{dx^2} + 3\dfrac{dy}{dx} - y = e^x + 2$

8. $(D^2 - 6D + 9)y = 6e^{3x} + 7e^{-2x} - 1$

ANSWERS

1. $y = c_1 e^x + c_2 e^{2x} + \dfrac{1}{2}e^{3x}$

2. $y = c_1 e^{2x} + (c_2 + c_3 x)e^{-x} + \dfrac{x}{9}e^{2x} - \dfrac{1}{4}e^x$

3. $y = c_1 e^x + (c_2 + c_3 x)e^{2x} + \dfrac{x^2}{2}e^{2x} + 2xe^x - \dfrac{1}{4}$

4. $y = c_1 e^{ax} + c_2 e^{-ax} + \dfrac{x}{2}\left(\dfrac{e^{ax} + e^{-ax}}{2}\right)$

5. $y=\left(c_1+c_2x+\dfrac{x^2}{2}\right)e^{-\frac{x}{2}}$

6. $y=e^{-x}(c_1\cos x+c_2\sin x)+\dfrac{e^x}{10}-\dfrac{e^{-x}}{2}$

7. $y=(c_1+c_2x+c_3x^2)e^x+\dfrac{x^3}{6}e^x-2$

8. $y=(c_1+c_2x)e^{3x}+3x^2e^{3x}+\dfrac{7}{25}e^{-2x}-\dfrac{1}{9}$

Case II : When $Q=\sin(ax+b)$ *or* $\cos(ax+b)$.

We know that

$$D\sin(ax+b)=a\cos(ax+b)$$

$$D^2\sin(ax+b)=(-a^2)\sin(ax+b)$$

$$D^3\sin(ax+b)=(-a^3)\cos(ax+b)$$

$$D^4\sin(ax+b)=(D^2)^2\sin(ax+b)=(-a^2)^2\sin(ax+b)$$

....

....

$\therefore \quad (D^2)^n\sin(ax+b)=(-a^2)^n\sin(ax+b)$

Hence $\quad f(D^2)\sin(ax+b)=f(-a^2)\sin(ax+b)$

Operating on both sides by $\dfrac{1}{f(D^2)}$, we have

$$\frac{1}{f(D^2)}f(D^2)\sin(ax+b)=\frac{1}{f(D^2)}f(-a^2)\sin(ax+b)$$

$$\Rightarrow \quad \sin(ax+b)=f(-a^2)\frac{1}{f(D^2)}\sin(ax+b)$$

$$\Rightarrow \quad \frac{1}{f(-a^2)}\sin(ax+b)=\frac{1}{f(D^2)}\sin(ax+b)$$

Hence **P.I.** $=\dfrac{1}{f(D^2)}\sin(ax+b)=\dfrac{1}{f(-a^2)}\sin(ax+b)$, **provided** $f(-a^2)\neq 0$

If $f(-a^2)=0$,the above method fails.

then P.I. $=\dfrac{1}{f(D^2)}\sin(ax+b)=x.\dfrac{1}{f'(-a^2)}\sin(ax+b)$, provided $f'(-a^2)\neq 0$

If $f'(-a^2)=0$, then P.I. = $\frac{1}{f(D^2)}\sin(ax+b)=x^2.\frac{1}{f''(-a^2)}\sin(ax+b)$, provided $f''(-a^2)\neq 0$ and so on.

Similarly, for $Q=\cos(ax+b)$

P.I. = $\frac{1}{f(D^2)}\cos(ax+b)=\frac{1}{f(-a^2)}\cos(ax+b)$, provided $f(-a^2)\neq 0$

If $f(-a^2)=0$,the above method fails.

then P.I. = $\frac{1}{f(D^2)}\cos(ax+b)=x.\frac{1}{f'(-a^2)}\cos(ax+b)$, provided $f'(-a^2)\neq 0$

If $f'(-a^2)=0$, then P.I. = $\frac{1}{f(D^2)}\cos(ax+b)=x^2.\frac{1}{f''(-a^2)}\cos(ax+b)$, provided $f''(-a^2)\neq 0$ and so on.

ILLUSTRATIVE EXAMPLES

Example 1. Solve : $\frac{d^2y}{dx^2}+4y=\cos(3x+1)$

Solution. We have $(D^2+4)y=\cos(3x+1)$

The auxiliary equation is

$$m^2+4=0$$

$$\Rightarrow \quad m=\pm 2i$$

$$\therefore \quad \text{C.F.}=c_1\cos 2x+c_2\sin 2x$$

$$\text{P.I.}=\frac{1}{D^2+4}\cos(3x+1)=\frac{1}{-3^2+4}\cos(3x+1)$$

$$=-\frac{1}{5}\cos(3x+1)$$

Hence, the complete solution is

$$y=c_1\cos 2x+c_2\sin 2x-\frac{1}{5}\cos(3x+1)$$

where c_1 and c_2 are arbitrary constants of integration.

Example 2. Solve : $\dfrac{d^2y}{dx^2}+4y=\sin 3x$

Solution. We have $(D^2+4)y=\sin 3x$

The auxiliary equation is

$$m^2+4 = 0$$

$$\Rightarrow \quad m = \pm 2i$$

$$\therefore \quad \text{C.F.} = c_1\cos 2x + c_2 \sin 2x$$

$$\text{P.I.} = \frac{1}{D^2+4}\sin 3x$$

$$= \frac{1}{-3^2+4}\sin 3x = -\frac{1}{5}\sin 3x$$

Hence, the complete solution is

$$y = c_1\cos 2x + c_2\sin 2x - \frac{1}{5}\sin 3x$$

where c_1 and c_2 are arbitrary constants of integration.

Example 3. Solve : $\dfrac{d^2y}{dx^2}+3\dfrac{dy}{dx}+2y=\cos x$

Solution. We have $(D^2+3D+2)y=\cos x$

The auxiliary equation is

$$m^2+3m+2 = 0$$

$$\Rightarrow \quad (m+1)(m+2) = 0$$

$$\Rightarrow \quad m = -1, -2$$

$$\therefore \quad \text{C.F.} = c_1e^{-x}+c_2e^{-2x}$$

$$\text{P.I.} = \frac{1}{D^2+3D+2}\cos x$$

$$= \frac{1}{-1^2+3D+2}\cos x = \frac{1}{3D+1}\cos x$$

$$= \frac{1}{3D+1}\cos x = \frac{3D-1}{9D^2-1}\cos x = \frac{3D-1}{9(-1^2)-1}\cos x$$

$$= \frac{(3D-1)}{-10}\cos x = -\frac{1}{10}(-3\sin x-\cos x)$$

$$= \frac{1}{10}(3\sin x+\cos x)$$

Hence, the complete solution is

$$y = c_1 e^{-x} + c_2 e^{-2x} + \frac{1}{10}(3\sin x + \cos x)$$

where c_1 and c_2 are arbitrary constants of integration.

Example 4. Solve : $(D^2 + 1)y = \sin x \sin 2x$

Solution. The auxiliary equation is

$$m^2 + 1 = 0$$

$$\Rightarrow \quad m = \pm i$$

$$\therefore \quad \text{C.F.} = c_1 \cos x + c_2 \sin x$$

$$\text{P.I.} = \frac{1}{D^2+1}\sin x \sin 2x = \frac{1}{2}\frac{1}{D^2+1}(2\sin x \sin 2x)$$

$$= \frac{1}{2}\frac{1}{D^2+1}(\cos x - \cos 3x)$$

$$= \frac{1}{2}\left[\frac{1}{D^2+1}\cos x - \frac{1}{D^2+1}\cos 3x\right]$$

(case of failure in Ist part)

$$= \frac{1}{2}\left[\frac{x}{2D}\cos x - \frac{1}{-3^2+1}\cos 3x\right]$$

$$= \frac{1}{2}\left[\frac{x}{2}\sin x + \frac{1}{8}\cos 3x\right] \quad \left(\frac{1}{D} \text{ means integration}\right)$$

Hence, the complete solution is

$$y = c_1 \cos x + c_2 \sin x + \frac{1}{2}\left[\frac{x}{2}\sin x + \frac{1}{8}\cos 3x\right]$$

where c_1 and c_2 are arbitrary constants of integration.

Example 5. Solve : $(D^2 + 4)y = e^x + \sin 2x$

Solution. The auxiliary equation is

$$m^2 + 4 = 0$$

$$\Rightarrow \quad m = \pm 2i$$

$$\therefore \quad \text{C.F.} = c_1 \cos 2x + c_2 \sin 2x$$

$$\text{P.I.} = \frac{1}{D^2+4}[e^x + \sin 2x]$$

$$= \left[\frac{1}{D^2+4}e^x + \frac{1}{D^2+4}\sin 2x\right]$$

$$= \left[\frac{1}{1^2+4} e^x + \frac{x}{2D} \sin 2x \right] \quad \text{(case of failure in IInd part)}$$

$$= \left[\frac{1}{5} e^x + \frac{x}{2} \left(\frac{-\cos 2x}{2} \right) \right] \quad \text{(means integration)}$$

$$= \frac{1}{5} e^x - \frac{x}{4} \cos 2x$$

Hence, the complete solution is

$$y = c_1 \cos 2x + c_2 \sin 2x + \frac{1}{5} e^x - \frac{x}{4} \cos 2x$$

where c_1 and c_2 are arbitrary constants of integration.

Example 6. Solve : $(D^2 - 4D + 3)y = 40 \cos x$

Solution. The auxiliary equation is

$$m^2 - 4m + 3 = 0$$

$$\Rightarrow \quad (m-1)(m-3) = 0$$

$$\Rightarrow \quad m = 1, 3$$

$$\therefore \quad \text{C.F.} = c_1 e^x + c_2 e^{3x}$$

$$\text{P.I.} = \frac{1}{D^2 - 4D + 3} 40 \cos x$$

$$= 40 \frac{1}{-1^2 - 4D + 3} \cos x$$

$$= 40 \frac{1}{-4D + 2} \cos x = 20 \frac{1}{1 - 2D} \cos x$$

$$= 20 \frac{1+2D}{1-4D^2} \cos x = 20 \frac{(1+2D)}{1-4(-1^2)} \cos x$$

$$= 20.\frac{1}{5}(\cos x - 2\sin x) = 4\cos x - 8\sin x$$

Hence, the complete solution is

$$y = c_1 e^x + c_2 e^{3x} + 4\cos x - 8\sin x$$

where c_1 and c_2 are arbitrary constants of integration.

EXERCISE 2.10

1. $\dfrac{d^2y}{dx^2}+4\dfrac{dy}{dx}+8y=\sin(2x+3)$
2. $(D^2+4D+5)y=65\sin 2x$
3. $(D^2+a^2)y=\sin ax$
4. $(D^2-4)y=\cos^2 x$
5. $(D^2-4D+4)y=e^{-4x}+5\cos 3x$
6. $(D^2+9)y=\cos 2x+\sin 2x$
7. $\dfrac{d^3y}{dx^3}+\dfrac{d^2y}{dx^2}-\dfrac{dy}{dx}-y=\cos 2x$
8. $(D^2-4D+3)y=\sin 3x\cos 2x$

ANSWERS

1. $y=e^{-2x}(c_1\cos 2x+c_2\sin 2x)+\dfrac{1}{20}[\sin(2x+3)-2\cos(2x+3)]$
2. $y=e^{-2x}(c_1\cos x+c_2\sin x)-8\cos 2x+\sin 2x$
3. $y=c_1\cos ax+c_2\sin ax-\dfrac{x}{2a}\cos ax$
4. $y=c_1e^{2x}+c_2e^{-2x}-\dfrac{1}{8}-\dfrac{1}{16}\cos 2x$
5. $y=(c_1+c_2x)e^{2x}+\dfrac{1}{36}e^{-4x}-\dfrac{5}{169}(12\sin 3x+5\cos 3x)$
6. $y=c_1\cos 3x+c_2\sin 3x+\dfrac{1}{5}(\cos 2x+\sin 2x)$
7. $y=c_1e^{x}+(c_2+c_3x)e^{-x}-\dfrac{1}{25}(2\sin 2x+\cos 2x)$
8. $y=c_1e^{x}+c_2e^{3x}+\dfrac{1}{884}(10\cos 5x-11\sin 5x)+\dfrac{1}{20}(\sin x+2\cos x)$

Case III : When $Q=x^m$, m being a positive integer.

In this case,

$$\text{P.I.} = \frac{1}{f(D)}x^m=[f(D)]^{-1}x^m$$

To evaluate P.I. , take out common the lowest degree term of D from $f(D)$, so that the first term becomes unity. Then the remaining factor in the denominator is of the form $1+\phi(D)$ or $1-\phi(D)$, which is taken in the numerator and it becomes $[1+\phi(D)]^{-1}$ or $[1-\phi(D)]^{-1}$.

Now expand $[1 \pm \phi(D)]^{-1}$ in ascending powers of D by using binomial theorem upto the term containing D^m. Thus, operate on x^m term by term.

Remark : If $Q = x^m$ expand $[1 \pm \phi(D)]^{-1}$ upto to the term D^m, since $D^m x^m = m!$, $D^{m+1}x^m = 0,\ D^{m+2}x^m = 0$, and all the other higher differential coefficient of x^m are zero.

ILLUSTRATIVE EXAMPLES

Example 1. Solve : $(D^2 - 4)y = x^2$

Solution. The auxiliary equation is

$$m^2 - 4 = 0$$

$$\Rightarrow \quad m = \pm 2$$

$$\therefore \quad \text{C.F.} = c_1 e^{2x} + c_2 e^{-2x}$$

$$\text{P.I.} = \frac{1}{D^2 - 4}x^2 = \frac{1}{-4\left(1 - \frac{D^2}{4}\right)}x^2 = -\frac{1}{4}\left(1 - \frac{D^2}{4}\right)^{-1} x^2$$

$$= -\frac{1}{4}\left(1 + \frac{D^2}{4} + \ldots\ldots\right)x^2 = -\frac{1}{4}\left(x^2 + \frac{D^2 x^2}{4}\right)$$

$$= -\frac{1}{4}\left(x^2 + \frac{2}{4}\right) = -\frac{1}{4}\left(x^2 + \frac{1}{2}\right)$$

Hence, the complete solution is

$$y = c_1 e^{2x} + c_2 e^{-2x} - \frac{1}{4}\left(x^2 + \frac{1}{2}\right)$$

Example 2. Solve : $(D^2 - 4D + 4)y = x^3$

Solution. The auxiliary equation is

$$m^2 - 4m + 4 = 0$$

$$\Rightarrow \quad (m-2)^2 = 0$$

$$\Rightarrow \quad m = 2, 2$$

$$\therefore \quad \text{C.F.} = (c_1 + c_2 x)e^{2x}$$

$$\text{P.I.} = \frac{1}{D^2 - 4D + 4}x^3 = \frac{1}{(D-2)^2} = \frac{1}{4\left(1 - \frac{D}{2}\right)^2}x^3$$

$$=\frac{1}{4}\left(1-\frac{D}{2}\right)^{-2}x^3=\frac{1}{4}\left(1+\frac{2D}{2}+\frac{3D^2}{4}+\frac{D^3}{2}+......\right)x^3$$

$$=\frac{1}{4}\left(x^3+Dx^3+\frac{3}{4}D^2x^3+\frac{1}{2}D^3x^3\right)$$

$$=\frac{1}{4}\left(x^3+3x^2+\frac{9}{2}x+\frac{1}{2}\right)$$

Hence, the complete solution is

$$y=(c_1+c_2x)e^{2x}+\frac{1}{4}\left(x^3+3x^2+\frac{9}{2}x+\frac{1}{2}\right)$$

Example 3. Solve : $(D^3-D^2-6D)y=1+x^2$

Solution. The auxiliary equation is

$$m^3-m^2-6m=0$$

$$\Rightarrow \quad m(m-3)(m+2)=0$$

$$\Rightarrow \quad m=0,-2,3$$

$$\therefore \quad \text{C.F.}=c_1+c_2e^{-2x}+c_3e^{3x}$$

$$\text{P.I.}=\frac{1}{D^3-D^2-6D}(1+x^2)=\frac{1}{-6D-D^2+D^3}(1+x^2)$$

$$=\frac{1}{-6D\left\{1+\left(\frac{D-D^2}{6}\right)\right\}}(1+x^2)$$

$$=-\frac{1}{6D}\left\{1+\left(\frac{D-D^2}{6}\right)\right\}^{-1}(1+x^2)$$

$$=-\frac{1}{6D}\left\{1-\left(\frac{D-D^2}{6}\right)+\left(\frac{D-D^2}{6}\right)^2+.......\right\}(1+x^2)$$

$$=-\frac{1}{6D}\left\{1-\frac{1}{6}D+\frac{7}{36}D^2\right\}(1+x^2)$$

$$=-\frac{1}{6D}\left\{(1+x^2)-\frac{1}{6}D(1+x^2)+\frac{7}{36}D^2(1+x^2)\right\}$$

$$=-\frac{1}{6D}\left\{(1+x^2)-\frac{1}{6}(0+2x)+\frac{7}{36}(0+2)\right\}$$

$$= -\frac{1}{6D}\left\{(1+x^2)-\frac{x}{3}+\frac{7}{18}\right\} = -\frac{1}{6D}\left\{x^2-\frac{x}{3}+\frac{25}{18}\right\}$$

$$= -\frac{1}{6}\left\{\frac{x^3}{3}-\frac{x^2}{6}+\frac{25}{18}x\right\} = -\frac{x}{18}\left\{x^2-\frac{x}{2}+\frac{25}{6}\right\}$$

(as $\frac{1}{D}f(x) = \int f(x)dx$)

Hence, the complete solution is

$$y = c_1 + c_2e^{-2x} + c_3e^{3x} - \frac{x}{18}\left\{x^2-\frac{x}{2}+\frac{25}{6}\right\}$$

Example 4. Solve : $(D^3+8)y = x^4+2x+1$

Solution. The auxiliary equation is

$$m^3+8 = 0$$

$$\Rightarrow \quad (m+2)(m^2-2m+4) = 0$$

$$\Rightarrow \quad m = -2,\ m = \frac{2\pm\sqrt{4-16}}{2}$$

$$\Rightarrow \quad m = -2,\ 1\pm i\sqrt{3}$$

$$\therefore \quad \text{C.F.} = c_1e^{-2x} + e^x\left(c_2\cos\sqrt{3}x + c_3\sin\sqrt{3}x\right)$$

$$\text{P.I.} = \frac{1}{D^3+8}(x^4+2x+1) = \frac{1}{8\left(1+\frac{D^3}{8}\right)}(x^4+2x+1)$$

$$= \frac{1}{8}\left\{1+\frac{D^3}{8}\right\}^{-1}(x^4+2x+1)$$

$$= \frac{1}{8}\left\{1-\frac{D^3}{8}+\left(\frac{D^3}{8}\right)^2+\ldots\ldots\right\}(x^4+2x+1)$$

$$= \frac{1}{8}\left\{1-\frac{D^3}{8}\right\}(x^4+2x+1)$$

$$= \frac{1}{8}\left\{(x^4+2x+1)-\frac{1}{8}D^3(x^4+2x+1)\right\}$$

$$=\frac{1}{8}\left\{(x^4+2x+1)-\frac{1}{8}(24x+0+0)\right\}=\frac{1}{8}(x^4-x+1)$$

Hence, the complete solution is

$$y=c_1e^{-2x}+e^x\left(c_2\cos\sqrt{3}x+c_3\sin\sqrt{3}x\right)+\frac{1}{8}(x^4-x+1)$$

Example 5. Solve : $(D^2-4D+4)y=x^2+e^x+\cos 2x$

Solution. The auxiliary equation is

$$m^2-4m+4=0$$

$$\Rightarrow \quad (m-2)^2=0$$

$$\Rightarrow \quad m=2,2$$

$$\therefore \quad \text{C.F.}=(c_1+c_2x)e^{2x}$$

$$\text{P.I.}=\frac{1}{D^2-4D+4}(x^2+e^x+\cos 2x)$$

Now $$\frac{1}{D^2-4D+4}x^2=\frac{1}{4}\left(1-D+\frac{D^2}{4}\right)^{-1}x^2=\frac{1}{4}\left(1-\left(D-\frac{D^2}{4}\right)\right)^{-1}x^2$$

$$=\frac{1}{4}\left\{1+\left(D-\frac{D^2}{4}\right)+\left(D-\frac{D^2}{4}\right)^2+\ldots\ldots\right\}x^2$$

$$=\frac{1}{4}\left\{1+\left(D-\frac{D^2}{4}\right)+\left(D^2+\frac{D^4}{16}-\frac{D^3}{2}\right)+\ldots\ldots\right\}x^2$$

$$=\frac{1}{4}\left\{1+D+\frac{3}{4}D^2\right\}x^2=\frac{1}{4}\left\{x^2+2x+\frac{3}{4}.2\right\}$$

$$=\frac{1}{4}\left\{x^2+2x+\frac{3}{2}\right\}$$

$$\frac{1}{D^2-4D+4}e^x=\frac{1}{1^2-4(1)+4}e^x=e^x$$

and $$\frac{1}{D^2-4D+4}\cos 2x=\frac{1}{2^2-4D+4}\cos 2x=-\frac{1}{4D}\cos 2x$$

$$=-\frac{1}{4}\int\cos 2x\,dx=-\frac{1}{4}.\frac{1}{2}\sin 2x=-\frac{1}{8}\sin 2x$$

$$\therefore \qquad \text{P.I.} = \frac{1}{4}\left\{x^2+2x+\frac{3}{2}\right\}+e^x-\frac{1}{8}\sin 2x$$

Hence, the complete solution is

$$y = (c_1+c_2x)e^{2x}+\frac{1}{4}\left\{x^2+2x+\frac{3}{2}\right\}+e^x-\frac{1}{8}\sin 2x.$$

EXERCISE 2.11

1. $\frac{d^2y}{dx^2}+\frac{dy}{dx}-2y=x$
2. $(D^2-6D+9)y=2x^2-x+3$
3. $(D^3+2D^2+D)y=x^2+x+e^{2x}$
4. $(D^2+5D+4)y=3-2x$
5. $(D^2+2D+1)y=2x+x^2$
6. $(D-1)^2(D+1)^2y=\sin^2\frac{x}{2}+e^x+x$
7. $(D^4+8D^2+16)y=16x+256$
8. $(D+3)(D-2)y=x$

ANSWERS

1. $y=c_1e^x+c_2e^{-2x}-\frac{1}{4}(2x+1)+\frac{1}{20}[\sin(2x+3)-2\cos(2x+3)]$
2. $y=(c_1+c_2x)e^{3x}+\frac{1}{9}\left(2x^2+\frac{5}{3}x+\frac{11}{3}\right)$
3. $y=c_1+(c_2+c_3x)e^{-x}+\frac{1}{3}x^3-\frac{3}{2}x^2+4x+\frac{1}{18}e^{2x}$
4. $y=c_1e^{-x}+c_2e^{-4x}+\frac{1}{8}(11-4x)$
5. $y=(c_1+c_2x)e^{-x}+x^2-2x+2$
6. $y=(c_1+c_2x)e^x+(c_3+c_4x)e^{-x}+\frac{1}{2}-\frac{1}{8}\cos x+\frac{x^2}{8}e^x+x$
7. $y=(c_1+c_2x)\cos 2x+(c_3+c_4x)\sin 2x+x+16$
8. $y=c_1e^{2x}+c_2e^{-3x}-\frac{1}{6}\left(x+\frac{1}{6}\right)$

Case IV : When $Q=e^{ax}V$, V where is a function of x.

Let V_1 be any function of x, then by successive differentiation, we have

$$D(e^{ax}V_1) = e^{ax}DV_1+ae^{ax}V_1=e^{ax}(D+a)V_1$$

$$D^2(e^{ax}V_1) = D\left[e^{ax}(D+a)V_1\right]=e^{ax}D(D+a)V_1+ae^{ax}(D+a)V_1$$

$$= e^{ax}(D^2 + aD)V_1 + ae^{ax}(D+a)V_1$$

$$= e^{ax}(D^2 + aD + a^2)V_1 = e^{ax}(D+a)^2 V_1$$

Similarly, $\quad D^3(e^{ax}V) = e^{ax} f(D+a)^3 V_1$

In general, $\quad D^n(e^{ax}V_1) = e^{ax}(D+a)^n V_1$

$\therefore \quad f(D)[e^{ax}V_1] = e^{ax} f(D+a)V_1$

$$\Rightarrow \quad e^{ax}V_1 = \frac{1}{f(D)}[e^{ax} f(D+a)V_1] \qquad \ldots (1)$$

Now put $f(D+a)V_1 = V$, so that

$$V_1 = \frac{1}{f(D+a)}V$$

Substituting these values in (1)

$$e^{ax}\frac{1}{f(D+a)}V = \frac{1}{f(D)}[e^{ax}V]$$

$$\Rightarrow \quad \frac{1}{f(D)}[e^{ax}V] = e^{ax}\frac{1}{f(D+a)}V$$

Remark : $\quad \frac{1}{f(D)}e^{ax} = \frac{1}{f(D)}e^{ax}.1 = e^{ax}\frac{1}{f(D+a)}.1$

ILLUSTRATIVE EXAMPLES

Example 1. Solve : $(D^2 - 2D + 1)y = x^2 e^{3x}$

Solution. The auxiliary equation is

$$m^2 - 2m + 1 = 0$$

$$\Rightarrow \quad m = 1, 1$$

$$\therefore \quad \text{C.F.} = (c_1 + c_2 x)e^x$$

$$\text{P.I.} = \frac{1}{D^2 - 2D + 1}x^2 e^{3x} = \frac{1}{(D-1)^2}x^2 e^{3x}$$

$$= e^{3x}\frac{1}{(D+3-1)^2}x^2 = e^{3x}\frac{1}{(D+2)^2}x^2$$

$$= e^{3x}.\frac{1}{4}\frac{1}{\left(1+\frac{D}{2}\right)^2}x^2 = \frac{e^{3x}}{4}\left(1+\frac{D}{2}\right)^{-2}x^2$$

$$= \frac{e^{3x}}{4}\left(1-2.\frac{D}{2}+3.\frac{D^2}{4}-.......\right)x^2$$

$$= \frac{e^{3x}}{4}\left(x^2 - Dx^2 + \frac{3}{4}D^2x^2\right) = \frac{e^{3x}}{4}\left(x^2 - 2x + \frac{3}{2}\right)$$

Hence, the complete solution is

$$y = (c_1 + c_2 x)e^x + \frac{e^{3x}}{4}\left(x^2 - 2x + \frac{3}{2}\right)$$

Example 2. Solve : $(D^2 - 2D + 4)y = e^x \cos x$

Solution. The auxiliary equation is

$$m^2 - 2m + 4 = 0$$

$$\Rightarrow \quad m = 1 \pm i\sqrt{3}$$

$$\therefore \quad \text{C.F.} = e^x\left(c_1 \cos\sqrt{3}x + c_2 \sin\sqrt{3}x\right)$$

$$\text{P.I.} = \frac{1}{D^2 - 2D + 4} e^x \cos x$$

$$= e^x \frac{1}{(D+1)^2 - 2(D+1) + 4} \cos x$$

$$= e^x \frac{1}{D^2 + 3} \cos x = e^x \frac{1}{-1^2 + 3} \cos x = \frac{1}{2} e^x \cos x$$

Hence, the complete solution is

$$y = e^x\left(c_1 \cos\sqrt{3}x + c_2 \sin\sqrt{3}x\right) + \frac{1}{2} e^x \cos x$$

Example 3. Solve : $(D^3 - 7D - 6)y = e^{2x}(x+1)$

Solution. The auxiliary equation is

$$m^3 - 7m - 6 = 0$$

$$\Rightarrow \quad (m+1)(m^2 - m - 6) = 0$$

$$\Rightarrow \quad (m+1)(m+2)(m-3) = 0$$

$$\Rightarrow \quad m = -1, -2, 3$$

$$\therefore \quad \text{C.F.} = c_1 e^{-x} + c_2 e^{-2x} + c_3 e^{3x}$$

$$\text{P.I.} = \frac{1}{D^3 - 7D - 6} e^{2x}(x+1)$$

$$= e^{2x} \frac{1}{(D+2)^3 - 7(D+2) - 6}(x+1)$$

$$= e^{2x} \frac{1}{D^3 + 6D^2 + 5D - 12}(x+1)$$

$$= e^{2x} \frac{1}{-12\left[1 - \dfrac{D^3 + 6D^2 + 5D}{12}\right]}(x+1)$$

$$= -\frac{e^{2x}}{12}\left(1 - \frac{D^3 + 6D^2 + 5D}{12}\right)^{-1}(x+1)$$

$$= -\frac{e^{2x}}{12}\left(1 - \frac{D^3 + 6D^2 + 5D}{12} + \left(\frac{D^3 + 6D^2 + 5D}{12}\right)^2 - ....\right)(x+1)$$

$$= -\frac{e^{2x}}{12}\left(1 - \frac{5D}{12}\right)(x+1) = -\frac{e^{2x}}{12}\left(x + 1 - \frac{5}{12}.1\right)$$

$$= -\frac{e^{2x}}{12}\left(x + \frac{7}{12}\right)$$

Hence, the complete solution is

$$y = c_1 e^{-x} + c_2 e^{-2x} + c_3 e^{3x} - \frac{e^{2x}}{12}\left(x + \frac{7}{12}\right)$$

Example 4. Solve : $(D^2 - 4)y = x \sinh x$

Solution. The auxiliary equation is

$$m^2 - 4 = 0 \Rightarrow m = 2, \ -2$$

$$\therefore \quad \text{C.F.} = c_1 e^{2x} + c_2 e^{-2x}$$

$$\text{P.I.} = \frac{1}{D^2 - 4} x \sinh x = \frac{1}{D^2 - 4} x\left(\frac{e^x - e^{-x}}{2}\right)$$

$$= \frac{1}{2}\left[\frac{1}{D^2 - 4} xe^x - \frac{1}{D^2 - 4} xe^{-x}\right]$$

$$= \frac{1}{2}\left[e^x \frac{1}{(D+1)^2 - 4} x - e^{-x} \frac{1}{(D-1)^2 - 4} x\right]$$

$$= \frac{1}{2}\left[e^x \frac{1}{D^2 + 2D - 3} x - e^{-x} \frac{1}{D^2 - 2D - 3} x\right]$$

$$= \frac{1}{2}\left[e^x \frac{1}{-3\left(1-\frac{2D}{3}-\frac{D^2}{3}\right)} x - e^{-x} \frac{1}{-3\left(1+\frac{2D}{3}-\frac{D^2}{3}\right)} x \right]$$

$$= \frac{1}{2}\left[-\frac{e^x}{3}\left\{1-\left(\frac{2D}{3}+\frac{D^2}{3}\right)\right\}^{-1} x + \frac{e^{-x}}{3}\left\{1+\left(\frac{2D}{3}-\frac{D^2}{3}\right)\right\}^{-1} x \right]$$

$$= -\frac{1}{6}\left[e^x \left\{1+\left(\frac{2D}{3}+\frac{D^2}{3}\right)-\ldots\ldots\right\} x - e^{-x}\left\{1-\left(\frac{2D}{3}-\frac{D^2}{3}\right)+\ldots\ldots\right\} x \right]$$

$$= -\frac{1}{6}\left[e^x \left(1+\frac{2D}{3}\right) x - e^{-x}\left(1-\frac{2D}{3}\right) x \right]$$

$$= -\frac{1}{6}\left[e^x \left(x+\frac{2}{3}\right) - e^{-x}\left(x-\frac{2}{3}\right) \right]$$

$$= -\frac{x}{3}\left(\frac{e^x - e^{-x}}{2}\right) - \frac{2}{9}\left(\frac{e^x + e^{-x}}{2}\right) = -\frac{x}{3}\sinh x - \frac{2}{9}\cosh x$$

Hence, the complete solution is

$$y = c_1 e^{2x} + c_2 e^{-2x} - \frac{x}{3}\sinh x - \frac{2}{9}\cosh x$$

Example 5. Solve : $(D^2 + 4D + 8)y = 12e^{-2x} \sin x \ \sin 3x$

Solution. The auxiliary equation is

$$m^2 + 4m + 8 = 0$$

$$\Rightarrow \quad m = -2 \pm 2i$$

$$\therefore \quad \text{C.F.} = e^{-2x}(c_1 \cos 2x + c_2 \sin 2x)$$

$$\text{P.I.} = \frac{1}{D^2 + 4D + 8} 12e^{-2x} \sin x \sin 3x$$

$$= 6.\frac{1}{D^2 + 4D + 8} e^{-2x}(2\sin x \sin 3x)$$

$$= 6.\frac{1}{D^2 + 4D + 8} e^{-2x}(\cos 2x - \cos 4x)$$

$$= 6e^{-2x} \frac{1}{(D-2)^2 + 4(D-2) + 8}(\cos 2x - \cos 4x)$$

$$= 6e^{-2x}\left[\frac{1}{D^2+4}\cos 2x - \frac{1}{D^2+4}\cos 4x\right]$$

$$= 6e^{-2x}\left[\frac{x}{2D}\cos 2x - \frac{1}{-4^2+4}\cos 4x\right]$$

(case of failure in Ist part)

$$= 6e^{-2x}\left[\frac{x}{2}\left(\frac{\sin 2x}{2}\right) - \frac{1}{-12}\cos 4x\right]$$

$$= 6e^{-2x}\left[\frac{x}{4}\sin 2x + \frac{1}{12}\cos 4x\right]$$

$$= \frac{e^{-2x}}{2}[3x\sin 2x + \cos 4x]$$

Hence, the complete solution is

$$y = e^{-2x}(c_1\cos 2x + c_2\sin 2x) + \frac{e^{-2x}}{2}[3x\sin 2x + \cos 4x]$$

Example 6. Solve : $(D^2+1)y = x^2\sin 2x$

Solution. The auxiliary equation is

$$m^2+1 = 0 \Rightarrow m = \pm i$$

$$\therefore \quad \text{C.F.} = c_1\cos x + c_2\sin x$$

$$\text{P.I.} = \frac{1}{D^2+1}x^2\sin 2x$$

$$= \text{Imaginary part of } \frac{1}{D^2+1}x^2e^{2ix} \qquad (\text{as } e^{2ix} = \cos 2x + i\sin 2x)$$

$$= \text{Imaginary part of } e^{2ix}\frac{1}{(D+2i)^2+1}x^2$$

$$= \text{Imaginary part of } e^{2ix}\frac{1}{D^2+4iD-3}x^2$$

$$= \text{Imaginary part of } e^{2ix}\frac{1}{-3\left[1-\left(\frac{D^2}{3}+\frac{4iD}{3}\right)\right]}x^2$$

$$= \text{Imaginary part of } \frac{-e^{2ix}}{3}\left[1-\left(\frac{D^2}{3}+\frac{4iD}{3}\right)\right]^{-1}x^2$$

$$= \text{Imaginary part of } \frac{-e^{2ix}}{3}\left[1+\left(\frac{D^2}{3}+\frac{4iD}{3}\right)+\left(\frac{D^2}{3}+\frac{4iD}{3}\right)^2+\ldots\right]x^2$$

$$= \text{Imaginary part of } \frac{-e^{2ix}}{3}\left[1+\frac{D^2}{3}+\frac{4iD}{3}+\frac{16i^2D^2}{9}\right]x^2$$

$$= \text{Imaginary part of } \frac{-e^{2ix}}{3}\left[1+\frac{D^2}{3}-\frac{16D^2}{9}+\frac{4iD}{3}\right]x^2$$

$$= \text{Imaginary part of } \frac{-e^{2ix}}{3}\left[x^2+\frac{2}{3}-\frac{16}{9}.2+\frac{4i}{3}.2x\right]$$

$$= \text{Imaginary part of } \frac{-1}{3}(\cos 2x+i\sin 2x)\left(x^2+\frac{8ix}{3}-\frac{26}{9}\right)$$

$$= -\frac{1}{3}\left(x^2\sin 2x+\frac{8}{3}x\cos 2x-\frac{26}{9}\sin 2x\right)$$

$$= -\frac{1}{3}\left(\frac{8}{3}x\cos 2x+\frac{1}{9}\left(9x^2-26\right)\sin 2x\right)$$

Hence, the complete solution is

$$y = c_1\cos x+c_2\sin x-\frac{1}{3}\left(\frac{8}{3}x\cos 2x+\frac{1}{9}\left(9x^2-26\right)\sin 2x\right)$$

EXERCISE 2.12

1. $(D^2-5D+6)y = x^3e^{2x}$
2. $(D^2-4D+3)y = 2xe^{3x}$
3. $(D^2-7D+10)y = e^{2x}\sin x$
4. $(D^2+1)y = e^x\sin x+xe^{2x}$
5. $(D^3+3D^2+3D+1)y = x^2e^{-x}$
6. $(D^2+4D+4)y = 2\sinh 2x$
7. $(D^2+4D-12)y = (x-1)e^{2x}$
8. $(D^2-3D+2)y = 2e^x\cos\frac{x}{2}$
9. $(D^2+4)y = x\sin^2 x$
10. $(D^2-2D+1)y = xe^x\sin x$

ANSWERS

1. $y = c_1e^{2x}+c_2e^{3x}-e^{2x}\left(\frac{x^4}{4}+x^3+3x^2+6x\right)$

2. $y = c_1e^{x}+c_2e^{3x}+e^{3x}\left(\frac{x^2}{2}-\frac{x}{2}+\frac{1}{4}\right)$

3. $y = c_1e^{2x}+c_2e^{5x}+\frac{1}{10}e^{2x}(3\cos x-\sin x)$

4. $y = c_1\cos x+c_2\sin x-\frac{e^x}{5}(2\cos x-\sin x)+\frac{e^{2x}}{25}(5x-4)$

5. $y=(c_1+c_2x+c_3x^2)e^{-x}+\frac{1}{60}x^5e^{-x}$

6. $y=(c_1+c_2x)e^{-x}+\frac{1}{16}e^{2x}-\frac{x^2}{2}e^{-2x}$

7. $y=c_1e^{2x}+c_2e^{-6x}+e^{2x}\left(\frac{x^2}{16}-\frac{9x}{64}\right)$

8. $y=c_1e^{x}+c_2e^{2x}-\frac{16}{5}e^{x}\left(\sin\frac{x}{2}+\frac{1}{2}\cos\frac{x}{2}\right)$

9. $y=c_1\cos 2x+c_2\sin 2x+\frac{1}{8}x(1+2\sin 2x)$

10. $y=(c_1+c_2x)e^{x}-e^{x}(2\cos x+x\sin x)$

Case V : When $Q=\phi(x)$, where $\phi(x)$ is any other function of x.

In this case we use the following formulae,

(*i*) $\frac{1}{D-a}.\phi(x)=e^{ax}\int e^{-ax}.\phi(x)dx$

(*ii*) $\frac{1}{D+a}.\phi(x)=e^{-ax}\int e^{ax}.\phi(x)dx$

ILLUSTRATIVE EXAMPLES

Example 1. Solve : $(D^2+9)y=\sec 3x$

Solution. Auxiliary equation is

$$m^2+9=0 \Rightarrow m=\pm 3i$$

$$\therefore \quad \text{C.F.} = c_1\cos 3x+c_2\sin 3x$$

$$\text{P.I.} = \frac{1}{D^2+9}\sec 3x=\frac{1}{(D+3i)(D-3i)}\sec 3x$$

$$=\frac{1}{6i}\left[\frac{1}{D-3i}-\frac{1}{D+3i}\right]\sec 3x$$

$$=\frac{1}{6i}\left[\frac{1}{D-3i}\sec 3x-\frac{1}{D+3i}\sec 3x\right]$$

Now,
$$\frac{1}{D-3i}\sec 3x = e^{3ix}\int e^{-3ix}\sec 3x dx$$

$$=e^{3ix}\int(\cos 3x-i\sin 3x)\sec 3x dx$$

$$= e^{3ix}\int(1 - i\tan 3x)dx = e^{3ix}\left(x + \frac{i}{3}\log\cos 3x\right)$$

$$\Rightarrow \qquad \frac{1}{D-3i}\sec 3x = e^{3ix}\left(x + \frac{i}{3}\log\cos 3x\right)$$

Similarly, $\quad \dfrac{1}{D+3i}\sec 3x = e^{-3ix}\left(x - \dfrac{i}{3}\log\cos 3x\right)$

$$\therefore \qquad \text{P.I.} = \frac{1}{6i}\left[e^{3ix}\left(x + \frac{i}{3}\log\cos 3x\right) - e^{-3ix}\left(x - \frac{i}{3}\log\cos 3x\right)\right]$$

$$= \frac{x}{3}\left(\frac{e^{3ix} - e^{-3ix}}{2i}\right) + \frac{1}{9}\left(\frac{e^{3ix} + e^{-3ix}}{2}\right)\log\cos 3x$$

$$= \frac{x}{3}\sin 3x + \frac{1}{9}\cos 3x.\log\cos 3x$$

Hence, the complete solution is

$$y = c_1\cos 3x + c_2\sin 3x + \frac{x}{3}\sin 3x + \frac{1}{9}\cos 3x.\log\cos 3x$$

Example 2. Solve : $(D^2 + 4)y = \tan 2x$

Solution. Auxiliary equation is

$$m^2 + 4 = 0 \Rightarrow m = \pm 2i$$

$$\therefore \qquad \text{C.F.} = c_1\cos 2x + c_2\sin 2x$$

$$\text{P.I.} = \frac{1}{D^2 + 4}\tan 2x = \frac{1}{(D+2i)(D-2i)}\tan 2x$$

$$= \frac{1}{4i}\left[\frac{1}{D-2i} - \frac{1}{D+2i}\right]\tan 2x$$

$$= \frac{1}{4i}\left[\frac{1}{D-2i}\tan 2x - \frac{1}{D+2i}\tan 2x\right]$$

Now, $\quad \dfrac{1}{D-2i}\tan 2x = e^{2ix}\displaystyle\int e^{-2ix}\tan 2x dx$

$$= e^{2ix}\int(\cos 2x - i\sin 2x)\frac{\sin 2x}{\cos 2x}dx$$

$$= e^{2ix}\int\left(\sin 2x - i.\frac{\sin^2 2x}{\cos 2x}\right)dx$$

$$= e^{2ix}\int\left(\sin 2x - i.\frac{1 - \cos^2 2x}{\cos 2x}\right)dx$$

$$= e^{2ix}\int[\sin 2x - i(\sec 2x - \cos 2x)]\,dx$$

$$= e^{2ix}\left[-\frac{\cos 2x}{2} - \frac{i}{2}\log(\sec 2x + \tan 2x) + \frac{i}{2}\sin 2x\right]$$

$$= \frac{e^{2ix}}{2}[-(\cos 2x - i\sin 2x) - i\log(\sec 2x + \tan 2x)]$$

$$= -\frac{e^{2ix}}{2}\left[e^{-2ix} + i\log(\sec 2x + \tan 2x)\right]$$

$\Rightarrow$ $$\frac{1}{D-2i}\tan 2x = -\frac{e^{2ix}}{2}\left[e^{-2ix} + i\log(\sec 2x + \tan 2x)\right]$$

Similarly, $$\frac{1}{D+2i}\tan 2x = -\frac{e^{-2ix}}{2}\left[e^{+2ix} - i\log(\sec 2x + \tan 2x)\right]$$

$\therefore$ $$\text{P.I.} = \frac{1}{4i}\left[-\frac{e^{2ix}}{2}\left\{e^{-2ix} + i\log(\sec 2x + \tan 2x)\right\} + \frac{e^{-2ix}}{2}\left\{e^{2ix} - i\log(\sec 2x + \tan 2x)\right\}\right]$$

$$= -\frac{1}{4i}\left[i\left(\frac{e^{2ix} - e^{-2ix}}{2}\right)\log(\sec 2x + \tan 2x)\right]$$

$$= -\frac{1}{4}\cos 2x.\log(\sec 2x + \tan 2x)$$

Hence, the complete solution is

$$y = c_1\cos 2x + c_2\sin 2x - \frac{1}{4}\cos 2x.\log(\sec 2x + \tan 2x)$$

Example 3. Solve : $(D^2 - 2D + 2)y = e^x \tan x$

Solution. Auxiliary equation is

$$m^2 - 2m + 2 = 0 \Rightarrow m = 1 \pm i$$

$\therefore$ $$\text{C.F.} = e^x(c_1\cos x + c_2\sin x)$$

$$\text{P.I.} = \frac{1}{D^2 - 2D + 2}e^x\tan x$$

$$= e^x\frac{1}{(D+1)^2 - 2(D+1) + 2}\tan x$$

$$= e^x\frac{1}{D^2+1}\tan x = e^x\left[\frac{1}{(D-i)(D+i)}\right]\tan x$$

$$= \frac{e^x}{2i}\left[\frac{1}{(D-i)} - \frac{1}{(D+i)}\right]\tan x$$

Now, $$\frac{1}{D-i}\tan x = e^{ix}\int e^{-ix}\tan x dx$$

$$= e^{ix}\int(\cos x - i\sin x)\frac{\sin x}{\cos x}dx$$

$$= e^{ix}\int\left(\sin x - i.\frac{\sin^2 x}{\cos x}\right)dx$$

$$= e^{ix}\int\left(\sin x - i.\frac{1-\cos^2 x}{\cos x}\right)dx$$

$$= e^{ix}\int[\sin x - i(\sec x - \cos x)]dx$$

$$= e^{ix}[-\cos x - i\log(\sec x + \tan x) + i\sin x]$$

$$= e^{ix}[-(\cos x - i\sin x) - i\log(\sec x + \tan x)]$$

$$= -e^{ix}\left[e^{-ix} + i\log(\sec x + \tan x)\right]$$

$\Rightarrow$ $$\frac{1}{D-i}\tan x = -e^{ix}\left[e^{-ix} + i\log(\sec x + \tan x)\right]$$

Similarly, $$\frac{1}{D+i}\tan x = -e^{-ix}\left[e^{ix} - i\log(\sec x + \tan x)\right]$$

$\therefore$ $$\text{P.I.} = \frac{e^x}{2i}\left[-e^{ix}\left\{e^{-ix} + i\log(\sec x + \tan x)\right\}\right.$$

$$\left. + e^{-ix}\left\{e^{ix} - i\log(\sec x + \tan x)\right\}\right]$$

$$= -\frac{e^x}{i}\left[i\left(\frac{e^{ix}+e^{-ix}}{2}\right)\log(\sec x + \tan x)\right]$$

$$= -e^x\cos x.\log(\sec x + \tan x)$$

Hence, the complete solution is

$$y = e^x(c_1\cos x + c_2\sin x) - e^x\cos x.\log(\sec x + \tan x)$$

EXERCISE 2.13

1. $(D^2+1)y = \operatorname{cosec} x$

2. $(D^2+4)y = \sec 2x$

3. $(D^2+2D+2)y = e^{-x}\sec^3 x$

4. $(D^2+1)y = x - \cot x$

5. $(D^2+a^2)y = \tan ax$

ANSWERS

1. $y = c_1 \cos x + c_2 \sin x - x\cos x + \sin x.\log \sin x$

2. $y = c_1 \cos 2x + c_2 \sin 2x + \frac{x}{2}\sin 2x + \frac{1}{4}\cos 2x.\log\cos 2x$

3. $y = e^{-x}\left(c_1 \cos x + c_2 \sin x + \frac{\sin x \tan x}{2}\right)$

4. $y = c_1 \cos x + c_2 \sin x + x - \sin x.\log(\cos ecx - \cot x)$

5. $y = c_1 \cos ax + c_2 \sin ax - \frac{1}{a^2}\cos ax.\log(\sec ax + \tan ax)$

2.23 HOMOGENEOUS LINEAR DIFFERENTIAL EQUATIONS

An equation of the form

$$x^n \frac{d^n y}{dx^n} + a_1 \; x^{n-1}\frac{d^{n-1} y}{dx^{n-1}} + a_2 \; x^{n-2}\frac{d^{n-2} y}{dx^{n-2}} + ... + a_{n-1} x\frac{dy}{dx} + a_n y = Q \qquad ...(1)$$

where a_i's are all constants and Q is a function of x, is called a homogeneous linear differential equation or Cauchy's homogeneous linear differential equation.

Method of Solving : To solve such types of equations, we reduce them to linear differential equations with constant coefficients by the substitution

$$x = e^z \text{ so that } z = \log x \text{ and } \frac{dz}{dx} = \frac{1}{x}$$

Now, $$\frac{dy}{dx} = \frac{dy}{dz}.\frac{dz}{dx} = \frac{dy}{dz}.\frac{1}{x}$$

or $$x\frac{dy}{dx} = \frac{dy}{dz} = Dy, \text{ where } \frac{d}{dz} \equiv D$$

Again, $$\frac{d^2 y}{dx^2} = \frac{d}{dx}\left(\frac{dy}{dx}\right) = \frac{d}{dx}\left(\frac{1}{x}.\frac{dy}{dz}\right) = -\frac{1}{x^2}\frac{dy}{dz} + \frac{1}{x}\frac{d}{dz}\left(\frac{dy}{dz}\right)\frac{dz}{dx}$$

$$= -\frac{1}{x^2}\frac{dy}{dz} + \frac{1}{x}\frac{d^2 y}{dz^2}\frac{dz}{dx} = -\frac{1}{x^2}\frac{dy}{dz} + \frac{1}{x^2}\frac{d^2 y}{dz^2}$$

$$\left(\because \frac{dz}{dx} = \frac{1}{x}\right)$$

$\Rightarrow$ $$x^2\frac{d^2 y}{dx^2} = \frac{d^2 y}{dz^2} - \frac{dy}{dz} = D^2 y - Dy = D(D-1)y$$

Similarly, $$x^3\frac{d^3 y}{dx^3} = D(D-1)(D-2)y$$

In general we have,

$$x^n \frac{d^n y}{dx^n} = D(D-1)(D-2)(D-3).......(D-n+1)y$$

Substituting these values in (1), we get a linear differential equations with constant coefficients, which can be solved by the methods discussed earlier.

ILLUSTRATIVE EXAMPLES

Example 1. Solve : $x^2 \frac{d^2 y}{dx^2} + 5x\frac{dy}{dx} + 4y = x \log x$

Solution. Put $x = e^z$ so that $z = \log x$ and $\frac{d}{dz} \equiv D$, then the given differential equation reduces to

$$[D(D-1)+5D+4]y = ze^z$$

$$\Rightarrow \qquad (D^2+4D+4)y = ze^z$$

which is a linear differential equations with constant coefficients.

Its auxiliary equation is

$$m^2+4m+4 = 0 \Rightarrow m = -2, \ -2$$

$$\therefore \qquad \text{C.F.} = (c_1+c_2 z)e^{-2z} = (c_1+c_2 \log x)x^{-2}$$

and

$$\text{P.I.} = \frac{1}{D^2+4D+4} ze^z$$

$$= \frac{1}{(D+2)^2} ze^z = e^z \frac{1}{(D+2+1)^2} z$$

$$= e^z \frac{1}{(D+3)^2} z = \frac{e^z}{9}\left(1+\frac{D}{3}\right)^{-2} z$$

$$= \frac{e^z}{9}\left(1-\frac{2D}{3}+\frac{D^2}{3}+.....\right)z$$

$$= \frac{e^z}{9}\left(1-\frac{2D}{3}\right)z = \frac{e^z}{9}\left(z-\frac{2}{3}\right) = \frac{x}{9}\left(\log x - \frac{2}{3}\right)$$

Hence, the complete solution is

$$y = (c_1+c_2 \log x)x^{-2} + \frac{x}{9}\left(\log x - \frac{2}{3}\right)$$

Example 2. Solve: $x^2\dfrac{d^2y}{dx^2}+x\dfrac{dy}{dx}-4y=x^2$

Solution. Put $x=e^z$ so that $z=\log x$ and $\dfrac{d}{dz}\equiv D$, then the given differential equation reduces to

$$[D(D-1)+D-4]y = e^{2z}$$

$$\Rightarrow \qquad (D^2-4)y = e^{2z}$$

which is a linear differential equations with constant coefficients.
Its auxiliary equation is

$$m^2-4 = 0 \Rightarrow m=2,\ -2$$

$$\therefore \qquad \text{C.F.} = c_1e^{2z}+c_2e^{-2z} = c_1x^2+c_2x^{-2}$$

and

$$\text{P.I.} = \frac{1}{D^2-4}e^{2z} = \frac{z}{2D}e^{2z} = \frac{z}{2}.\frac{1}{D}e^{2z}$$

$$= \frac{z}{2}.\frac{e^{2z}}{2} = \frac{x^2}{4}\log x$$

Hence, the complete solution is

$$y = c_1x^2+c_2x^{-2}+\frac{x^2}{4}\log x$$

Example 3. Solve : $x^2\dfrac{d^2y}{dx^2}-x\dfrac{dy}{dx}+4y=\cos(\log x)+x\sin(\log x)$

Solution. Put $x=e^z$ so that $z=\log x$ and $\dfrac{d}{dz}\equiv D$, then the given differential equation reduces to

$$[D(D-1)-D+4]y = \cos z+e^z\sin z$$

$$\Rightarrow \qquad (D^2-2D+4)y = \cos z+e^z\sin z$$

which is a linear differential equations with constant coefficients.
Its auxiliary equation is

$$m^2-2m+4 = 0$$

$$\Rightarrow \qquad m = \frac{2\pm\sqrt{4-16}}{2} = 1\pm i\sqrt{3}$$

$$\therefore \qquad \text{C.F.} = e^z\left(c_1\cos\sqrt{3}z+c_2\sin\sqrt{3}z\right)$$

$$= x\left(c_1\cos(\sqrt{3}\log x)+c_2\sin(\sqrt{3}\log x)\right)$$

and

$$\text{P.I.} = \frac{1}{D^2-2D+4}(\cos z + e^z \sin z)$$

$$= \frac{1}{D^2-2D+4}\cos z + \frac{1}{D^2-2D+4} e^z \sin z$$

$$= \frac{1}{-1^2-2D+4}\cos z + e^z \frac{1}{(D+1)^2-2(D+1)+4}\sin z$$

$$= \frac{1}{3-2D}\cos z + e^z \frac{1}{D^2+3}\sin z$$

$$= \frac{3+2D}{9-4D^2}\cos z + e^z \frac{1}{-1^2+3}\sin z$$

$$= \frac{1}{9-4(-1^2)}(3\cos z - 2\sin z) + \frac{e^z}{2}\sin z$$

$$= \frac{1}{13}(3\cos z - 2\sin z) + \frac{e^z}{2}\sin z$$

$$= \frac{1}{13}[3\cos(\log x) - 2\sin(\log x)] + \frac{x}{2}(\log x)$$

Hence, the complete solution is

$$y = x\left(c_1 \cos(\sqrt{3}\log x) + c_2 \sin(\sqrt{3}\log x)\right) + \frac{1}{13}[3\cos(\log x) - 2\sin(\log x)] + \frac{x}{2}(\log x)$$

Example 4. Solve : $x^3 \dfrac{d^3y}{dx^3} + 2x^2 \dfrac{d^2y}{dx^2} + 2y = 10\left(x + \dfrac{1}{x}\right)$

Solution. Put $x = e^z$ so that $z = \log x$ and $\dfrac{d}{dz} \equiv D$, then the given differential equation reduces to

$$[D(D-1)(D-2) + 2D(D-1) + 2]y = 10(e^z + e^{-z})$$

$$\Rightarrow \quad (D^3 - D^2 + 2)y = 10(e^z + e^{-z})$$

which is a linear differential equations with constant coefficients.
Its auxiliary equation is

$$m^3 - m^2 + 2 = 0$$

$$\Rightarrow \quad (m+1)(m^2 - 2m + 2) = 0$$

$$\Rightarrow \quad m = -1,\ 1 \pm i$$

$\therefore$ $$\text{C.F.} = c_1e^{-z} + e^z(c_2\cos z + c_3\sin z)$$

$$= \frac{c_1}{x} + x\left(c_2\cos(\log x) + c_3\sin(\log x)\right)$$

and $$\text{P.I.} = 10\frac{1}{D^3 - D^2 + 2}(e^z + e^{-z})$$

$$= 10\left(\frac{1}{D^3 - D^2 + 2}e^z + \frac{1}{D^3 - D^2 + 2}e^{-z}\right)$$

$$= 10\left(\frac{1}{1^3 - 1^2 + 2}e^z + \frac{z}{3D^2 - 2D}e^{-z}\right)$$

$$= 10\left(\frac{1}{2}e^z + \frac{z}{3(-1)^2 - 2(-1)}e^{-z}\right)$$

$$= 10\left(\frac{1}{2}e^z + \frac{z}{5}e^{-z}\right) = 10\left(\frac{1}{2}x + \frac{1}{5x}\log x\right)$$

$$= 5x + \frac{2}{x}\log x$$

Hence, the complete solution is

$$y = \frac{c_1}{x} + x\left(c_2\cos(\log x) + c_3\sin(\log x)\right) + 5x + \frac{2}{x}\log x$$

Example 5. Solve : $x^2\dfrac{d^2y}{dx^2} + x\dfrac{dy}{dx} = 2x^2 + 5x^2\log x$

Solution. Put $x = e^z$ so that $z = \log x$ and $\dfrac{d}{dz} \equiv D,$ then the given differential equation reduces to

$$[D(D-1) + D]y = 2e^{2z} + 5ze^{2z}$$

$\Rightarrow$ $$D^2y = 2e^{2z} + 5ze^{2z}$$

which is a linear differential equations with constant coefficients.

Its auxiliary equation is

$$m^2 = 0 \Rightarrow m = 0,\ 0$$

$\therefore$ $$\text{C.F.} = e^{0z}(c_1 + c_2z) = c_1 + c_2\log x$$

and $$\text{P.I.} = \frac{1}{D^2}(2e^{2z} + 5ze^{2z}) = 2.\frac{1}{D^2}e^{2z} + 5.\frac{1}{D^2}ze^{2z}$$

$$= 2.\frac{1}{4}e^{2z} + 5e^{2z}\frac{1}{(D+2)^2}z$$

$$= \frac{1}{2}e^{2z} + 5e^{2z}\frac{1}{4\left(1+\frac{D}{2}\right)^2}z$$

$$= \frac{1}{2}e^{2z} + \frac{5}{4}e^{2z}\left(1+\frac{D}{2}\right)^{-2}z$$

$$= \frac{1}{2}e^{2z} + \frac{5}{4}e^{2z}\left(1-2.\frac{D}{2}+3.\frac{D^2}{4}+....\right)z$$

$$= \frac{1}{2}e^{2z} + \frac{5}{4}e^{2z}(1-D)z = \frac{1}{2}e^{2z} + \frac{5}{4}e^{2z}(z-1)$$

$$= \frac{1}{2}x^2 + \frac{5}{4}x^2(\log x - 1) = \frac{5}{4}x^2 \log x - \frac{3}{4}x^2$$

Hence, the complete solution is

$$y = c_1 + c_2 \log x + \frac{5}{4}x^2 \log x - \frac{3}{4}x^2$$

2.24 LEGENDRE'S LINEAR DIFFERENTIAL EQUATION

An equation of the form

$$(a+bx)^n \frac{d^n y}{dx^n} + a_1 (a+bx)^{n-1}\frac{d^{n-1}y}{dx^{n-1}} + ...+ a_{n-1}(a+bx)\frac{dy}{dx} + a_n y = Q \quad ...(1)$$

where a_i's are all constants and Q is a function of x, is called a homogeneous linear differential equation or Cauchy's homogeneous linear differential equation.

Method of Solving : To solve such types of equations, we transform them to linear differential equations with constant coefficients by the substitution

$$a + bx = e^z \text{ so that } z = \log(a+bx) \text{ and } \frac{dz}{dx} = \frac{b}{a+bx}$$

Then, we get, $(a+bx)\frac{dy}{dx} = bDy$, where $D \equiv \frac{d}{dz}$;

$$(a+bx)^2 \frac{d^2y}{dx^2} = b^2 D(D-1)y;$$

Similarly, $(a+bx)^3 \frac{d^3y}{dx^3} = b^3 D(D-1)(D-2)y$

In general we have,

$$(a+bx)^n \frac{d^n y}{dx^n} = b^n D(D-1)(D-2)(D-3).......(D-n+1)y$$

Substituting these values in (1), we get a linear differential equations with constant coefficients, which can be solved by the methods discussed earlier.

ILLUSTRATIVE EXAMPLES

Example 1. Solve : $(1+x)^2 \dfrac{d^2y}{dx^2} - 4(1+x)\dfrac{dy}{dx} + 6y = x$

Solution. Put $1+x = e^z$ so that $z = \log(1+x)$ and $\dfrac{d}{dz} \equiv D$, then the given differential equation reduces to

$$\left[1^2.D(D-1) - 4.1.D + 6\right]y = e^z - 1$$

$$\Rightarrow \qquad (D^2 - 5D + 6)y = e^z - 1$$

which is a linear differential equations with constant coefficients.
Its auxiliary equation is

$$m^2 - 5m + 6 = 0 \Rightarrow m = 2,\ 3$$

$$\therefore \qquad \text{C.F.} = c_1e^{2z} + c_2e^{3z} = c_1(1+x)^2 + c_2(1+x)^3$$

and

$$\text{P.I.} = \frac{1}{D^2 - 5D + 6}(e^z - 1)$$

$$= \frac{1}{D^2 - 5D + 6}.e^z - \frac{1}{D^2 - 5D + 6}.e^{0z}$$

$$= \frac{1}{1^2 - 5.1 + 6}.e^z - \frac{1}{0^2 - 5.0 + 6}.e^{0z}$$

$$= \frac{1}{2}e^z - \frac{1}{6}e^{0z} = \frac{1}{6}(3e^z - 1)$$

$$= \frac{1}{6}[3(1+x) - 1] = \frac{1}{6}(3x + 2)$$

Hence, the complete solution is

$$y = c_1(1+x)^2 + c_2(1+x)^3 + \frac{1}{6}(3x+2).$$

Example 2. Solve : $(3+2x)^2 \dfrac{d^2y}{dx^2} - 6(3+2x)\dfrac{dy}{dx} + 16y = 8\sin\log(3+2x)$

Solution. Put $3+2x = e^z$ so that $z = \log(3+2x)$ and $\dfrac{d}{dz} \equiv D$, then the given differential equation reduces to

$$[2^2.D(D-1) - 6.2.D + 16]y = 8\sin z$$

$$\Rightarrow \qquad (4D^2 - 16D + 16)y = \sin z$$

which is a linear differential equation with constant coefficients.

Its auxiliary equation is

$$4m^2 - 16m + 16 = 0 \Rightarrow m = 2,\ 2$$

$\therefore$ $$\text{C.F.} = (c_1 + c_2 z)e^{2z} = [c_1 + c_2 \log(3+2x)](3+2x)^2$$

and $$\text{P.I.} = \frac{1}{4D^2 - 16D + 16} 8\sin z = 8.\frac{1}{4D^2 - 16D + 16}\sin z$$

$$= 8.\frac{1}{4(-1^2) - 16D + 16}\sin z$$

$$= 8.\frac{1}{12 - 16D}\sin z = 2.\frac{1}{3 - 4D}\sin z$$

$$= 2.\frac{(3+4D)}{9 - 16D^2}\sin z = \frac{2}{9 - 16(-1^2)}[3\sin z + 4\cos z]$$

$$= \frac{2}{25}[3\sin\log(3+2x) + 4\cos\log(3+2x)]$$

Hence, the complete solution is

$$y = [c_1 + c_2 \log(3+2x)](3+2x)^2 + \frac{2}{25}[3\sin\log(3+2x) + 4\cos\log(3+2x)]$$

EXERCISE 2.14

1. $x^2 \frac{d^2y}{dx^2} + x\frac{dy}{dx} + y = \sin(\log x^2)$

2. $x^2 \frac{d^2y}{dx^2} - x\frac{dy}{dx} - 3y = x^2 \log x$

3. $x^3 \frac{d^3y}{dx^3} + 3x^2 \frac{d^2y}{dx^2} + x\frac{dy}{dx} + y = x + \log x$

4. $x^2 \frac{d^2y}{dx^2} + x\frac{dy}{dx} + y = \log x \sin(\log x)$

5. $x^3 \frac{d^3y}{dx^3} + 2x^2 \frac{d^2y}{dx^2} + 3x\frac{dy}{dx} - 3y = x + x^2$

6. $x^2 \frac{d^2y}{dx^2} + x\frac{dy}{dx} - y = x^2 e^x + x^7 + 5$

7. $x^2 \dfrac{d^2y}{dx^2} - 2x\dfrac{dy}{dx} - 4y = x^4$

8. $(2+3x)^2 \dfrac{d^2y}{dx^2} + 3(2+3x)\dfrac{dy}{dx} - 36y = 3x^2 + 4x + 1$

9. $(a+x)^2 \dfrac{d^2y}{dx^2} + (a+x)\dfrac{dy}{dx} + y = 4\cos\log(a+x)$

10. $(3+x)^2 \dfrac{d^2y}{dx^2} - 4(3+x)\dfrac{dy}{dx} + 6y = x + 3$

ANSWERS

1. $y = c_1\cos(\log x) + c_2\sin(\log x) - \dfrac{1}{3}\sin(\log x^2)$

2. $y = c_1x^3 + c_2x^{-1} - \dfrac{x^2}{9}(3\log x + 2)$

3. $y = c_1x^{-1} + \sqrt{x}\left[c_2\cos\left(\dfrac{\sqrt{3}}{2}\log x\right) + c_3\sin\left(\dfrac{\sqrt{3}}{2}\log x\right)\right] + \dfrac{x}{2} + \log x$

4. $y = c_1\cos(\log x) + c_2\sin(\log x) - \dfrac{1}{4}(\log x)^2\cos(\log x) + \dfrac{1}{4}\log x\sin(\log x)$

5. $y = c_1x + c_2\cos(\sqrt{3}\log x) + c_3\sin(\sqrt{3}\log x) + \dfrac{x^2}{7} + \dfrac{1}{4}x\log x$

6. $y = c_1x + \dfrac{c_2}{x} + \dfrac{e^x}{x^2} - \dfrac{e^x}{x^3} + \dfrac{x^7}{48} - 5$

7. $y = c_1x^4 + c_2x^{-1} + \dfrac{x^4}{5}\log x$

8. $y = c_1(2+3x)^2 + c_2(2+3x)^{-2} + \dfrac{(2+3x)^2\log(2+3x) + 1}{108}$

9. $y = c_1\cos\log(a+x) + c_2\sin\log(a+x) + 2\log(a+x)\sin\log(a+x)$

10. $y = c_1(3+x)^2 + c_2(3+x)^3 + \dfrac{x+3}{2}$

2.25 LINEAR DIFFERENTIAL EQUATIONS OF SECOND ORDER

The differential equation of the form

$$\frac{d^2y}{dx^2} + P\frac{dy}{dx} + Qy = R \qquad \text{... (1)}$$

where P, Q are R functions of x alone; is called a linear differential equation of second order.

Note : The coefficient of $\frac{d^2y}{dx^2}$ should always be unity.

We shall discuss the following method for solving the linear differential equation of second order :

2.26 METHOD OF VARIATION OF PARAMETERS FOR SOLVING LINEAR DIFFERENTIAL EQUATIONS OF SECOND ORDER

Steps for Solution

1. Find C.F. of the given differential equation.
2. Let the parts of C.F.be u and v.
3. Consider $y = Au + Bv$ as the complete solution of the differential equation.
4. Where A and B are found by

   $$A = \int \frac{-Rv}{W} dx + c_1 \text{ and } B = \int \frac{Ru}{W} dx + c_2$$

   where c_1 and c_2 are the arbitrary constants of integration and

   $$W = uv_1 - u_1 v$$

   v_1 = first derivative of v, and u_1 = first derivative of u.
5. Write $y = Au + Bv$ as the complete solution.

ILLUSTRATIVE EXAMPLES

Example 1. Solve : $\frac{d^2y}{dx^2} + y = \operatorname{cosec} x$

Solution. Auxiliary equation is

$$m^2 + 1 = 0 \Rightarrow m = \pm i$$

$\therefore$ $$\text{C.F.} = c_1 \cos x + c_2 \sin x$$

Here $$u = \cos x \text{ and } v = \sin x \text{ and } R = \operatorname{cosec} x$$

Let $y = Au + Bv = A\cos x + B\sin x$ be the complete solution of the given equation

and $W = uv_1 - u_1 v = \cos x.\cos x - (-\sin x).\cos x = \cos^2 x + \sin^2 x = 1$

Now, $$A = \int \frac{-Rv}{W} dx + c_1 = \int \frac{-\operatorname{cosec} x.\sin x}{1} dx + c_1$$

$$= \int \frac{-1}{1} dx + c_1 = -\int 1.dx + c_1 = -x + c_1$$

$$B = \int \frac{Ru}{W} dx + c_2 = \int \frac{\operatorname{cosec} x.\cos x}{1} dx + c_2$$

$$= \int \frac{\cot x}{1} dx + c_2 = \int \cot x dx + c_2 = \log \sin x + c_2$$

Hence, complete solution is,

$$y = Au + Bv = (-x + c_1)\cos x + (\log \sin x + c_2)\sin x$$

$$= c_1 \cos x + c_2 \sin x - x\cos x + \sin x.\log \sin x$$

Example 2. Solve : $\dfrac{d^2y}{dx^2} - 3\dfrac{dy}{dx} + 2y = \dfrac{e^x}{1+e^x}$

Solution. Auxiliary equation is

$$m^2 - 3m + 2 = 0 \Rightarrow m = 1,\ 2$$

$\therefore$ $$\text{C.F.} = c_1 e^x + c_2 e^{2x}$$

Here $u = e^x$ and $v = e^{2x}$ and $R = \dfrac{e^x}{1+e^x}$

Let $y = Au + Bv = Ae^x + Be^{2x}$ be the complete solution of the given equation

and $W = uv_1 - u_1 v = e^x.2e^{2x} - e^x.e^{2x} = e^{3x}$

Now, $$A = \int \frac{-Rv}{W}\,dx + c_1 = \int \frac{-\dfrac{e^x}{1+e^x}.e^{2x}}{e^{3x}}\,dx + c_1$$

$$= \int \frac{-\dfrac{e^{3x}}{1+e^x}}{e^{3x}}\,dx + c_1 = \int \frac{-1}{1+e^x}\,dx + c_1$$

$$= \int \frac{-e^{-x}}{e^{-x}+1}\,dx + c_1 = \log(1+e^{-x}) + c_1$$

$$= \log\left(\frac{e^x+1}{e^x}\right) + c_1 = \log(e^x+1) - \log e^x + c_1$$

$$B = \int \frac{Ru}{W}\,dx + c_2 = \int \frac{\dfrac{e^x}{1+e^x}.e^x}{e^{3x}}\,dx + c_2$$

$$= \int \frac{\dfrac{e^{2x}}{1+e^x}}{e^{3x}}\,dx + c_2 = \int \frac{1}{e^x(1+e^x)}\,dx + c_2$$

$$= \int \left[\frac{1}{e^x} - \frac{1}{1+e^x}\right] dx + c_2$$

$$= \int \left[e^{-x} - \frac{e^{-x}}{e^{-x}+1}\right] dx + c_2 = -e^{-x} + \log(1+e^{-x}) + c_1$$

$$= -e^{-x} + \log(e^x+1) - \log e^x + c_2$$

Hence, complete solution is,

$$y = Au + Bv = \left[\log(e^x+1) - \log e^x + c_1\right]e^x + \left[-e^{-x} + \log(e^x+1) - \log e^x + c_2\right]e^{2x}$$

$$= (e^x + e^{2x})\log(1+e^x) + e^x(c_1 - 1 - x) + e^{2x}(c_2 - x) \quad [\because \log e^x = x]$$

Example 3. Solve : $x^2\dfrac{d^2y}{dx^2} + 2x\dfrac{dy}{dx} - 12y = x^3 \log x$

Solution. Put $x = e^z$ so that $z = \log x$ and $\dfrac{d}{dz} \equiv D,$ then the given differential equation reduces to

$$[D(D-1) + 2D - 12]y = ze^{3z}$$

$$\Rightarrow \quad (D^2 + D - 12)y = ze^z$$

which is a linear differential equations with constant coefficients.
Auxiliary equation is

$$m^2 + m - 12 = 0 \Rightarrow m = 3, -4$$

$$\therefore \quad \text{C.F.} = c_1e^{3z} + c_2e^{-4z} = c_1x^3 + \frac{c_2}{x^4}$$

Here $u = x^3$ and $v = \dfrac{1}{x^4}$

and making the coefficient of $\dfrac{d^2y}{dx^2}$ as unity, we get $R = x\log x$

Let $y = Au + Bv = Ax^3 + \dfrac{B}{x^4}$ be the complete solution of the given equation

and $W = uv_1 - u_1v = x^3.(-4x^{-5}) - x^{-4}.3x^2 = -4x^{-2} - 3x^{-2} = -7x^{-2}$

Now,

$$A = \int \frac{-Rv}{W}dx + c_1 = \int \frac{-x\log x.\frac{1}{x^4}}{-7x^{-2}}dx + c_1$$

$$= \int \frac{-x\log x.\frac{1}{x^4}}{-7x^{-2}}dx + c_1 = \frac{1}{7}\int \frac{\log x}{x}dx + c_1$$

$$= \frac{1}{7}\frac{(\log x)^2}{2} + c_1 = \frac{(\log x)^2}{14} + c_1$$

$$B = \int \frac{Ru}{W}dx + c_2 = \int \frac{x\log x.x^3}{-7x^{-2}}dx + c_2$$

$$= -\frac{1}{7}\int x^6 \log x dx + c_2 = -\frac{1}{7}\left[\log x.\frac{x^7}{7} - \int \frac{1}{x}.\frac{x^7}{7}\right] + c_2$$

$$= -\frac{1}{7}\left[\frac{x^7 \log x}{7} - \int \frac{x^6}{7}\right] + c_2$$

$$= -\frac{1}{7}\left[\frac{x^7 \log x}{7} - \frac{x^7}{7.7}\right] + c_2 = \frac{x^7}{49}\left[\frac{1}{7} - \log x\right] + c_2$$

Hence, complete solution is

$$y = Au + Bv = \left[\frac{(\log x)^2}{14} + c_1\right]x^3$$

$$+ \left[\frac{x^7}{49}\left(\frac{1}{7} - \log x\right) + c_2\right]x^{-4}$$

$$= \left[c_1 + \frac{1}{343}\right]x^3 + c_2 x^{-4} + \frac{x^3}{98}\log x(7\log x - 2)$$

Example 4. Solve : $\frac{d^2y}{dx^2} - 2\frac{dy}{dx} + 2y = e^x \tan x$

Solution. Auxiliary equation is

$$m^2 - 2m + 2 = 0 \Rightarrow m = 1 \pm i$$

$\therefore$ $$\text{C.F.} = e^x(c_1 \cos x + c_2 \sin x)$$

Here $u = e^x \cos x$ and $v = e^x \sin x$ and $R = e^x \tan x$

Let $y = Au + Bv = Ae^x \cos x + Be^x \sin x$ be the complete solution of the given equation

and $$W = uv_1 - u_1 v = e^x \cos x.(e^x \cos x + e^x \sin x)$$

$$-e^x \sin x.(e^x \cos x - e^x \sin x)$$

$$= e^{2x}(\cos^2 x + \cos x.\sin x - \sin x.\cos x + \sin^2 x) = e^{2x}$$

Now, $$A = \int \frac{-Rv}{W}dx + c_1 = \int \frac{-e^x \tan x.e^x \sin x}{e^{2x}}dx + c_1$$

$$= \int \frac{-\sin^2 x}{\cos x}dx + c_1 = -\int \frac{1 - \cos^2 x}{\cos x}dx + c_1$$

$$= -\int [\sec x - \cos x]dx + c_1$$

$$= -\int \sec x dx + \int \cos x dx + c_1$$

$$= -\log(\sec x + \tan x) + \sin x + c_1$$

$$B = \int \frac{Ru}{W} dx + c_2$$

$$= \int \frac{e^x \tan x . e^x \cos x}{e^{2x}} dx + c_2$$

$$= \int \sin x dx + c_2 = -\cos x + c_2$$

Hence, complete solution is

$$y = Au + Bv = [-\log(\sec x + \tan x) + \sin x + c_1] e^x \cos x$$

$$+[-\cos x + c_2] e^x \sin x$$

$$= e^x (c_1 \cos x + c_2 \sin x) - e^x \cos x . \log(\sec x + \tan x).$$

Example 5. Solve : $\dfrac{d^2y}{dx^2} + 4y = \sec 2x$

Solution. Auxiliary equation is

$$m^2 + 4 = 0 \Rightarrow m = \pm 2i$$

$\therefore$ $\quad$ C.F. $= c_1 \cos 2x + c_2 \sin 2x$

Here $u = \cos 2x$ and $v = \sin 2x$ and $R = \sec 2x$

Let $y = Au + Bv = A\cos 2x + B\sin 2x$ be the complete solution of the given equation and

$$W = uv_1 - u_1 v = \cos 2x . 2\cos 2x - (-2\sin 2x)\sin 2x$$

$$= 2(\cos^2 2x + \sin^2 2x) = 2$$

Now, $\quad A = \int \dfrac{-Rv}{W} dx + c_1 = \int \dfrac{-\sec 2x . \sin 2x}{2} dx + c_1$

$$= -\int \frac{\tan 2x}{2} dx + c_1 = -\frac{1}{2.2} \log \sec 2x + c_1$$

$$= \frac{1}{4} \log \cos 2x + c_1$$

$$B = \int \frac{Ru}{W} dx + c_2 = \int \frac{\sec 2x . \cos 2x}{2} dx + c_2$$

$$= \frac{1}{2} \int 1 . dx + c_2 = \frac{x}{2} + c_2$$

Hence, complete solution is

$$y = Au + Bv = \left[\frac{1}{4}\log\cos 2x + c_1\right]\cos 2x + \left[\frac{x}{2} + c_2\right]\sin 2x$$

$$= c_1 \cos 2x + c_2 \sin 2x + \frac{1}{2}x\sin 2x$$

$$+ \frac{1}{4}\cos 2x.\log\cos 2x$$

Example 6. Solve : $\frac{d^2y}{dx^2} + y = x$

Solution. Auxiliary equation is

$$m^2 + 1 = 0 \Rightarrow m = \pm i$$

$\therefore$ $\quad$ C.F. $= c_1 \cos x + c_2 \sin x$

Here $u = \cos x$ and $v = \sin x$ and $R = x$

Let $y = Au + Bv = A\cos x + B\sin x$ be the complete solution of the given equation

and $\quad W = uv_1 - u_1v = \cos x.\cos x - (-\sin x)\sin x$

$$= (\cos^2 x + \sin^2 x) = 1$$

Now, $\quad A = \int \frac{-Rv}{W}dx + c_1 = \int \frac{-x.\sin x}{1}dx + c_1$

$$= -\int x.\sin x dx + c_1 = -\left[x(-\cos x) - \int 1.(-\cos x)\right] + c_1$$

$$= x.\cos x - \sin x + c_1$$

$$B = \int \frac{Ru}{W}dx + c_2 = \int \frac{x.\cos x}{1}dx + c_2$$

$$= \int x.\cos x dx + c_2 = \left[x.\sin x - \int 1.\sin x\, dx\right] + c_2$$

$$= x.\sin x + \cos x + c_2$$

Hence, complete solution is

$$y = Au + Bv = [x.\cos x - \sin x + c_1]\cos x$$

$$+ [x.\sin x + \cos x + c_2]\sin x$$

$$= c_1 \cos x + c_2 \sin x + x$$

EXERCISE 2.15

1. $\frac{d^3y}{dx^3} + \frac{dy}{dx} = \tan x$

2. $\frac{d^2y}{dx^2} + y = \sec x$ (*I.U. 2007, 08*)

3. $\frac{d^2y}{dx^2} - y = \frac{2}{1+e^x}$

4. $\frac{d^2y}{dx^2} - \frac{dy}{dx} = e^x \sin x$

5. $x^2\dfrac{d^2y}{dx^2}+x\dfrac{dy}{dx}-y=x^2e^x$

6. $\dfrac{d^2y}{dx^2}+y=\sec x\tan x$

7. $\dfrac{d^2y}{dx^2}-3\dfrac{dy}{dx}+2y=\sin x$

8. $x^2\dfrac{d^2y}{dx^2}-2x\dfrac{dy}{dx}-4y=32(\log x)^2$

9. $\dfrac{d^2y}{dx^2}+y=x\cos x$

ANSWERS

1. $y=c_1+c_2\cos x+c_3\sin x+\log\sec x+1-\sin x\log(\sec x+\tan x)$

2. $y=c_1\cos ax+c_2\sin ax+\dfrac{1}{a}x\sin ax+\dfrac{1}{a^2}\cos ax.\log\cos ax$

3. $y=e^x\left[-e^{-x}+\log(1+e^{-x})+c_1\right]+e^{-x}\left[-\log(1+e^x)+c_2\right]$

4. $y=c_1+c_2e^{2x}-\dfrac{e^x}{2}\sin x$

5. $y=c_1x+\dfrac{c_2}{x}+e^x-\dfrac{e^x}{x}$

6. $y=c_1\cos x+c_2\sin x+x\cos x-\sin x+\sin x\log\sec x$

7. $y=c_1e^x+c_2e^{2x}+\dfrac{1}{10}(3\cos x+\sin x)$

8. $y=c_1x^4+\dfrac{c_2}{x}-8(\log x)^2+12\log x+13$

9. $y=c_1\cos x+\left(c_2-\dfrac{1}{8}\right)\sin x+\dfrac{x^2}{4}\sin x+\dfrac{x}{4}\cos x$

UNIT – III

Partial Differential Equation and Geometry

3.1 PARTIAL DIFFERENTIAL EQUATION

A differential equation containing partial derivatives of a function of two or more independent variables is called a partial differential equation. *e.g.*,

$$y\frac{\partial z}{\partial x}+x\frac{\partial z}{\partial y} = 0$$

$$\frac{\partial^2 z}{\partial x^2}+\frac{\partial^2 z}{\partial y^2} = 0$$

$$\frac{\partial^2 z}{\partial x^2} = a^2\frac{\partial^2 z}{\partial y^2}$$

are the partial differential equations.

The independent variables will be denoted by x and y and the dependent variable by z. The partial differential coefficients are denoted as follows.

$$\frac{\partial z}{\partial x} = p,\ \frac{\partial z}{\partial y} = q,\ \frac{\partial^2 z}{\partial x^2} = r,\ \frac{\partial^2 z}{\partial x \partial y} = s,\ \frac{\partial^2 z}{\partial y^2} = t$$

3.2 ORDER AND DEGREE OF PARTIAL DIFFERENTIAL EQUATION

Order of a partial differential equation is the order of the highest ordered derivative present in the equation.

Degree of partial differential equation is the greatest exponent (power) of the highest ordered derivative present in the equation, when it has been made free from radical signs and fractional powers.

3.3 SOLUTION OF PARTIAL DIFFERENTIAL EQUATION

The solution of a partial differential equation in a region D is a function having partial derivatives which satisfy the differential equation at every point in D.

The general solution of a partial differential equation contains arbitrary constants or arbitrary functions or both.

3.4 FORMATION OF PARTIAL DIFFERENTIAL EQUATION

A partial differential equation is formed by two methods :

1. By eliminating arbitrary constants
2. By eliminating arbitrary functions

ILLUSTRATIVE EXAMPLES

1. By eliminating arbitrary constants

Example 1. Form a partial differential equation from the following equations by eliminating the arbitrary constants :

(*i*) $Z = ax + a^2y^2 + b$ (*ii*) $Z = (x + a)(y + b)$

(*iii*) $\frac{x^2}{a^2} + \frac{y^2}{b^2} + \frac{z^2}{c^2} = 1$ (*iv*) $x^2 + y^2 + (z - c)^2 = a^2$

Solution. (*i*) $z = ax + a^2y^2 + b$ ... (1)

Diff. equation (1) partially w.r.t. x and y

$$\frac{\partial z}{\partial x} = p = a$$

$$\frac{\partial z}{\partial y} = q = 2a^2y$$

Eliminating a between these result, we get $q = 2p^2y$

(*ii*) $Z = (x + a)(y + b)$ ... (1)

Differentiating equation (1) partially w.r.t. x, we get

$$\frac{\partial z}{\partial x} = p = y + b \quad \text{... (2)}$$

Differentiating equation (1) partially w.r.t. y, we get

$$\frac{\partial z}{\partial y} = q = x + a \quad \text{... (3)}$$

Multiplying equations (2) and (3) we get

$$pq = (y + b)(x + a)$$

$$p\,q = z$$

(*iii*) $\frac{x^2}{a^2} + \frac{y^2}{b^2} + \frac{z^2}{c^2} = 1$ ... (1)

Differentiating equation (1) partially w.r.t x and y we have

$$\frac{2x}{a^2} + \frac{2z}{c^2}\frac{\partial z}{\partial x} = 0$$

or $c^2x + a^2z\frac{\partial z}{\partial x} = 0$ ... (2)

and $\frac{2y}{b^2} + \frac{2z}{c^2}\frac{\partial z}{\partial y} = 0$

or $c^2y + b^2z\frac{\partial z}{\partial y} = 0$ ... (3)

Again differentiating equation (2) partially w.r.t. x we have

$$c^2 + a^2\left(\frac{\partial z}{\partial x}\right)^2 + a^2 z\frac{\partial^2 z}{\partial x^2} = 0$$

or
$$\frac{c^2}{a^2} + \left(\frac{\partial z}{\partial x}\right)^2 + z\frac{\partial^2 z}{\partial x^2} = 0$$

Substituting $\frac{c^2}{a^2} = \frac{-z}{x}\frac{\partial z}{\partial x}$ from equation (2), we have

$$\frac{-z}{x}\frac{\partial z}{\partial x} + \left(\frac{\partial z}{\partial x}\right)^2 + z\frac{\partial^2 z}{\partial x^2} = 0$$

or
$$xz\frac{\partial^2 z}{\partial x^2} + x\left(\frac{\partial z}{\partial x}\right)^2 - z\frac{\partial z}{\partial x} = 0$$

Which is a partial differential equation of the second order

(*iv*) $x^2 + y^2 + (z-c)^2 = a^2$ ...(1)

Differentiating equation (1) partially w.r.t. x, we get

$$2x + 2(z-c)\frac{\partial z}{\partial x} = 0$$

$$x + (z-c)p = 0 \quad ...(2)$$

Differentiating equation (1) partially w.r.t. y, we get

$$2y + 2(z-c)\frac{\partial z}{\partial y} = 0$$

$$y + (z-c)q = 0 \quad ...(3)$$

From equations (2) and (3), we get

$$(z-c) = \frac{-x}{p}$$

Putting this value of $z - c$ in equation (3), we get

$$y - \frac{x}{p}q = 0$$

$$yp - xq = 0$$

2. By eliminating arbitrary functions

Example 2. Form the partial differential equation from $z = f(x^2 - y^2)$

Solution. $z = f(x^2 - y^2)$ ... (1)

Differentiating equation (1) partially w.r.t. x and y, we get

$$p = \frac{\partial z}{\partial x} = f'(x^2 - y^2)2x \quad ... (2)$$

$$q = \frac{\partial z}{\partial y} = f'(x^2 - y^2)(-2y) \quad ...(3)$$

Dividing equation (2) by (3), we get

$$\frac{p}{q} = \frac{-x}{y} \text{ or } py = -qx$$

$$py + qx = 0$$

Example 3. Form the partial differential equation from $z = x + y + f(xy)$

Solution.

$$z = x + y + f(xy) \qquad \text{... (1)}$$

Differentiating equation (1) partially w.r.t. x and y, we get

$$p = \frac{\partial z}{\partial x} = 1 + f'(xy).y$$

$$\Rightarrow \quad p - 1 = yf'(xy) \qquad \text{... (2)}$$

$$q = \frac{\partial z}{\partial y} = 1 + f'(xy).x$$

$$\Rightarrow \quad q - 1 = xf'(xy) \qquad \text{... (3)}$$

Dividing equation (2) by (3), we get

$$\frac{p-1}{q-1} = \frac{y}{x}$$

$$\Rightarrow \quad px - qy = x - y$$

EXERCISE 3.1

Form the partial differential equations.

1. $z = ax + by + ab$
2. $az + b = a^2x + y$
3. $2z = \frac{x^2}{a^2} + \frac{y^2}{b^2}$
4. $(x - h)^2 + (y - k)^2 + z^2 = a^2$
5. $ax^2 + by^2 + z^2 = 1$
6. $x^2 + y^2 = (z - c)^2 \tan^2 \alpha$

ANSWERS

1. $z = px + qy + pq$ **2** $pq = 1$
3. $2z = px + qy$ **4.** $z^2(p^2 + q^2 + 1) = a^2$
5. $z(px + qy) = z^2 - 1$ **6.** $yp - xq = 0$

3.5 DEFINITIONS

The solution $f(x, y, z, a, b) = 0$ of a first order partial differential equation which contains two arbitrary constant is called a complete solution or complete integral. If in this solution, we put $b = \phi(a)$ and find the envelope of the family of surface $f(x, y, z, a, \phi(a)) = 0$, we get a solution involving an arbitrary function ϕ. This is called the general solution or general integral.

A solution obtained from the complete integral by giving particular values to the arbitrary constant is called a particular solution or particular integral.

3.6 SOLUTION OF EQUATION BY DIRECT INTEGRATION

The equations which contain only one partial derivative can be solved by direct integration. However, in place of the constants of integration we use arbitrary functions of the variable kept constant.

ILLUSTRATIVE EXAMPLES

Example 1. Solve $\frac{\partial^2 z}{\partial x \partial y} = x^2 y$ subject to the conditions

$$z(x, 0) = x^2 \text{ and } z(1, y) = \cos y$$

Solution.

$$\frac{\partial^2 z}{\partial x \partial y} = x^2 y$$

$$\frac{\partial}{\partial x}\left(\frac{\partial z}{\partial y}\right) = x^2 y$$

Integrating w.r.t. x, we get

$$\frac{\partial z}{\partial y} = \frac{x^3}{3} y + f(y)$$

Integrating w.r.t. y, we get

$$z = \frac{x^3}{3} \cdot \frac{y^2}{2} + \int f(y) dy + g(x)$$

$$z = \frac{x^3 y^2}{6} + F(y) + g(x) \qquad \dots (1)$$

[where $\int f(y) dy = F(y)$]

Now putting $z = x^2$ and $y = 0$ in equation (1), we get

$$x^2 = 0 + F(0) + g(x)$$

$$g(x) = x^2 - F(0)$$

Put in equation (1), we get

$$z = \frac{x^3 y^2}{6} + F(y) + x^2 - F(0) \qquad \dots (2)$$

Again $\qquad z(1, y) = \cos y$

Putting $x = 1$ and $z = \cos y$ in equation (2), we get

$$\cos y = \frac{y^2}{6} + F(y) + 1 - F(0)$$

Putting the value of $F(y)$ in equation (2), we obtain

$$z = \frac{1}{6} x^3 y^2 + \cos y - \frac{1}{6} y^2 - 1 + F(0) + x^2 - F(0)$$

$$z = \frac{1}{6} x^3 y^2 + \cos y - \frac{1}{6} y^2 - 1 + x^2$$

Example 2. Solve $\dfrac{\partial^3 z}{\partial x^2 \partial y} = \cos(2x+3y)$

Solution. $\dfrac{\partial^3 z}{\partial x^2 \partial y} = \cos(2x+3y)$

Integrating w.r.t. x, we get

$$\frac{\partial^2 z}{\partial x \partial y} = \frac{1}{2}\sin(2x+3y) + f(y)$$

Again Integrating w.r.t. x, we get

$$\frac{\partial z}{\partial y} = -\frac{1}{4}\cos(2x+3y) + x\int f(y)dx + g(y)$$

$$= -\frac{1}{4}\cos(2x+3y) + x\,\phi(y) + g(y)$$

Integrating w.r.t. y, we get

$$z = -\frac{1}{12}\sin(2x+3y) + x\int \phi(y)dy + \int g(y)dy$$

$$z = -\frac{1}{12}\sin(2x+3y) + x\,\phi_1(y) + \phi_2(y)$$

EXERCISE 3.2

Solve the following equations :

1. $\dfrac{\partial^2 z}{\partial x \partial y} = \dfrac{y}{x} + 2$

2. $\dfrac{\partial^2 z}{\partial x \partial y} = e^y \cos x$

3. $\dfrac{\partial^2 z}{\partial x^2} = a^2 z$, given that when $x = 0$, $\dfrac{\partial z}{\partial x} = a \sin y$ and $\dfrac{\partial z}{\partial y} = 0$

4. $\dfrac{\partial^2 u}{\partial y dx} = 4x \sin 3xy$

ANSWERS

1. $z = \dfrac{y^2}{2}\log x + 2xy + f(y) + \phi(x)$
2. $z = e^y \sin x + f(y) + \phi(x)$
3. $z = \sin x + e^y \cos x$
4. $u = \dfrac{-4}{9y}\sin 3xy + f(x) + \phi(y)$

3.7 LINEAR PARTIAL DIFFERENTIAL EQUATIONS OF THE FIRST ORDER

A differential equation involving first order partial derivatives p and q only is called a partial differential equation of the first order. If p and q both occurs in the first degree only and are not multiplied together then it is called linear partial differential equation of the first order.

3.8 LAGRANGE'S LINEAR EQUATION

The partial differential equation of the form $Pp + Qq = R$ where P, Q and R are functions of x, y, z and $p = \frac{\partial z}{\partial x}$, $q = \frac{\partial z}{\partial y}$ is the standard form of a linear partial differential equation of the first order and is called Lagrange's linear equation.

Working Rule

Step I: Write down the auxiliary equations

$$\frac{dx}{P} = \frac{dy}{Q} = \frac{dz}{R}$$

Step II : Solve the above auxiliary equations
Let the two solutions be $u = c_1$ and $v = c_2$

Step III: Then $F(u, v) = 0$ or $u = \phi(v)$ is the required solution of

$$Pp + Qq = R$$

ILLUSTRATIVE EXAMPLES

Example 1. Solve $y^2p - xyq = x(z - 2y)$
Solution. $y^2p - xyq = x(z - 2y)$
The auxiliary equations are

$$\frac{dx}{y^2} = \frac{dy}{-xy} = \frac{dz}{x(z-2y)} \qquad \text{... (1)}$$

Taking first two members of the equations

$$\frac{dx}{y} = \frac{dy}{-x} \Rightarrow xdx = -ydy$$

Integrating both sides, we get

$$\frac{x^2}{2} = \frac{-y^2}{2} + \frac{c_1}{2}$$

$$\Rightarrow \quad x^2 + y^2 = c_1 \qquad \text{... (2)}$$

Taking last two equations of (1), we have

$$-\frac{dy}{y} = \frac{dz}{z-2y}$$

$$-zdy + 2y\,dy = ydz$$

or

$$2y\,dy = ydz + zdy$$

Integrating, we get

$$y^2 = yz + c_2$$

$$y^2 - yz = c_2 \quad \text{... (3)}$$

From (2) and (3), we get

$$x^2 + y^2 = f(y^2 - yz)$$

Example 2. Solve $(x^2 - yz)\,p + (y^2 - zx)q = z^2 - xy$

Solution. The Auxiliary equations are

$$\frac{dx}{x^2 - yz} = \frac{dy}{y^2 - zx} = \frac{dz}{z^2 - xy}$$

$$\frac{dx - dy}{(x-y)(x+y+z)} = \frac{dy - dz}{(y-z)(x+y+z)} = \frac{dz - dx}{(z-x)(x+y+z)}$$

Taking first two members, we get

$$\frac{dx - dy}{(x-y)} = \frac{dy - dz}{y - z}$$

Integrating both sides, we get

$\log (x - y) = \log (y - z) + \log C_1$

$$\log\left(\frac{x-y}{y-z}\right) = \log C_1$$

$$\frac{x-y}{y-z} = C_1 \quad \text{... (1)}$$

Similarly taking last two equations, we get

$$\frac{y-z}{z-x} = C_2 \quad \text{... (2)}$$

From equations (1) and (2), we get

$$f\left(\frac{x-y}{y-z}, \frac{y-z}{z-x}\right) = 0$$

3.9 METHOD OF MULTIPLIERS

Let the auxiliary equations be

$$\frac{dx}{P} = \frac{dy}{Q} = \frac{dz}{R}$$

l, m, n may be constants or function of x, y, z then, we have

$$\frac{dx}{P} = \frac{dy}{Q} = \frac{dz}{R} = \frac{l dx + m dy + n dz}{lP + mQ + nR}$$

l, m, n are chosen in such a way that

$$lP + mQ + nR = 0$$

Thus $ldx + mdy + ndz = 0$

Solve this differential equation if the solution is $u = c_1$ similarly choose another set of multipliers l_1, m_1, n_1 and if the second solution is $v = c_2$

$\therefore$ Required solution is $f(u, v) = 0$

ILLUSTRATIVE EXAMPLES

Example 1. Solve the differential equation

$$x^2(y-z)p + y^2(z-x)q = z^2(x-y)$$

Solution. The auxiliary equations are

$$\frac{dx}{x^2(y-z)} = \frac{dy}{y^2(z-x)} = \frac{dz}{z^2(x-y)}$$

Using $\frac{1}{x^2}, \frac{1}{y^2}, \frac{1}{z^2}$ as multipliers, we get

$$\text{each fraction} = \frac{\frac{1}{x^2}dx + \frac{1}{y^2}dy + \frac{1}{z^2}dz}{0}$$

$$\frac{dx}{x^2} + \frac{dy}{y^2} + \frac{dz}{z^2} = 0$$

Integrating, we get

$$\frac{1}{x} + \frac{1}{y} + \frac{1}{z} = a \qquad \text{... (1)}$$

Again using $\frac{1}{x}, \frac{1}{y}, \frac{1}{z}$ as multipliers, we get

$$\text{each fraction} = \frac{\frac{1}{x}dx + \frac{1}{y}dy + \frac{1}{z}dz}{0}$$

$$\frac{dx}{x} + \frac{dy}{y} + \frac{dz}{z} = 0$$

Integrating, we get

$$\log x + \log y + \log z = \log b$$

$$\log(xyz) = \log b \quad \Rightarrow \quad xyz = b \qquad \text{... (2)}$$

From (1) and (2) the solution is given by

$$f\left(\frac{1}{x} + \frac{1}{y} + \frac{1}{z}, xyz\right) = 0$$

Example 2. Solve the differential equation

$$\frac{y-z}{yz}p = \frac{z-x}{zx}q = \frac{x-y}{xy}$$

Solution. $\frac{y-z}{yz}p = \frac{z-x}{zx}q = \frac{x-y}{xy}$

Multiplying by xyz, we get

$$x(y-z)p + y(z-x)q = z(x-y)$$

$$\frac{dx}{x(y-z)} = \frac{dy}{y(z-x)} = \frac{dz}{z(x-y)}$$

$$= \frac{dx+dy+dz}{x(y-z)+y(z-x)+z(x-y)} \quad \text{... (1)}$$

$$= \frac{dx+dy+dz}{0}$$

$\therefore \quad dx + dy + dz = 0$

Integrating, we get $\quad x + y + z = a$ ... (2)

Again equation (1) can be written as

$$\frac{\frac{dx}{x}}{y-z} = \frac{\frac{dy}{y}}{z-x} = \frac{\frac{dz}{z}}{x-y} = \frac{\frac{dx}{x}+\frac{dy}{y}+\frac{dz}{z}}{(y-z)+(z-x)+(x-y)}$$

$$= \frac{\frac{dx}{x}+\frac{dy}{y}+\frac{dz}{z}}{0}$$

$$\therefore \quad \frac{dx}{x}+\frac{dy}{y}+\frac{dz}{z} = 0$$

Integrating, we get

$$\log x + \log y + \log z = \log b$$

$$\log (xyz) = \log b$$

$$xyz = b \quad \text{... (3)}$$

From (2) and (3) the solution is given by $f(x + y + z, xyz) = 0$

EXERCISE 3.3

Solve the following equations :

1. $xp + yq = 3z$
2. $yzp + zxq = xy$
3. $x^2p + y^2q = (x + y) z$
4. $x\frac{\partial u}{dx} + y\frac{\partial u}{dy} + z\frac{\partial u}{\partial z} = xyz$
5. $(y + z) p + (z + x) q = x + y$
6. $p \tan x + q \tan y = \tan z$
7. $xp - yq + x^2 - y^2 = 0$
8. $x^2p + y^2q + z^2 = 0$

ANSWERS

1. $f\left(x/y, \frac{x^3}{z}\right) = 0$
2. $f(x^2 - y^2, x^2 - z^2) = 0$
3. $f\left(\frac{xy}{z}, \frac{1}{y} - \frac{1}{x}\right) = 0$
4. $f\left(\frac{x}{y}, \frac{y}{z}, xyz - 3x\right) = 0$

5. $f\left[\dfrac{x-y}{y-z}, (x-y)^2(x+y+z)\right]=0$ **6.** $f\left[\dfrac{\sin x}{\sin y}, \dfrac{\sin y}{\sin z}\right]=0$

7. $f\left[xy, \dfrac{x^2}{2}+\dfrac{y^2}{2}+z\right]=0$ **8.** $f\left[\dfrac{1}{y}-\dfrac{1}{x}, \dfrac{1}{y}+\dfrac{1}{z}\right]=0$

3.10 NON-LINEAR PARTIAL DIFFERENTIAL EQUATION OF THE FIRST ORDER

A partial differential equation which involves first order partial derivatives p and q with degree higher than one and the products of p and q is called a non-linear partial differential equation of the first order. The complete solution of such an equation involves only two arbitrary constants.

I. Equation of the form $f(p, q) = 0$

i.e. equations containing only p and q and no x, y, z.

Method: Let the required solution be

$$z = ax + by + c \qquad \text{... (1)}$$

$$\frac{\partial z}{\partial x} = a, \frac{\partial z}{\partial y} = b$$

on putting these values in $f(p, q) = 0$ we get

$$f(a, b) = 0$$

From this find the value of b in term of a and substitute the value of b in (1) that will be the required solution.

ILLUSTRATIVE EXAMPLES

Example 1. Solve $pq = p + q$

Solution. The equation is of the form $f(p, q) = 0$ the complete solution is

$$z = ax + by + c \qquad \text{... (1)}$$

where $$ab = a+b \Rightarrow b = \frac{a}{a-1}$$

∴ From equation (1) the complete solution is

$$z = ax + \frac{a}{a-1}y + c$$

Example 2. Solve $p^2 + q^2 = 1$

Solution. The equation is of the form $f(p, q) = 0$

The complete solution is $z = ax + by + c$ ... (1)

where $$a^2 + b^2 = 1 \Rightarrow b = \sqrt{1-a^2}$$

∴ From equation (1) the complete solution is

$$z = ax + \sqrt{1-a^2}\,y + c$$

Example 3. Solve $(y-x)(qy-px) = (p-q)^2$

Solution. Let $x + y = X$ and $xy = Y$

So, that
$$p = \frac{\partial z}{\partial x} = \frac{\partial z}{\partial X} \cdot \frac{\partial X}{\partial x} + \frac{\partial z}{\partial Y}\frac{\partial Y}{\partial x}$$

i.e.
$$p = \frac{\partial z}{\partial X} + \frac{\partial z}{\partial Y} y$$

and
$$q = \frac{\partial z}{\partial y} = \frac{\partial z}{\partial X}\frac{\partial X}{\partial y} + \frac{\partial z}{\partial Y}\frac{\partial Y}{\partial y}$$

i.e.
$$q = \frac{\partial z}{\partial X} + \frac{\partial z}{\partial Y} .x$$

$\therefore$
$$(qy - px) = (y-x)\frac{\partial z}{\partial X} \text{ and } p - q = (y-x)\frac{\partial z}{\partial Y}$$

$\therefore$ The given equation can be written as $\dfrac{\partial z}{\partial X} = \left(\dfrac{\partial z}{\partial Y}\right)^2$

or
$$P = Q^2, \text{ where } P = \frac{\partial z}{\partial X} \text{ and } Q = \frac{\partial z}{\partial Y}$$

It is of the form $f(P, Q) = 0$

Its complete solution is $z = aX + bY + c$

Where
$$a = b^2 \Rightarrow b = \sqrt{a}$$

$\therefore$ The complete solution is $z = a(x+y) + \sqrt{a}\ xy + c$

EXERCISE 3.4

Solve the following partial differential equations :

1. $p^2 - q^2 = 1$
2. $p^2 + p = q^2$
3. $p^3 - q^3 = 0$
4. $p^2 + q^2 = 2$

ANSWERS

1. $z = ax + \sqrt{a^2 - 1}\ y + c$
2. $z = ax + \sqrt{a^2 + a}\ y + c$
3. $z = a(x+y) + c$
4. $z = ax + \sqrt{2 - a^2}\ y + c$

II. Equation of the form $z = px + qy + f(p, q)$

Its complete solution is $z = ax + by + f(a, b)$ obtained by writing a for p and b for q.

ILLUSTRATIVE EXAMPLES

Example 1. Solve $z = px + qy + 2\sqrt{pq}$

Solution. $z = px + qy + 2\sqrt{pq}$

Its solution is $z = ax + by + 2\sqrt{ab}$

Example 2. Solve $z = px + qy + \sqrt{1+p^2+q^2}$

Solution. $z = px + qy + \sqrt{1+p^2+q^2}$

Its solution is $z = ax + by + \sqrt{1+a^2+b^2}$

Example 3. Solve $z = px + qy + p^3 + q^3$

Solution. $z = px + qy + p^3 + q^3$

Its solution is $z = ax + by + a^3 + b^3$

III. Equations of the form *f* (*z*, *p*, *q*) = 0

i.e. equations not containing x and y.

Working rule:

1. Assume $u = x + ay$ be the solution replace, $p = \frac{dz}{du}$ and $q = a\frac{dz}{du}$ in the given equation.
2. Solve the resulting ordinary differential equation in z and u.
3. Replace u by $x + ay$.

ILLUSTRATIVE EXAMPLES

Example 1. Solve $p\,(1+q^2) = q\,(z-a)$

Solution. Let $u = x + by$

So, that
$$p = \frac{dz}{du} \text{ and } q = b\frac{dz}{du}$$

Substituting these values of p and q in the given equation, we have

$$\frac{dz}{du}\left[1+b^2\left(\frac{dz}{du}\right)^2\right] = b\,\frac{dz}{du}\,(z-a)$$

$$1 + b^2\left(\frac{dz}{du}\right)^2 = b\,(z-a)$$

$$b^2\left(\frac{dz}{du}\right)^2 = bz - ab - 1$$

$$\frac{dz}{du} = \frac{1}{b}\sqrt{bz-ab-1}$$

$$\int \frac{b\,dz}{\sqrt{bz-ab-1}} = \int du + c$$

$$2\sqrt{bz-ab-1} = u + c$$

$$4(bz-ab-1) = (u+c)^2$$

$$4(bz-ab-1) = (x+by+c)^2$$

Example 2. Solve $z^2(p^2+q^2+1) = a^2$

Solution. Let $u = x + by$, so that $p = \frac{dz}{du}$ and $q = b\frac{dz}{du}$

Substituting these values of p and q in the given equation we have

$$z^2\left[\left(\frac{dz}{du}\right)^2 + b^2\left(\frac{dz}{du}\right)^2 + 1\right] = a^2$$

$$z^2(1+b^2)\left(\frac{dz}{du}\right)^2 = a^2 - z^2$$

$$z\sqrt{1+b^2}\,\frac{dz}{du} = \pm\sqrt{a^2-z^2}$$

$$\pm\sqrt{1+b^2}\;\frac{z}{\sqrt{a^2-z^2}}\,dz = du$$

Integrating, we get

$$\pm\sqrt{1+b^2}\;\sqrt{a^2-z^2} = u + c$$

$$(1+b^2)(a^2-z^2) = (x+by+c)^2$$

EXERCISE 3.5

Solve the following partial differential equations :

1. $x^2p^2 + y^2q^2 = z$
2. $z^2(p^2x^2 + q^2) = 1$
3. $z^2(p^2 + q^2 + 1) = 1$

ANSWERS

1. $2\sqrt{z} = \frac{\log x + a \log y}{\sqrt{1+a^2}} + c$

2. $\sqrt{1+a^2}\; z^2 = \pm 2(\log x + ay) + c$

3. $(1-z^2)^{\frac{1}{2}} = -\frac{x+ay}{\sqrt{1+a^2}} + c$

IV. Equations of the form $f_1(x, p) = f_2(y, q)$

In these equations z is absent and the terms containing x and p can be written on one side and the terms containing y and q can be written on the other side.
Method: Let $f_1(x, p) = f_2(y, q) = a$
Solving these equations for p and q. Let $p = F_1(x)$ and $q = F_2(y)$

Since $$dz = \frac{\partial z}{\partial x}dx + \frac{\partial z}{\partial y}dy = pdx + q\,dy$$

$\therefore$ $$dz = F_1(x)dx + F_2(y)\,dy$$

Integrating both sides, we get

$$z = \int F_1(x)dx + \int F_2(y)dy + c$$

ILLUSTRATIVE EXAMPLES

Example 1. Solve $p^2 - q^2 = x - y$

Solution. $$p^2 - q^2 = x - y$$
$$p^2 - x = q^2 - y$$
Let $$p^2 - x = q^2 - y = a$$
$$p^2 = x + a \text{ and } q^2 = y + a$$
$$p = \sqrt{x+a} \text{ and } q = \sqrt{y+a}$$

Substituting these values of p and q in $dz = pdx + qdy$, we get

$$dz = \sqrt{x+a}\,dx + \sqrt{y+a}\,dy$$

Integrating $$z = \frac{2}{3}(x+a)^{3/2} + \frac{2}{3}(y+a)^{3/2} + b$$

Example 2. Solve $p - x^2 = q + y^2$

Solution. $$p - x^2 = q + y^2$$
Let $$p - x^2 = q + y^2 = a$$
then $p = x^2 + a$ and $q = a - y^2$
Substituting these values of p and q in $dz = pdx + qdy$, we get
$$dz = (x^2 + a)\,dx + (a - y^2)dy$$

Integrating, we get $z = \left(\frac{x^3}{3} + ax\right) + \left(ay - \frac{y^3}{3}\right) + b$

EXERCISE 3.6

Solve the following partial differential equations :

1. $yp = 2yx + \log q$
2. $p^2 + q^2 = z^2(x + y)$
3. $p^2 - q^2 = x - y$
4. $z(p^2 - q^2) = x - y$
5. $z^2(p^2 + q^2) = x^2 + y^2$

ANSWERS

1. $z - x^2 + ax + \frac{1}{a}e^{ay} + b$ **2.** $\log z = \frac{2}{3}(a+x)^{3/2} + \frac{2}{3}(y-a)^{3/2} + b$

3. $z = \frac{2}{3}(x+a)^{3/2} + \frac{2}{3}(y+a)^{3/2} + c$

4. $z^{3/2} = (x+a)^{3/2} + (y-a)^{3/2} + b$

5. $z^2 = x\sqrt{x^2+a} + y\sqrt{y^2-a} + a\log\frac{x+\sqrt{x^2+a}}{y+\sqrt{y^2-a}} + c$

3.11. CHARPITS METHOD

This is the general method for solving non-linear partial differential equations of the first order.
To solve such equations, we obtained auxilary equations by

$$\frac{dp}{\frac{\partial f}{\partial x}+p\frac{\partial f}{\partial z}} = \frac{dq}{\frac{\partial f}{\partial y}+q\frac{\partial f}{\partial z}} = \frac{dz}{-p\frac{\partial f}{\partial p}-q\frac{\partial f}{\partial q}} = \frac{dx}{-\frac{\partial f}{\partial p}} = \frac{dy}{-\frac{\partial f}{\partial q}} = \frac{dF}{0}$$

Solve auxiliary equations and integrating we get the required complete solution.

ILLUSTRATIVE EXAMPLES

Example 1. Solve $(p^2+q^2)\,y = qz$

Solution. $f = (p^2+q^2)\,y - qz = 0$ ... (1)

$$\therefore \frac{\partial f}{\partial x} = 0,\ \frac{\partial f}{\partial y} = p^2+q^2,\ \frac{\partial f}{\partial z} = -q,\ \frac{\partial f}{\partial p} = 2py,\ \frac{\partial f}{\partial q} = 2qy - z$$

Charpits auxiliary equations are

$$\frac{dp}{\frac{\partial f}{\partial x}+p\frac{\partial f}{\partial z}} = \frac{dq}{\frac{\partial f}{\partial y}+q\frac{\partial f}{\partial z}} = \frac{dz}{-p\frac{\partial f}{\partial p}-q\frac{\partial f}{\partial q}} = \frac{dx}{-\frac{\partial f}{\partial p}} = \frac{dy}{-\frac{\partial f}{\partial q}} = \frac{dF}{0}$$

$$\frac{dp}{-pq} = \frac{dq}{p^2} = \frac{dz}{-qz} = \frac{dx}{-2py} = \frac{dy}{-2qy+z} = \frac{dF}{0}$$

From first two members we have $pdp + qdq = 0$

Integrating, we get $p^2 + q^2 = a^2$ ... (2)

Putting in equation (1), $q = \frac{a^2 y}{z}$

$\therefore$ From equation (2)

$$p = \sqrt{a^2 - q^2}$$

$$= \sqrt{a^2 - \frac{a^4y^2}{z^2}} = \frac{a}{z}\sqrt{z^2 - a^2y^2}$$

$$\therefore \quad dz = pdx + qdy = \frac{a}{z}\sqrt{z^2-a^2y^2}\,dx + \frac{a^2y}{z}\,dy$$

$$zdz - a^2ydy = a\sqrt{z^2 - a^2y^2}\ dx$$

$$\frac{\frac{1}{2}d(z^2 - a^2y^2)}{\sqrt{z^2 - a^2y^2}} = adx$$

Integrating, we get $\sqrt{z^2 - a^2y^2} = ax + b$

or $\qquad z^2 = (ax + b)^2 + a^2y^2$

Example 2. Solve $px + qy = pq$

Solution. $\qquad f(x, y, z, p, q) = 0$ is $px + qy - pq = 0$ $\qquad$ ... (1)

$$\frac{\partial f}{\partial x} = p,\ \frac{\partial f}{\partial y} = q,\ \frac{\partial f}{\partial z} = 0,\ \frac{\partial f}{\partial p} = x - q,\ \frac{\partial f}{\partial q} = y - p$$

Charpits auxiliary equations are

$$\frac{dx}{-\frac{\partial f}{\partial p}} = \frac{dy}{-\frac{\partial f}{\partial q}} = \frac{dz}{-p\frac{\partial f}{\partial p} - q\frac{\partial f}{\partial q}} = \frac{dp}{\frac{\partial f}{\partial x} + p\frac{\partial f}{\partial z}} = \frac{dq}{\frac{\partial f}{\partial y} + q\frac{\partial f}{\partial z}} = \frac{dF}{0}$$

$$\frac{dx}{-(x-q)} = \frac{dy}{-(y-p)} = \frac{dz}{-p(x-q) - q(y-p)} = \frac{dp}{p} = \frac{dq}{q} = \frac{dF}{0}$$

From fourth and fifth members, we have

$$\frac{dp}{p} = \frac{dq}{q} \Rightarrow \log p = \log q + \log a \Rightarrow p = aq$$

Putting $p = aq$ in the given equation (1), we get

$$aqx - aq^2 = -qy$$

$$\Rightarrow \qquad q = \frac{ax + y}{a}$$

$$\therefore \qquad p = aq = ax + y \qquad \text{... (2)}$$

$$\frac{dx}{-x + q} = \frac{dy}{-y + p} = \frac{dz}{2pq - px - qy}$$

$$\frac{pdx}{-px + pq} = \frac{qdy}{-qy + pq} = \frac{dz}{2pq - px - qy}$$

Now $\qquad dz = pdx + qdy$

$$dz = (ax + y)\,dx + \frac{(ax + y)}{a}dy$$

$$adz = a(y + ax)\,dx + (y + ax)\,dy$$

$$adz = (y + ax)\,(adx + dy)$$

Integrating, we get $\qquad az = \dfrac{(ax + y)^2}{2} + b$

3.12 PARTIAL DIFFERENTIAL EQUATION OF THE SECOND ORDER

A differential equation is said to be a partial differential equation of the second order when it includes at least one of the partial derivatives *r*, *s*, *t* but none of the higher order.

3.13 LINEAR PARTIAL DIFFERENTIAL EQUATION WITH CONSTANT COEFFICIENTS

A partial differential equation in which the dependent variable and its derivatives occur only in the first degree and are not multiplied together is called a linear partial differential equation.
General form of such an equation is

$$A_0 \frac{\partial^n z}{\partial x^n} + A_1 \frac{\partial^n z}{\partial x^{n-1} \partial y} + \ldots + A_n \frac{\partial^n z}{\partial y^n} + B_0 \frac{\partial^{n-1} z}{\partial x^{n-1}} + B_1 \frac{\partial^{n-1} z}{\partial x^{n-2} \partial y} + \ldots$$

$$+ B_{n-1} \frac{\partial^{n-1} z}{\partial y^{n-1}} + C_0 \frac{\partial z}{\partial x} + C_1 \frac{\partial z}{\partial y} + P_0 Z = F(x, y) \qquad \ldots(1)$$

Where the coefficients $A_0, A_1 \ldots A_n, B_0, B_1 \ldots B_{n-1}\ C_0, C_1, P_0$ are constants or function of x and y.
If these coefficients are constant then equation (1) is called a linear partial differential equation with constant coefficients. If all the derivatives appearing in equation (1) are of the same order then the differential equation is called a linear homogeneous partial differential equation with constant coefficients while if all the derivatives are not of same order then it is called a non-homogeneous linear partial differential equation with constant coefficients.

3.14. LINEAR HOMOGENEOUS PARTIAL DEFERENTIAL EQUATION OF THE N^{TH} ORDER WITH CONSTANT COEFFICIENTS

An equation of the form

$$a_0 \frac{\partial^n z}{\partial x^n} + a_1 \frac{\partial^n z}{\partial x^{n-1} \partial y} + a_2 \frac{\partial^n z}{\partial x^{n-2} \partial y^2} + \ldots + a_n \frac{\partial^n z}{\partial y^n} = F(x, y) \qquad \ldots (1)$$

Where $a_0, a_1, a_2, \ldots, a_n$ are constant is called a homogeneous linear partial differential equation of the nth order with constant coefficients.
Here, all the partial derivatives are of the nth order.

Putting $$D \equiv \frac{\partial}{\partial x} \text{ and } D' \equiv \frac{\partial}{\partial y} \qquad \ldots (1)$$

becomes

$$(a_0 D^n + a_1 D^{n-1} D' + a_2 D^{n-2} D'^2 + \ldots + a_n D'^n) z = F(x, y)$$

$$\Rightarrow \qquad f(D, D') z = F(x, y) \qquad \ldots (2)$$

The complete solution of (2) is $z =$ C. F. + P. I.

3.15. RULES FOR FINDING THE COMPLEMENTARY FUNCTION (C. F.)

Consider the equation

$$a_0 \frac{\partial^2 z}{\partial x^2} + a_1 \frac{\partial^2 z}{\partial x dy} + a_2 \frac{\partial^2 z}{dy^2} = 0$$

$$(a_0 D^2 + a_1 DD' + a_2 D'^2)\, z = 0 \qquad ...(1)$$

Ist Step: Put $D = m$, $D' = 1$ in equation (1), we get

$$a_0 m^2 + a_1 m + a_2 = 0 \qquad ... (2)$$

This is the auxiliary equation

IInd Step: Solve the auxiliary equation

Case I: If the roots of the auxiliary equation are real and distinct say m_1, m_2 $(m_1 \neq m_2)$ then

$$\text{C. F.} = f_1 (y + m_1 x) + f_2 (y + m_2 x)$$

Case II: If the roots are equal say $m_1 = m_2 = m$ then

$$\text{C. F.} = f_1 (y + mx) + x f_2 (y + mx)$$

ILLUSTRATIVE EXAMPLES

Example 1. Solve the following equations :

(*i*) $\dfrac{\partial^2 z}{\partial x^2} + 4\dfrac{\partial^2 z}{\partial x \partial y} - 5\dfrac{\partial^2 z}{\partial y^2} = 0$

(*ii*) $\dfrac{\partial^3 z}{\partial x^3} - 3\dfrac{\partial^3 z}{\partial x^2 \partial y} + 2\dfrac{\partial^3 z}{\partial x \partial y^2} = 0$

(*iii*) $\dfrac{\partial^4 z}{\partial x^4} - \dfrac{\partial^4 z}{\partial y^4} = 0$

Solution. (*i*) $\dfrac{\partial^2 z}{\partial x^2} + 4\dfrac{\partial^2 z}{\partial x \partial y} - 5\dfrac{\partial^2 z}{\partial y^2} = 0$

$$(D^2 + 4DD' - 5D'^2)z = 0$$

The Auxiliary equation is [Putting $D = m$, $D' = 1$]

$$\begin{aligned} m^2 + 4m - 5 &= 0 \\ m^2 + 5m - m - 5 &= 0 \\ m\,(\text{m} + 5) - 1(m + 5) &= 0 \\ (m - 1)\,(m + 5) &= 0 \\ m &= 1, -5 \\ \text{C. F.} &= f_1 (y + x) + f_2 (y - 5x) \\ \text{P.I.} &= 0 \end{aligned}$$

Hence, complete solution is

$$z = \text{C. F} + \text{P. I} = f_1 (y + x) + f_2 (y - 5x)$$

(*ii*) $$\frac{\partial^3 z}{\partial x^3} - 3\frac{\partial^3 z}{\partial x^2 \partial y} + 2\frac{\partial^3 z}{\partial x \partial y^2} = 0$$

$$(D^3 - 3D^2D' + 2DD'^2)\, z = 0$$

The auxiliary equation is

$$m^3 - 3m^2 + 2m = 0$$

$$m(m^2 - 3m + 2) = 0$$

$$m\,(m - 1)\,(m - 2) = 0$$

$$m = 0, 1, 2$$

$$\text{C. F} = f_1\,(y) + f_2\,(y + x) + f_3\,(y + 2x)$$

$$\text{P.I.} = 0$$

Hence, complete solution is $z = \text{C.F} + \text{P.I}$

$$z = f_1\,(y) + f_2\,(y + x) + f_3\,(y + 2x)$$

(*iii*) $$\frac{\partial^4 z}{\partial x^4} - \frac{\partial^4 z}{\partial y^4} = 0$$

$$(D^4 - D'^4)\, z = 0$$

The auxiliary equation is

$$(m^4 - 1) = 0$$

$$(m^2 - 1)(m^2 + 1) = 0$$

$$m = \pm 1, \pm i$$

$$\text{C.F.} = f_1(y + x) + f_2(y - x) + f_3(y + ix) + f_4(y - ix)$$

$$\text{P.I.} = 0$$

Hence, complete solution is $z = \text{C.F.} + \text{P.I.}$

$$z = f_1(y + x) + f_2(y - x) + f_3(y + ix) + f_4(y - ix)$$

Example 2. Solve $4\frac{\partial^2 z}{\partial x^2} - 12\frac{\partial^2 z}{\partial x \partial y} + 9\frac{\partial^2 z}{\partial y^2} = 0$

Solution. $(4D^2 - 12DD' + 9D'^2)\, z = 0$

The auxiliary equation is

$$4m^2 - 12m + 9 = 0$$

$$(2m - 3)^2 = 0 \Rightarrow m = \frac{3}{2}, \frac{3}{2}$$

$$\text{C.F.} = f_1\left(y + \frac{3}{2}x\right) + xf_2\left(y + \frac{3}{2}x\right)$$

$$\text{P.I.} = 0$$

Hence, complete solution is $z = \text{C.F.} + \text{P.I.}$

$$z = f_1\left(y + \frac{3}{2}x\right) + xf_2\left(y + \frac{3}{2}x\right)$$

Example 3. Solve $r - 4s + 4t = 0$

Solution. $r - 4s + 4t = 0$

or $$\frac{\partial^2 z}{\partial x^2} - 4\frac{\partial^2 z}{\partial x \partial y} + 4\frac{\partial^2 z}{\partial y^2} = 0$$

$$(D^2 - 4DD' + 4D'^2)\, z = 0$$

The Auxiliary equation is

$$m^2 - 4m + 4 = 0$$

$$(m-2)^2 = 0 \Rightarrow m = 2, 2$$

$$\text{C. F.} = f_1(y+2x) + x f_2(y+2x)$$

$$\text{P.I.} = 0$$

Hence, complete solution is $z = \text{C. F.} + \text{P.I.}$

$$z = f_1(y+2x) + x f_2(y+2x)$$

Example 4. Solve $(D^3 - 6D^2D' + 12DD'^2 - 8D'^3)z = 0$

Solution. $(D^3 - 6D^2D' + 12DD'^2 - 8D'^3)z = 0$

The Auxiliary equation is

$$m^3 - 6m^2 + 12m - 8 = 0$$

$$(m-2)^3 = 0 \quad \Rightarrow \quad m = 2, 2, 2$$

$$\text{C. F.} = f_1(y+2x) + x f_2(y+2x) + x^2 f_3(y+2x)$$

$$\text{P.I.} = 0$$

Hence, complete solution is $z = \text{C. F.} + \text{P. I}$

$$z = f_1(y+2x) + x f_2(y+2x) + x^2 f_3(y+2x)$$

EXERCISE 3.7

Solve the following equations :

1. $(D^3 - 6D^2D' + 11DD'^2 - 6D'^3)z = 0$

2. $\dfrac{\partial^2 z}{\partial x^2} + \dfrac{\partial^2 z}{\partial x \partial y} = 0$

3. $\dfrac{\partial^3 z}{\partial x^3} + 3\dfrac{\partial^3 z}{\partial x^2 \partial y} + 2\dfrac{\partial^3 z}{\partial x \partial y^2} = 0$

4. $\dfrac{\partial^3 z}{\partial x^3} - 7\dfrac{\partial^3 z}{\partial x \partial y^2} + 6\dfrac{\partial^3 z}{\partial y^3} = 0$

5. $r + 2s + t = 0$

6. $\dfrac{\partial^4 z}{\partial x^4} + \dfrac{\partial^4 z}{\partial y^4} = 0$

ANSWERS

1. $z = f_1(y+x) + f_2(y+2x) + f_3(y+3x)$

2. $z = f_1(y) + f_2(y-x)$

3. $z = f_1(y) + f_2(y-x) + f_3(y-2x)$

4. $z = f_1(y+x) + f_2(y+2x) + f_3(y-3x)$

5. $z = f_1(y-x) + x f_2(y-x)$

6. $z = f_1(y+z_1x) + f_2(y+\overline{z_1}\,x) + f_3(y+z_2x) + f_4(y+\overline{z_2}\,x)$

3.16. RULES FOR FINDING PARTICULAR INTEGRAL (P. I.)

P. I. of the equation $f(D, D')\, z = F(x, y)$ is given by

$$\frac{1}{f(D,D')} F(x, y)$$

Case I: When $F(x, y) = e^{ax + by}$

$$\text{P.I.} = \frac{1}{f(D,D')} e^{ax+by} = \frac{e^{ax+by}}{f(a,b)}$$

[Replacing $D = a$, $D' = b$]; provided $f(a, b) \neq 0$

If $f(a, b) = 0$, then

$$\text{P.I.} = x \frac{1}{\frac{\partial}{\partial D}[f(D,D')]} e^{ax+by}$$

provided $f'(a, b) \neq 0$

Case II: When $F(x, y) = \sin(ax + by)$

$$\text{P.I.} = \frac{1}{f(D,D')} \sin(ax+by)$$

$$\text{P.I.} = \frac{1}{f(D^2, DD', D'^2)} \sin(ax+by)$$

$$= \frac{1}{f(-a^2, -ab, -b^2)} \sin(ax+by)$$

provided $f(-a^2, -ab - b^2) \neq 0$

Replacing $D^2 = -a^2$, $DD' = -ab$, $D'^2 = -b^2$

similarly, when $F(x, y) = \cos(ax + by)$

$$\text{P. I.} = \frac{1}{f(D^2, DD', D'^2)} \cos(ax+by)$$

$$= \frac{1}{f(-a^2, -ab, -b^2)} \cos(ax+by)$$

provided $f(-a^2, -ab - b^2) \neq 0$

Case III: When $F(x, y) = x^m y^n$

$$\text{P.I.} = \frac{1}{f(D,D^1)} x^m y^n = [f(D, D')]^{-1} x^m y^n$$

expand $[f(D, D')]^{-1}$ in ascending power of D or D' and operate on $x^m y^n$ term by them

Case IV: When $F(x, y)$ is not always of the form given above. Then this method is applicable to all cases, where $F(x, y)$ is not of the form given above.

$$\text{P. I.} = \frac{1}{f(D, D')} F(x, y)$$

Resolve $\frac{1}{f(D, D')}$ into partial fractions

$$\text{P. I.} = \frac{1}{(D - mD^1)} F(x, y) = \int F(x_1, c - mx)dx$$

where c is replaced by $y + mx$ after integration.

ILLUSTRATIVE EXAMPLES

Case I:

Example 1. Solve $\frac{\partial^2 z}{\partial x^2} - 7\frac{\partial^2 z}{\partial x \partial y} + 12\frac{\partial^2 z}{\partial y^2} = e^{x-y}$

Solution. $\frac{\partial^2 z}{\partial x^2} - 7\frac{\partial^2 z}{\partial x \partial y} + 12\frac{\partial^2 z}{\partial y^2} = e^{x-y}$

$$(D^2 - 7DD' + 12D'^2)\, z = e^{x-y}$$

where $D \equiv \frac{\partial}{\partial x},\ D' \equiv \frac{\partial}{\partial y}$

The auxiliary equation is $m^2 - 7m + 12 = 0$

$$(m - 3)(m - 4) = 0$$

$$m = 3, 4$$

$$\text{C. F.} = f_1(y + 3x) + f_2(y + 4x)$$

$$\text{P. I.} = \frac{1}{(D^2 - 7DD' + 12D'^2)} e^{x-y}$$

$$\text{P. I.} = \frac{1}{(1)^2 - 7 \times -1 + 12 \times (-1)^2} e^{x-y}$$

[Replacing $D = 1$, $D' = -1$]

$$\text{P.I.} = \frac{1}{1 + 7 + 12} e^{x-y} = \frac{1}{20} e^{x-y}$$

Hence, complete solution is

$$z = \text{C. F.} + \text{P. I.}$$

$$= f_1(y + 3x) + f_2(y + 4x) + \frac{1}{20} e^{x-y}$$

Example 2. Solve the linear partial differential equation

$$\frac{\partial^2 z}{\partial x^2}+2\frac{\partial^2 z}{\partial x\partial y}+\frac{\partial^2 z}{\partial y^2} = e^{3x+2y}$$

Solution. $$\frac{\partial^2 z}{\partial x^2}+2\frac{\partial^2 z}{\partial x\partial y}+\frac{\partial^2 z}{\partial y^2} = e^{3x+2y}$$

$$(D^2 + 2DD' + D'^2)\, z = e^{3x+2y}$$

where $$D = \frac{\partial}{\partial x},\ D' = \frac{\partial}{\partial y}$$

Auxiliary equation is

$$m^2 + 1$$

$$\text{C. F.} = f_1\,(y-x) + x\ f_2\,(y-x)$$

$$\text{P. I.} = \frac{1}{(D^2+2DD'+D'^2)}e^{3x+2y}$$

$$\text{P.I.} = \frac{1}{(3)^2+2\times 3\times 2+(2)^2)}e^{3x+2y}$$

$$\text{P.I.} = \frac{1}{9+12+4}e^{3x+2y} = \frac{1}{25}e^{3x+2y}$$

Hence, complete solution is

$$z = C.f. + \text{P. I.}$$

$$= f_1\,(y-x) + x\ f_2\,(y-x) + \frac{1}{25}e^{3x+2y}$$

Example 3. Solve the linear partial differential equation

$$4\frac{\partial^2 z}{\partial x^2}-4\frac{\partial^2 z}{\partial x\partial y}+\frac{\partial^2 z}{\partial y^2} = e^{x+2y}$$ *(I.U. 2007, 08)*

Solution. $$4\frac{\partial^2 z}{\partial x^2}-4\frac{\partial^2 z}{\partial x\partial y}+\frac{\partial^2 z}{\partial y^2} = e^{x+2y}$$

$$(4D^2 - 4DD' + D'^2)\, z = e^{x+2y},$$

Auxiliary equation is

$$4m^2 - 4m + 1 = 0$$

$$(2m-1)^2 = 0 \quad \Rightarrow \quad m = \frac{1}{2}, \frac{1}{2}$$

$$\text{C.F.} = f_1\,(y + \frac{x}{2}) + x\ f_2\,(y+\frac{x}{2})$$

$\Rightarrow$ $$\text{C. F.} = f_1\,(2y+x) + x\ f_2\,(2y+x)$$

$$\text{P.I.} = \frac{1}{(4D^2-4DD'+D'^2)}e^{x+2y}$$

On putting $D = 1$, $D' = 2$ the denominator becomes zero so, we differentiate the denominator w.r.t. D and multiply by x, we get

$$\text{P.I.} = x\,\frac{1}{8D-4D'}e^{x+2y}$$

Again on putting $D = 1$, $D' = 2$ the denominator $8D - 4D' \doteq 0$, So we differentitate the denominator w.r.t. D and multiply by x, we get

$$\text{P. I.} = x^2\,\frac{1}{8}e^{x+2y}$$

$$= \frac{x^2}{8}e^{x+2y}$$

Hence, the complete solution is

$$z = \text{C. F.} + \text{P. I.}$$

$$= f_1\,(2y+x) + x\,f_2\,(2y+x) + \frac{x^2}{8}e^{x+2y}$$

Case II:

Example 4. Solve the linear partial differential equation

$$\frac{\partial^2 z}{\partial x^2}+2\frac{\partial^2 z}{\partial x\partial y}+\frac{\partial^2 z}{\partial y^2} = \sin\,(2x+3y)$$

Solution. $\dfrac{\partial^2 z}{\partial x^2}+2\dfrac{\partial^2 z}{\partial x\partial y}+\dfrac{\partial^2 z}{\partial y^2} = \sin\,(2x+3y)$

$$(D^2 + 2DD' + D'^2)\,z = \sin\,(2x+3y), \quad \text{where } D = \frac{\partial}{\partial x},\ D' = \frac{\partial}{\partial y}$$

Auxiliary equation

$$m^2 + 2m + 1 = 0$$

$$(m+1)^2 = 0$$

$$m = -1, -1$$

$$\text{C.F.} = f_1\,(y-x) + x\,f_2\,(y-x)$$

$$\text{P. I.} = \frac{1}{(D^2+2DD'+D'^2)}\sin(2x+3y)$$

$$= \frac{1}{(-4+2(-6)-9)}\sin(2x+3y) \quad \left[\begin{array}{c}\text{Replacing}\\ D^2 = -2^2 = -4\\ D'^2 = -3^2 = -9\\ DD' = -2\times 3 = -6\end{array}\right]$$

$$= -\frac{1}{25}\sin(2x+3y)$$

Hence, complete solution is

$$z = \text{C. F.} + \text{P. I.}$$

$$= f_1\,(y-x) + x f_2\,(y-x) - \frac{1}{25}\sin(2x+3y)$$

Example 5. Solve $(D^2 + DD' - 6D'^2)\,z = \cos(2x + y)$

Solution. $(D^2 + DD' - 6D'^2)\,z = \cos(2x + y)$

Auxiliary equation

$$m^2 + m - 6 = 0$$

$$m = 2, -3$$

$$\text{C. F.} = f_1(y + 2x) + f_2(y - 3x)$$

$$\text{P. I.} = \frac{1}{(D^2 + DD' - 6D'^2)}\cos(2x + y)$$

$$= x.\frac{1}{(2D + D')}\cos(2x + y) = x.\frac{D}{2D^2 + DD'}\cos(2x + y)$$

$$= x.\frac{D}{2\times -4 - 2}\cos(2x + y) = -\frac{x}{10}D\cos(2x + y)$$

$$\text{P. I.} = \frac{2x}{10}\sin(2x + y) = \frac{x}{5}\sin(2x + y)$$

Hence, complete solution is

$$z = \text{C. F.} + \text{P. I.}$$

$$= f_1(y + 2x) + f_2(y - 3x) + \frac{x}{5}\sin(2x + y)$$

Example 6. Solve $\dfrac{\partial^3 z}{\partial x^3} - 4\dfrac{\partial^3 z}{\partial x^2 \partial y} + 4\dfrac{\partial^3 z}{\partial x \partial y^2} = 2\sin(3x + 2y)$

Solution. $\dfrac{\partial^3 z}{\partial x^3} - 4\dfrac{\partial^3 z}{\partial x^2 \partial y} + 4\dfrac{\partial^3 z}{\partial x \partial y^2} = 2\sin(3x + 2y)$

$$(D^3 - 4D^2D' + 4DD'^2)\,z = 2\sin(3x + 2y)$$

Auxiliary equation

$$m^3 - 4m^2 + 4m = 0$$

$$m(m^2 - 4m + 4) = 0$$

$$m(m - 2)^2 = 0$$

$$m = 0, 2, 2$$

$$\text{C.F.} = f_1(y) + f_2(y + 2x) + x f_3(y + 2x)$$

$$\text{P. I.} = \frac{1}{(D^3 - 4D^2D' + 4DD'^2)}2\sin(3x + 2y)$$

$$= 2\frac{1}{D(D^2 - 4DD' + 4DD'^2)}\sin(3x + 2y)$$

$$= 2\frac{1}{D[-9 - 4(-6) + 4(-4)]}\sin(3x + 2y)$$

$$= -\frac{2}{D}\sin(3x + 2y) = -\frac{2}{3}[-\cos(3x + 2y)]$$

$$= \frac{2}{3}\cos(3x + 2y)$$

Hence, complete solution is

$$z = \text{C. F.} + \text{P. I.}$$

$$= f_1(y) + f_2(y+2x) + x f_3(y+2x) + \frac{2}{3}\cos(3x+2y)$$

Case III:

Example 7. Solve the linear partial differential equation

$$\frac{\partial^2 z}{\partial x^2} + 3\frac{\partial^2 z}{\partial x \partial y} + 2\frac{\partial^2 z}{\partial y^2} = x + y$$

Solution. $\frac{\partial^2 z}{\partial x^2} + 3\frac{\partial^2 z}{\partial x \partial y} + 2\frac{\partial^2 z}{\partial y^2} = x + y$

$$(D^2 + 3DD' + 2D'^2)\, z = x + y$$

Auxiliary equation

$$m^2 + 3m + 2 = 0$$

$$(m+1)(m+2) = 0$$

$$m = -1, -2$$

$$\text{C. F.} = f_1(y-x) + f_2(y-2x)$$

$$\text{P. I.} = \frac{1}{(D^2 + 3DD' + 2D'^2)}(x+y)$$

$$= \frac{1}{D^2}\left[1 + \frac{3D'}{D} + \frac{2D'^2}{D^2}\right]^{-1}(x+y)$$

$$= \frac{1}{D^2}\left[1 - \frac{3D'}{D} - \ldots\right](x+y)$$

$$= \frac{1}{D^2}\left[x + y - \frac{3}{D}(1)\right] = \frac{1}{D^2}[x + y - 3x]$$

$$\text{P. I.} = \frac{1}{D^2}(y - 2x) = \frac{x^2}{2}y - \frac{x^3}{3}$$

Hence, the complete solution is

$$z = \text{C. F.} + \text{P. I.}$$

$$z = f_1(y-x) + f_2(y-2x) + \frac{x^2}{2}y - \frac{x^3}{3}$$

Example 8. Solve $\frac{\partial^3 z}{\partial x^3} - \frac{\partial^3 z}{\partial y^3} = x^3 y^3$

Solution. $\frac{\partial^3 z}{\partial x^3} - \frac{\partial^3 z}{\partial y^3} = x^3 y^3$

$$(D^3 - D'^3)\, z = x^3 y^3 \text{ where } D \equiv \frac{\partial}{\partial x},\ D' \equiv \frac{\partial}{\partial y}$$

Auxiliary equation

$$m^3 - 1 = 0$$

$$m = 1, w, w^2$$

where w is the one of the cube roots of unity

$$\text{C.F.} = f_1(y+x) + f_2(y+wx) + f_3(y+w^2x)$$

$$\text{P. I.} = \frac{1}{(D^3 - D'^3)}(x^3 - y^3)$$

$$= \frac{1}{D^3}\left[1 - \frac{D'^3}{D^3}\right]^{-1}(x^3y^3)$$

$$= \frac{1}{D^3}\left[1 + \frac{D'^3}{D^3} + \ldots\right](x^3y^3)$$

$$= \frac{1}{D^3}\left[x^3y^3 + \frac{1}{D^3}D'^3(x^3y^3)\right]$$

$$= \frac{1}{D^3}\left[x^3y^3 + \frac{1}{D^3}(6x^3)\right]$$

$$= \frac{1}{D^3}\left[x^3y^3\right] + \frac{1}{D^6}(6x^3)$$

$$= \frac{x^6y^3}{6.5.4} + \frac{6x^9}{4.5.6.7.8.9} + \frac{x^6y^3}{120} + \frac{x^9}{10080}$$

Hence, the complete solution is

$$z = \text{C. F.} + \text{P. I.}$$

$$z = f_1(y+x) + f_2(y+wx) + f_3(y+w^2x) + \frac{x^6y^3}{120} + \frac{x^9}{10080}$$

Case IV:

Example 9. Solve $\frac{\partial^2 z}{\partial x^2} + \frac{\partial^2 z}{\partial x \partial y} - 6\frac{\partial^2 z}{\partial y^2} = y \cos x$

Solution. $\frac{\partial^2 z}{\partial x^2} + \frac{\partial^2 z}{\partial x \partial y} - 6\frac{\partial^2 z}{\partial y^2} = y \cos x$

$$(D^2 + DD' - 6D'^2)\, z = y \cos x$$

Auxiliary equation is

$$m^2 + m - 6 = 0$$

$$\Rightarrow \quad (m+3)(m-2) = 0$$

$$m = 2, -3$$

$$\text{C.F.} = f_1(y+2x) + f_2(y-3x)$$

$$\text{P.I.} = \frac{1}{(D^2 + DD' - 6D'^2)} y \cos x$$

$$= \frac{1}{(D - 2D')(D + 3D')}(y \cos x)$$

$$= \frac{1}{(D-2D^1)}\int (c+3x)\cos x dx \quad \text{Put } y = c + 3x$$

$$= \frac{1}{(D-2D^1)}[(c+3x)\sin x + 3\cos x]$$

$$= \frac{1}{(D-2D^1)}[y\sin x + 3\cos x] \quad \text{Put } c + 3x = y$$

$$= \int [(c-2x)\sin x + 3\cos x]dx$$

$$= (c - 2x)(-\cos x) - 2\sin x + 3\sin x$$

$$\text{P. I.} = -y\cos x + \sin x \qquad \text{put } c - 2x = y$$

Hence, complete solution is

$$z = \text{C. F.} + \text{P. I.}$$

$$z = f_1(y + 2x) + f_2(y - 3x) + \sin x - y\cos x$$

EXERCISE 3.8

Solve the following equations :

1. $\dfrac{\partial^2 z}{\partial x^2} - 5\dfrac{\partial^2 z}{\partial x \partial y} + 6\dfrac{\partial^2 z}{\partial y^2} = e^{x+y}$

2. $\dfrac{\partial^2 z}{\partial x^2} - 4\dfrac{\partial^2 z}{\partial x \partial y} + 4\dfrac{\partial^2 z}{\partial y^2} = e^{2x+y}$

3. $\dfrac{\partial^3 z}{\partial x^3} - 3\dfrac{\partial^3 z}{\partial x^2 \partial y} + 4\dfrac{\partial^3 z}{\partial y^3} = e^{x+2y}$

4. $(D^2 - 2DD' + D'^2)\, z = e^{x+2y}$

5. $\dfrac{\partial^2 z}{\partial x^2} - 2\dfrac{\partial^2 z}{\partial x \partial y} + \dfrac{\partial^2 z}{\partial y^2} = \sin x$

6. $\dfrac{\partial^2 z}{\partial x^2} - \dfrac{\partial^2 z}{\partial x \partial y} = \cos(x + 2y)$

7. $(2D^2 - 5DD' + 2D'^2)\, z = 5\sin(2x + y)$

8. $\dfrac{\partial^2 z}{\partial x^2} + \dfrac{\partial^2 z}{\partial x \partial y} - 6\dfrac{\partial^2 z}{\partial y^2} = \cos(3x + y)$

9. $\dfrac{\partial^2 z}{\partial x^2} - \dfrac{\partial^2 z}{\partial x \partial y} - 6\dfrac{\partial^2 z}{\partial y^2} = xy$

10. $\dfrac{\partial^2 z}{\partial x^2} + 3\dfrac{\partial^2 z}{\partial x \partial y} + 2\dfrac{\partial^2 z}{\partial y^2} = 12\,xy$

11. $\dfrac{\partial^2 z}{\partial x^2}+\dfrac{\partial^2 z}{\partial x \partial y}-2\dfrac{\partial^2 z}{\partial y^2}=(y-1)\,e^x$

12. $(D-D')\,(D+2D')\,z=(y+1)\,e^x$

13. $(D^2-DD'-2D'^2)\,z=(y-1)\,e^x$

ANSWERS

1. $z=f_1(y+2x)+f_2(y+3x)+\dfrac{1}{2}\,e^{x+y}$

2. $z=f_1(y+2x)+xf_2(y+2x)+\dfrac{x^2}{2}\,e^{2x+y}$

3. $z=f_1(y-x)+f_2(y+2x)+xf_3(y+2x)+\dfrac{1}{27}\,e^{x+2y}$

4. $z=f_1(y+x)+xf_2(y+x)+e^{x+2y}$

5. $z=f_1(y+x)+x\,f_2(y+x)-\sin x$

6. $z=f_1(y)+f_2(y+x)+\cos(x+2y)$

7. $z=f_1(y+2x)+f_2(2y+x)-\dfrac{5}{3}x\cos(2x+y)$

8. $z=f_1(y-3x)+f_2(y+2x)-\dfrac{1}{6}\cos(3x+y)$

9. $z=f_1(y-2x)+f_2(y+3x)+\dfrac{x^3y}{6}+\dfrac{x^4}{24}$

10. $z=f_1(y-x)+f_2(y-2x)+2x^3y-\dfrac{3}{2}x^4$

11. $z=f_1(y+x)+f_2(y-2x)+(y-2)\,e^x$

12. $z=f_1(y+x)+f_2(y-2x)+y\,e^x$

13. $z=f_1(y-x)+f_2(y+2x)+y\,e^x$

3.17. NON-HOMOGENOUS LINEAR PARTIAL DIFFERENTIAL EQUATION WITH CONSTANT COEFFICIENTS

The linear partial differential equations which are not homogeneous are called Non-homogeneous linear partial differential equations

For example, $3\dfrac{\partial^2 z}{\partial x^2}+2\dfrac{\partial^2 z}{\partial x \partial y}+\dfrac{\partial^2 z}{\partial y^2}+\dfrac{\partial z}{\partial x}+\dfrac{\partial z}{\partial y}+z=0$

$$\phi\,(D, D') = F\,(x, y) \qquad \text{... (1)}$$

i.e., if the polynomial $\phi\,(D, D')$ in D, D' is not homogeneous then (1) is called a non-homogeneous linear partial differential equation. Its complete solution is $z =$ C. F. + P.I.

3.18. METHODS FOR FINDING OUT C. F.

To find the C. F., we resolve $\phi(D, D')$ into linear factors of the form $D - mD' - a$. Now consider the equation

$$(D - mD' - a) z = 0 \qquad \ldots(2)$$

$\therefore$ The complete solution of (2) is $z = e^{ax} f(y + mx)$

Similarly the solution of $(D - mD' - a)^2 z = 0$ is

$$z = e^{ax} f_1 (y + mx) + xe^{ax} f_2 (y + mx)$$

ILLUSTRATIVE EXAMPLES

Example 1. Solve $(D + D' - 1)(D + 2D' - 2) z = 0$

Solution. $(D + D' - 1)(D + 2D' - 2) z = 0$

$$\text{C. F.} = e^x f_1 (y - x) + e^{2x} f_2 (y - 2x)$$

$$\text{P. I.} = 0$$

Hence, complete solution is $z = \text{C. F.} + \text{P. I.}$

$$z = e^x f_1 (y - x) + e^{2x} f_2 (y - 2x)$$

where f_1, f_2 are arbitrary functions.

Example 2. Solve $(D + D' - 2)(D + 4D' - 3) z = 0$

Solution. $(D + D' - 2)(D + 4D' - 3) z = 0$

$$\text{C. F.} = e^{2x} f_1 (y - x) + e^{3x} f_2 (y - 4x)$$

$$\text{P. I.} = 0$$

Hence, complete solution is $z = \text{C. F.} + \text{P. I.}$

$$z = e^{2x} f_1 (y - x) + e^{3x} f_2 (y - 4x)$$

Where f_1, f_2 are arbitrary functions.

Example 3. Solve $r + 2s + t + 2p + 2q + z = 0$

Solution. $r + 2s + t + 2p + 2q + z = 0$

$\Rightarrow$ $[D^2 + 2DD' + D'^2 + 2D + 2D' + 1] z = 0$

$\Rightarrow$ $[(D + D')^2 + 2(D + D') + 1] z = 0$

$\Rightarrow$ $(D + D' + 1)^2 z = 0$

$$\text{C. F.} = e^{-x} f_1(y - x) + x\, e^{-x} f_2 (y - x)$$

$$\text{P. I.} = 0$$

Hence, the complete solution is $z = \text{C. F.} + \text{P. I}$

$$z = e^{-x} f_1 (y - x) + x\, e^{-x} f_2 (y - x)$$

Example 4. Solve $(D + 1)(D + D' - 1) z = 0$

Solution. $(D + 1)(D + D' - 1) z = 0$

$$\text{C. F.} = e^{-x} f_1(y) + e^{x} f_2 (y - x)$$

$$\text{P.I.} = 0$$

Hence, the complete solution is $z = \text{C. F.} + \text{P. I}$

$$z = e^{-x} f_1 (y) + e^{x} f_2 (y - x)$$

where f_1, f_2 are arbitrary functions.

EXERCISE 3.9

Solve the following partial differential equations :

1. $(D-D'+1)(D+2D'-2)z=0$
2. $(D-D')(D+D'-3)z=0$
3. $(D-D'-1)(D-D'-2)z=0$
4. $(D^2+DD'+D'-1)z=0$
5. $t+s+q=0$

ANSWERS

1. $z=e^{-x}f_1(y+x)+e^{2x}f_2(y-2x)$
2. $z=f_1(y+x)+e^{3x}f_2(y-x)$
3. $z=e^{x}f_1(y+x)+e^{2x}f_2(y+x)$
4. $z=e^{-x}f_1(y)+e^{x}f_2(y-x)$
5. $z=f_1(x)+e^{-x}f_2(y-x)$

3.19. P.I. OF NON-HOMOGENEOUS LINEAR PARTIAL DIFFERENTIAL EQUATION WITH CONSTANT COEFFICIENTS

Let the given equation be $\phi(D, D')z = F(x, y)$ then

$$\text{P. I.} = \frac{1}{\phi(D,D')} F(x, y)$$

Case I: When $F(x, y) = e^{ax+by}$ and $\phi(a, b) \neq 0$

$$\text{P. I.} = \frac{1}{\phi(D,D')} e^{ax+by}$$

$$\text{P. I.} = \frac{1}{\phi(a,b)} e^{ax+by}$$

[Replacing D by a and D' by b]

Case II: When $F(x, y) = \sin(ax+by)$

$$\text{P. I.} = \frac{1}{\phi(D,D')} \sin(ax+by)$$

$$= \frac{1}{\phi(D^2, DD', D'^2)} \sin(ax+by)$$

$$\text{P. I.} = \frac{1}{\phi(-a^2,-ab,-b^2)} \sin(ax+by) \quad \begin{bmatrix} \text{Replacing } D^2 = -a^2 \\ DD' = -ab \\ D'^2 = -b^2 \end{bmatrix}$$

Where $\phi(-a^2, -ab, -b^2) \neq 0$

similarly when $F(x, y) = \cos(ax+by)$

$$\text{P. I.} = \frac{1}{\phi(D, D')} \cos(ax + by)$$

$$= \frac{1}{\phi(D^2, DD', D'^2)} \cos(ax + by)$$

$$= \frac{1}{\phi(-a^2, -ab, -b^2)} \cos(ax + by) \quad \begin{bmatrix} \text{Replacing} \\ D^2 = -a^2 \\ DD' = -ab \\ D'^2 = -b^2 \end{bmatrix}$$

Where $\phi(-a^2, -ab, -b^2) \neq 0$

Case III: When $F(x, y) = x^m y^n$ [where m and n are +ve integer]

$$\text{P. I.} = \frac{1}{\phi(D, D')} x^m y^n = [\phi(D, D')]^{-1}(x^m y^n)$$

Expand $[\phi(D, D')]^{-1}$ in ascending powers of $\frac{D'}{D}$ (when $m > n$) or $\frac{D}{D'}$ (when $m < n$) or D or D' as the case may be

Case IV: When $F(x, y) = e^{ax + by} V$, where V is a function of x and y, then we have

$$\text{P. I.} = \frac{1}{\phi(D, D')} e^{ax + by} V = e^{ax + by} \frac{1}{\phi(D + a, D' + b)} V$$

which can be evaluated by using any one of the previous cases discussed.

ILLUSTRATIVE EXAMPLES

Case I:

Example 1. Solve $D(D - 2D' - 3) z = e^{x + 2y}$

Solution. $D(D - 2D' - 3) z = e^{x + 2y}$

$$\text{C. F.} = f_1(y) + e^{3x} f_2(y + 2x)$$

$$\text{P. I.} = \frac{1}{D(D - 2D' - 3)} e^{x+2y} = \frac{e^{x+2y}}{1\{1 - 2 \times 2 - 3\}} = -\frac{1}{6} e^{x+2y}$$

Hence, complete solutions is $z = \text{C. F.} + \text{P. I.}$

$$z = f_1(y) + e^{3x} f_2(y + 2x) - \frac{1}{6} e^{x+2y}$$

where f_1, f_2 are arbitrary functions.

Example 2. Solve $(D - D' - 2)(D - D' - 3) z = e^{3x - 2y}$

Solution. $(D - D' - 2)(D - D' - 3) z = e^{3x - 2y}$

$$\text{C.F.} = e^{2x} f_1(y + x) + e^{3x} f_2(y + x)$$

$$\text{P.I.} = \frac{1}{(D - D' - 2)(D - D' - 3)} e^{3x - 2y}$$

$$\text{P. I.} = \frac{e^{3x-2y}}{[3-(-2)-2][3-(-2)-3]} = \frac{1}{6}e^{3x-2y}$$

Hence, complete solution is $z = \text{C. F.} + \text{P. I.}$

$$z = e^{2x} f_1(y+x) + e^{3x} f_2(y+x) - \frac{1}{6}e^{3x-2y}$$

Case II:

Example 3. Solve $(D^2 - DD' + D' - 1)z = \sin(x + 2y)$

Solution. $(D^2 - DD' + D' - 1)z = \sin(x + 2y)$

$$\Rightarrow \quad [(D+1)(D-1) - D'(D-1)]z = \sin(x+2y)$$

$$(D-1)(D-D'+1)z = \sin(x+2y)$$

$$\therefore \quad \text{C. F.} = e^x f_1(y) + e^{-x} f_2(y+x)$$

$$\text{P. I.} = \frac{1}{(D-1)(D-D'+1)}\sin(x+2y)$$

$$= \frac{1}{(D^2 - DD' + D' - 1)}\sin(x+2y)$$

$$= \frac{1}{-1+2+D'-1}\sin(x+2y)$$

$$= \frac{1}{D'}\sin(x+2y)$$

$$\text{P. I.} = \frac{-\cos(x+2y)}{2}$$

Hence, complete solution is

$$z = \text{C. F.} + \text{P. I.}$$

$$z = e^x f_1(x) + e^{-x} f_2(y+x) - \frac{\cos(x+2y)}{2}$$

where f_1, f_2 are arbitrary functions.

Example 4. Solve $(D + 1)(D + D' - 1)z = \sin(x + 2y)$

Solution. $(D + 1)(D + D' - 1)z = \sin(x + 2y)$

$$\text{C.F.} = e^{-x} f_1(y) + e^x f_2(y-x)$$

$$\text{P.I.} = \frac{1}{(D+1)(D+D'-1)}\sin(x+2y)$$

$$= \frac{1}{D^2 + DD' + D' - 1}\sin(x+2y)$$

$$= \frac{1}{-1-2+D'-1}\sin(x+2y)$$

$$= \frac{1}{D'-4}\sin(x+2y)$$

$$\text{P.I.} = \frac{(D'+4)}{(D'^2-16)}\sin(x+2y)$$

$$= \frac{D'+4}{-4-16)}\sin(x+2y) = -\frac{1}{20}\,(D'+4)\sin(x+2y)$$

$$= -\frac{1}{20}\,[D'\sin(x+2y) + 4\sin(x+2y)\,]$$

$$\text{P.I.} = -\frac{1}{20}\,[2\cos(x+2y) + 4\sin(x+2y)]$$

Hence, complete solution is

$$z = \text{C. F.} + \text{P. I.}$$

$$z = e^{-x}f_1(y) + e^x f_2(y-x) - \frac{1}{20}\,[2\cos(x+2y) + 4\sin(x+2y)]$$

Case III:

Example 5. Solve $\quad s + p - q = z + xy$

Solution. $\quad s + p - q = z + xy$

The given equation is

$$(DD' + D - D' - 1)\,z = xy$$

$$(D-1)(D'+1)\,z = xy$$

$\therefore$
$$\text{C.F.} = e^x f_1(y) + e^y f_2(x)$$

$$\text{P.I.} = \frac{1}{(D-1)(D'+1)}(xy)$$

$$= -[(1-D)^{-1}(1+D')^{-1}]\,(xy)$$

$$= -[(1+D+D^2+\ldots)(1-D'+\ldots)]\,(xy)$$

$$= -(1 + D - D' - DD')\,(xy)$$

[Leaving higher powers]

$$= -(xy + y - x - 1)$$

$$\text{P.I.} = x + 1 - xy - y$$

Hence, the complete solution is $z = \text{C. F.} + \text{P. I.}$

$$z = e^x f_1(y) + e^{-y} f_2(x) + x + 1 - xy - y$$

where f_1, f_2 are arbitrary functions.

Example 6. Solve $(D^2 - DD' + D)\,z = 1$

Solution. $\quad (D^2 - DD' + D)\,z = 1$

$$D(D - D' + 1)\,z = 1$$

$\therefore$
$$\text{C.F.} = f_1(y) + e^{-x} f_2(y+x)$$

$$\text{P.I.} = \frac{1}{D(D-D'+1)}(1)$$

$$= \frac{1}{D}[1 + (D - D')]^{-1}\,(1)$$

$$= \frac{1}{D}\,[1 - D + D']\,(1)$$

$$\text{P. I.} = \frac{1}{D}(1) = x$$

Hence, complete solution is $z = \text{C. F.} + \text{P. I.}$

$$z = f_1(y) + e^{-x} f_2(y+x) + x$$

Case IV.

Example 7. Solve $(D - 3D' - 2)^3 z = 2e^{2x} \sin(3x + y)$

Solution. $(D - 3D' - 2)^3 z = 2e^{2x} \sin(3x + y)$

$$\text{C.F.} = e^{2x} f_1(y + 3x) + xe^{2x} f_2(y + 3x) + x^2 e^{2x} f_3(y + 3x)$$

$$\text{P.I.} = \frac{1}{(D - 3D' - 2)^3} 2e^{2x} \sin(3x + y)$$

$$= 2e^{2x} \frac{1}{(D + 2 - 3D' - 2)^3} \sin(3x + y)$$

$$= 2e^{2x} \frac{1}{(D - 3D')^3} \sin(3x + y)$$

$$= 2e^{2x} \frac{x^3}{6} \sin(3x + y)$$

$$\text{P.I.} = \frac{1}{3} x^3 e^{2x} \sin(3x + y)$$

Hence, complete solution is $z = \text{C. F.} + \text{P. I.}$

$$z = e^{2x} f_1(y + 3x) + x e^{2x} f_2(y + 3x) + x^2 e^{2x} f_3(y + 3x) + \frac{1}{3} x^3 e^{2x} \sin(3x + y)$$

where f_1, f_2 and f_3, are arbitrary functions.

Example 8. Solve $(D - 3D' - 2) z = e^{2x} \tan(y + 3x)$

Solution. $(D - 3D' - 2) z = e^{2x} \tan(y + 3x)$

$$\text{C. F.} = e^{2x} f_1(y + 3x)$$

$$\text{P. I.} = \frac{1}{(D - 3D' - 2)} e^{2x} \tan(y + 3x)$$

$$= e^{2x} \frac{1}{(D + 2 - 3D' - 2)} \tan(y + 3x)$$

$$= e^{2x} \frac{1}{(D - 3D')} \tan(y + 3x)$$

$$= e^{2x} \left(\frac{x}{1}\right) \tan(y + 3x)$$

$$\text{P.I.} = x e^{2x} \tan(y + 3x)$$

Hence, complete solution is $z = \text{C. F.} + \text{P. I.}$

$$z = e^{2x} f_1(y + 3x) + x e^{2x} \tan(y + 3x)$$

EXERCISE 3.10

Solve the following equations :

1. $(D^2 - D'^2 - 3D + 3D')\,z = e^{x-2y}$
2. $(D^2 - D'^2 + D - D')\,z = e^{2x+3y}$
3. $(D - D' - 1)(D - D' - 2)\,z = e^{2x-y}$
4. $(D - D' - 1)(D + D' - 2)\,z = \sin(2x + 3y)$
5. $(D^2 - DD' + D' - 1)\,z = \cos(x + 2y)$
6. $(D^2 + DD' + D' - 1)\,z = \sin(x + 2y)$
7. $(D + D' - 1)\,(D + 2D' - 3)\,z = 2x + 3y$
8. $\dfrac{\partial^2 z}{\partial x^2} - \dfrac{\partial^2 z}{\partial x \partial y} + \dfrac{\partial z}{\partial y} = x^2 + y^2$
9. $r - 4s + 4t + p - 2q = e^{x+y}$
10. $(D - D' - 1)(D + D' - 2)\;z = e^{2x-y}$
11. $(D^2 - DD' + D' - 1)\,z = e^x + \cos(x + 2y)$

ANSWERS

1. $z = f_1(y + x) + e^{3x} f_2(y - x) - \dfrac{1}{12}\, e^{x-2y}$
2. $z = f_1(y + x) + e^{-x} f_2(y - x) - \dfrac{1}{6}\, e^{2x+3y}$
3. $z = e^x f_1(y + x) + e^{2x} f_2(y - x) - \dfrac{1}{2}\, e^{2x-y}$
4. $z = e^x f_1(y + x) + e^{2x} f_2(y + x) + \dfrac{1}{10}\,[\sin(2x + 3y) - 3\cos(2x + 3y)]$
5. $z = e^x f_1(y) + e^{-x} f_2(y + x) + \dfrac{1}{2}\sin(x + 2y)$
6. $z = e^{-x} f_1(y) + e^x f_2(y - x) - \dfrac{1}{10}\,[2\sin(x + 2y) + \cos(x + 2y)]$
7. $z = e^x f_1(y - x) + e^{3x} f_2(y - 2x) + \dfrac{2}{3}x + y + \dfrac{23}{9}$
8. $z = f_1(y) + e^{-x} f_2(y + x) + \dfrac{x^3}{3} - x^2 + xy^2 + 6x$
9. $z = f_1(y + 2x) + e^{-x} f_2(y + 2x) - x\,e^{x+y}$
10. $z = e^x f_1(y + x) + e^{2x} f_2(y - x) - \dfrac{1}{2}\, e^{2x-y}$
11. $z = e^x f_1(y) + e^{-x} f_2(y + x) + \dfrac{1}{2}\sin(x + 2y) + \dfrac{1}{2}x\,e^x$

3.20. CLASSIFICATION OF LINEAR PARTIAL DIFFERENTIAL EQUATION OF SECOND ORDER

Consider the differential equation of second order.

$$A\frac{\partial^2 u}{\partial x^2}+B\frac{\partial^2 u}{\partial x\partial y}+C\frac{\partial^2 u}{\partial y^2}+f(x,y,u,p,q)=0 \qquad \dots (1)$$

where A, B, C are constant or continuous functions of x and y.
Now equation (1) is

(*i*) elliptic if $B^2 - 4AC < 0$

(*ii*) hyperbolic if $B^2 - 4AC > 0$

(*iii*) parabolic if $B^2 - 4AC = 0$

(*i*) **Elliptic Equations:** Laplace's equation : $\frac{\partial^2 u}{\partial x^2}+\frac{\partial^2 u}{\partial y^2}=0$

Poisson's equation: $\frac{\partial^2 u}{\partial x^2}+\frac{\partial^2 u}{\partial y^2}=f(x,y)$ is elliptic in nature.

(*ii*) **Hyperbolic Equations:** The wave equation $\frac{\partial^2 u}{\partial t^2}=c^2\frac{\partial^2 u}{\partial x^2}$ is hyperbolic equation.

(*iii*) **Parabolic Equation:** The one-dimensional heat conduction equation $\frac{\partial u}{\partial t}=c^2\frac{\partial^2 u}{\partial x^2}$ is an example of parabolic differential equation.

ILLUSTRATIVE EXAMPLES

Example 1. Classify the following equations:

(*i*) $2\frac{\partial^2 u}{\partial x^2}+6\frac{\partial^2 u}{\partial x\partial y}+3\frac{\partial^2 u}{\partial y^2}=0$

(*ii*) $2\frac{\partial^2 u}{\partial x^2}+4\frac{\partial^2 u}{\partial x\partial y}+3\frac{\partial^2 u}{\partial y^2}=0$

(*iii*) $4\frac{\partial^2 u}{\partial x^2}+4\frac{\partial^2 u}{\partial x\partial t}+\frac{\partial^2 u}{\partial t^2}=0$

Solution. (*i*) $2\frac{\partial^2 u}{\partial x^2}+6\frac{\partial^2 u}{\partial x\partial y}+3\frac{\partial^2 u}{\partial y^2}=0$

Here $A = 2, B = 6, C = 3$

$$B^2 - 4AC = (6)^2 - 4 \times 2 \times 3 = 36 - 24 = +12 > 0$$

$$B^2 - 4AC > 0$$

$\therefore$ The given equation is hyperbolic.

(*ii*) $2\frac{\partial^2 u}{\partial x^2}+4\frac{\partial^2 u}{\partial x \partial y}+3\frac{\partial^2 u}{\partial y^2}=0$

Here $A = 2, B = 4, C = 3$

$$B^2 - 4AC = (4)^2 - 4 \times 2 \times 3 = 16 - 24 = -8 < 0$$

$$B^2 - 4AC < 0$$

$\therefore$ The given equation is an elliptic.

(*iii*) $4\frac{\partial^2 u}{\partial x^2}+4\frac{\partial^2 u}{\partial x \partial t}+\frac{\partial^2 u}{\partial t^2}=0$

Here $A = 4, B = 4, C = 1$

$$B^2 - 4AC = (4)^2 - 4 \times 4 \times 1 = 16 - 16 = 0$$

$$B^2 - 4AC = 0$$

$\therefore$ The given equation is parabolic.

Example 2. Classify the partial differential equation

$$X^2\frac{\partial^2 u}{\partial t^2}+3\frac{\partial^2 u}{\partial x \partial t}+x\frac{\partial^2 u}{\partial x^2}+17\frac{\partial u}{\partial t}=100\,u$$

Solution. $X^2\frac{\partial^2 u}{\partial t^2}+3\frac{\partial^2 u}{\partial x \partial t}+x\frac{\partial^2 u}{\partial x^2}+17\frac{\partial u}{\partial t}=100\,u$

Here $A = x^2, B = 3, C = x$

$$B^2 - 4AC = (3)^2 - 4 \times x^2 x = 9 - 4x^3$$

The given equation is elliptic if $9 - 4x^3 < 0$

The given equation is parabolic if $9 - 4x^3 = 0$

The given equation is hyperbolic if $9 - 4x^3 > 0$

Example 3. Classify the partial differential equation

$$4\frac{\partial^2 u}{\partial x^2}-4\frac{\partial^2 u}{\partial x \partial t}+\frac{\partial^2 u}{\partial t^2}=0$$

Solution. $4\frac{\partial^2 u}{\partial x^2}-4\frac{\partial^2 u}{\partial x \partial t}+\frac{\partial^2 u}{\partial t^2}=0$

Here $A = 4, B = -4, C = 1$

$$B^2 - 4AC = (-4)^2 - 4 \times 4 \times 1 = 16 - 16 = 0$$

$$B^2 - 4AC = 0$$

$\therefore$ The given equation is parabolic.

Example 4. Classify the partial differential equation

$$x^2 \frac{\partial^2 u}{\partial t^2} - \frac{\partial^2 u}{\partial x^2} + u = 0$$

Solution. $x^2 \frac{\partial^2 u}{\partial t^2} - \frac{\partial^2 u}{\partial x^2} + u = 0$

Here $A = -1, B = 0, C = x^2$

$B^2 - 4AC = -\ 4 \times -1 \times x^2 = 4x^2$

$\therefore$ The given equation is hyperbolic if $4x^2 > 0$ *i.e.* $x > 0$

parabolic if $4x^2 = 0$ *i.e.* $x = 0$

Since $4x^2$ being a square cannot be negative hence given equation cannot be elliptic.

EXERCISE 3.11

Classify the following partial differential equation :

1. $\frac{\partial^2 u}{\partial x^2} - 2\frac{\partial^2 u}{\partial x \partial y} + \frac{\partial^2 u}{\partial y^2} = 0$

2. $t\frac{\partial^2 u}{\partial t^2} + 3\frac{\partial^2 u}{\partial x \partial t} + x\frac{\partial^2 u}{\partial x^2} + 17\frac{\partial u}{\partial x} = 0$

3. $\frac{\partial^2 u}{\partial t^2} + t\frac{\partial^2 u}{\partial x \partial t} + x\frac{\partial^2 u}{\partial x^2} + 2\frac{\partial u}{\partial t} + \frac{\partial u}{\partial x} + 6u = 0$

ANSWERS

1. Parabolic
2. Hyperbolic if $xt < \frac{9}{4}$, parabolic if $xt = \frac{9}{4}$, elliptic if $xt > \frac{9}{4}$
3. Elliptic if $t^2 - 4x < 0$
   Parabolic if $t^2 - 4x = 0$
   Hyperbolic if $t^2 - 4x > 0$

3.21. EQUATION OF A STRAIGHT LINE IN DIFFERENT FORMS

1. The equation of a straight line parallel to x-axis is $y = b$.

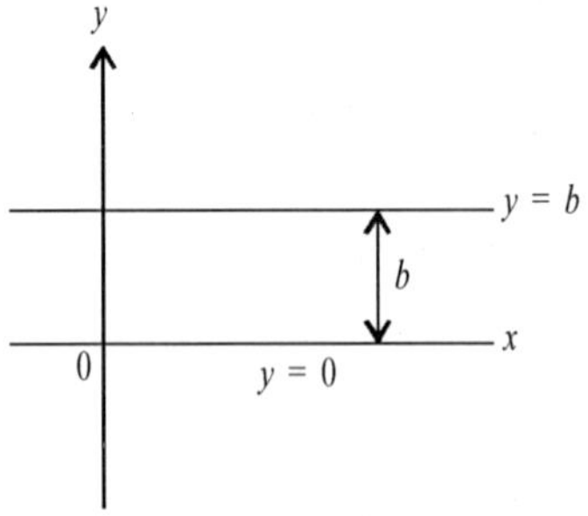

Remarks:

(*i*) The line $y = b$ lies above or below the x-axis according as b is positive or negative.

(*ii*) The equation of x-axis is $y = 0$.

2. The equation of a straight line parallel to y-axis is $x = a$

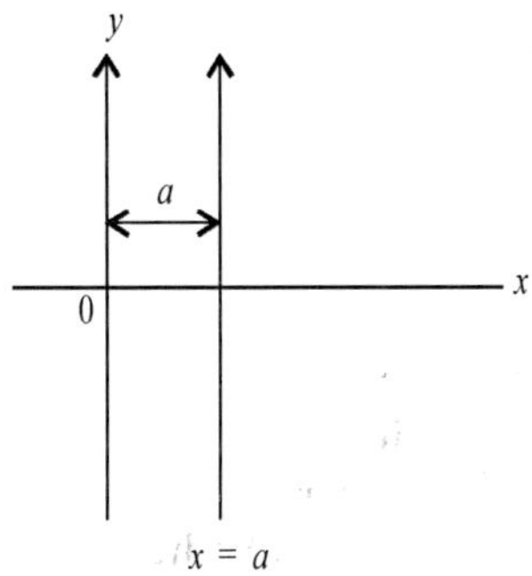

Remarks:

(*i*) The line $x = a$ lies to the right or left of y-axis according as a is positive or negative.

(*ii*) The equation of y-axis is $x = 0$.

3. Slope intercept form : The equation of a straight line which makes an angle θ to the x-axis in anti-clockwise direction and cuts off an intercept c from y-axis is

$$y = mx + c$$

where $m = \tan\theta$, and $\tan\theta$ is the slope or gradient of the line.

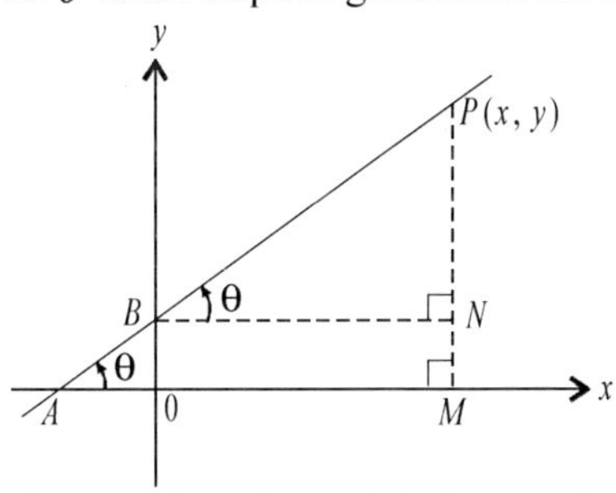

4. Point-slope form : The equation of a straight line passing through a fixed point say (x_1, y_1) and having a given slope m is given by

$$y - y_1 = m(x - x_1)$$

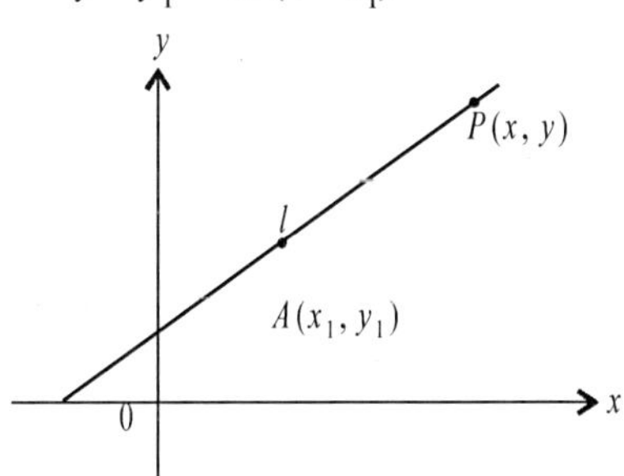

5. Two-point form : The equation of a (non-vertical) straight line passing through two fixed points. $A(x_1, y_1)$ and $B(x_2, y_2)$ is given by

$$y \quad y_1 = \frac{y_2 - y_1}{x_2 - x_1}(x - x_1)$$

6. Intercept form : The equation of a straight line which cuts off intercepts a and b from the axes is

$$\frac{x}{a}+\frac{y}{b}=1$$

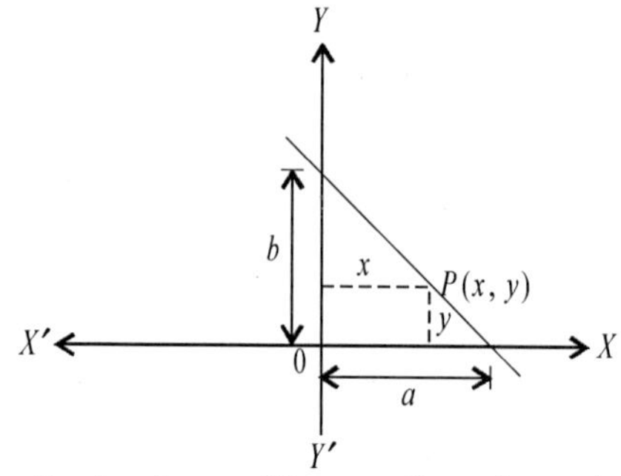

7. Normal or perpendicular form : The equation of a straight line in terms of the perpendicular p, let fall from the origin upon it, and the angle θ that this perpendicular makes with the x-axis is

$$x \cos\theta + y \sin\theta = p$$

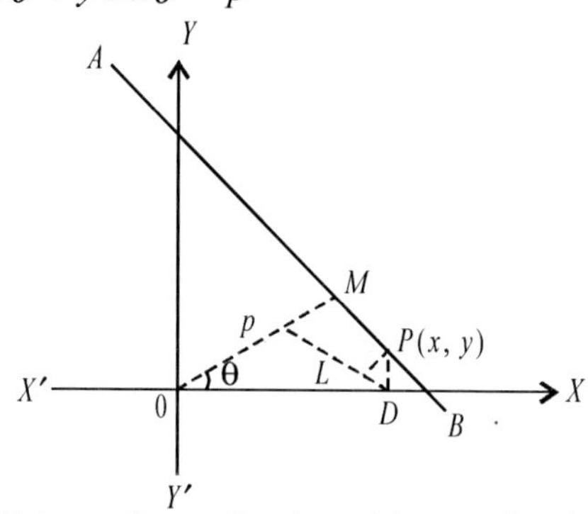

8. Symmetrical or distance form : Let (x_1, y_1) be any fixed point and (x, y) be any other point whose distance from (x_1, y_1) is r. Then the equation to the line is

$$\frac{x-x_1}{\cos\theta}=\frac{y-y_1}{\sin\theta}=r$$

where θ is the angle of inclination.

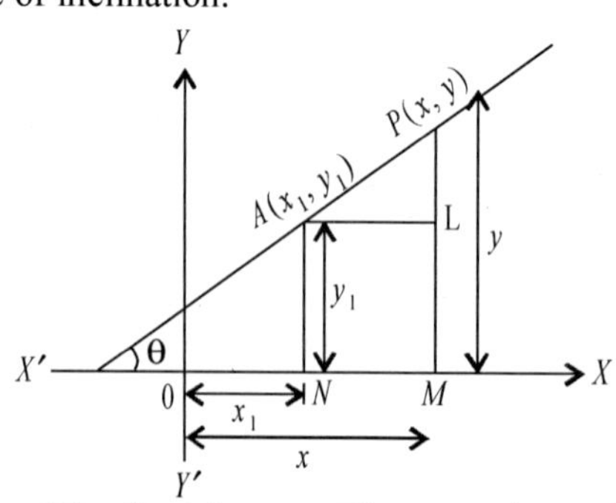

9. General equation of the first degree : The general equation of the first degree $Ax + By + C = 0$ represent a straight line.

ILLUSTRATIVE EXAMPLES

Example 1. Find the equation of a straight line parallel to x-axis at a distance 5 units below it.

Solution. We know that the equation of a straight line parallel to x-axis is $y = b$.
Here $b = -5$

$\therefore$ The equation of the required line is $y = -5$

$\Rightarrow \qquad y + 5 = 0$

Example 2. Find the equation of a straight line parallel to y-axis and passing through the point $(-3, 2)$

Solution. We know that the equation of a straight line parallel to y-axis is

$$x = a \qquad ...(1)$$

Since (1) passes through the point $(-3, 2)$, we get $-3 = a$ *i.e.* $a = -3$

From (1) $x = -3 \Rightarrow x + 3 = 0$

Example 3. Write the equation of a straight line with gradient $-\frac{4}{5}$ and cutting off an intercept of 3 units on y-axis.

Solution. Here $m = -\frac{4}{5}$, $c = 3$

$\therefore$ The equation of the line is $y = \frac{4}{5}x + 3 \qquad [y = mx + c]$

$\Rightarrow \qquad 4x + 5y - 15 = 0$

Example 4. Find the equation of the straight line which cuts an intercepts $\sqrt{2}$ from negative direction of y-axis and makes an angle $\tan^{-1}\frac{1}{\sqrt{2}}$ with x-axis.

Solution. Here $\theta = \tan^{-1}\frac{1}{\sqrt{2}} \Rightarrow \tan\theta = \frac{1}{\sqrt{2}}$ *i.e.* $m = \frac{1}{\sqrt{2}}$, $c = -\sqrt{2}$

The equation of the line is $y = mx + c$

$$y - \frac{1}{\sqrt{2}}x - \sqrt{2} \Rightarrow x - \sqrt{2}y = 2$$

Example 5. Find the equation of a straight line whose inclination is $\frac{5\pi}{6}$ and which cuts off an intercept of 4 units on negative direction of y-axis.

Solution. Let m be the slope of the line then

$m = \tan\frac{5\pi}{6} = \tan 150° = \tan(180° - 30°) = -\tan 30° = \frac{1}{\sqrt{3}}$

Also $c = y$-intercept $= -4$

$\therefore$ The equation of the line is $y = \frac{-1}{\sqrt{3}}x + (-4)$

$\Rightarrow \qquad x + \sqrt{3}y + 4\sqrt{3} = 0$

Example 6. Find the equation to the straight line which passes through the point (3, 4) and cuts intercepts, the sum of which length is 14.

Solution. Let the equation of the line is

$$\frac{x}{a}+\frac{y}{b}=1 \qquad \dots (1)$$

According to the question $a + b = 14 \Rightarrow a = 14 - b$

$\therefore$ The equation (1) becomes

$$\frac{x}{14-b}+\frac{y}{b}=1$$

$$\Rightarrow \quad bx + (14 - b)\,y = b(14 - b) \qquad \dots (2)$$

Since the line passes through the point (3, 4), then

$$\Rightarrow \quad 3b + (14 - b) \times 4 = b\,(14 - b)$$

$$\Rightarrow \quad 3b + 56 - 4b = 14b - b^2$$

$$\Rightarrow \quad b^2 - 15b + 56 = 0$$

$$\Rightarrow \quad (b - 7)\,(b - 8) = 0$$

$$\Rightarrow \quad b = 7, 8$$

From $a + b = 14$, we have $a = 7, 6$. Putting these values of a and b in equation (1), the equations of the required lines are

$$\frac{x}{7}+\frac{y}{7}=1 \Rightarrow x + y = 7$$

and

$$\frac{x}{6}+\frac{y}{8}=1 \Rightarrow 4x + 3y = 24$$

Example 7. Find the equation of the line through the points (4, – 3) and (0, 1).

Solution. Using two-points form, the equation of the line is

$$y-(-3) = \frac{1-(-3)}{0-4}(x - 4)$$

$$\Rightarrow \quad y + 3 = -1(x - 4)$$

$$\Rightarrow \quad x + y - 1 = 0$$

Example 8. Find the equation of the line passing through the points p (2, – 3) and making an angle of 120° with the positive direction of x-axis.

Solution. Here inclination of the line = 120°,

$\therefore$ Its slope $= m = \tan 120° = \tan (180° - 60°) = -\tan 60° = -\sqrt{3}$

The line passes through the point p (2, –3) the equation of the line is

$$y - (-3) = -\sqrt{3}\,(x - 2)$$

$$\sqrt{3}\,x + y + 3 - 2\sqrt{3} = 0$$

Example 9. The perpendicular drawn from the origin to a straight line, makes an angle of 60° with x-axis. The area of triangle the straight line makes with axes is $54\sqrt{3}$ square units. Find the equation to the straight line.

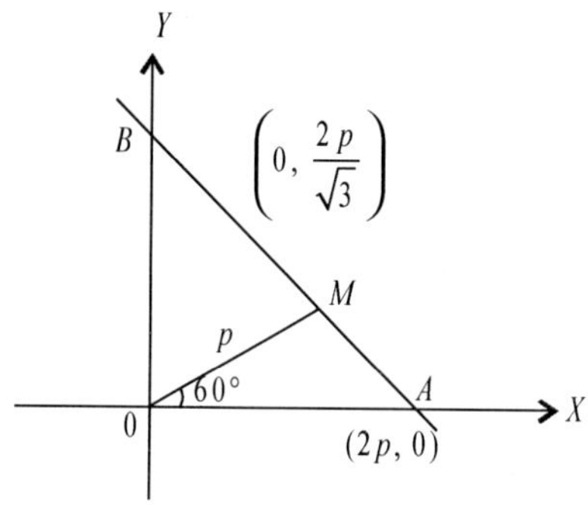

Solution. Let the equation of the straight line be

$$x \cos 60° + y \sin 60° = p$$

$$\Rightarrow \quad \frac{x}{2} + \frac{\sqrt{3}}{2} y = p$$

$$\Rightarrow \quad x + \sqrt{3}y = 2p \qquad \ldots (1)$$

$$\Rightarrow \quad \frac{x}{2p} + \frac{y}{\frac{2p}{\sqrt{3}}} = 1$$

which shows that the intercepts on x-axis is $2p$ and on y-axis is $\frac{2p}{\sqrt{3}}$ *i.e.*

$$OA = 2p \text{ and } OB = \frac{2p}{\sqrt{3}}$$

$$\therefore \text{ Area of } \quad \Delta OAB = \frac{1}{2}. OA. OB = \frac{1}{2}. 2p. \frac{2p}{\sqrt{3}} = \frac{2p^2}{\sqrt{3}}$$

According to the question

$$\frac{2p^2}{\sqrt{3}} = 54\sqrt{3}$$

$$\Rightarrow \quad p^2 = 27 \times 3$$

$$r = \pm 9, \qquad \text{put in equation (1), we get}$$

$$x + \sqrt{3}y = \pm 18$$

Example 10. Find the equation of a straight line on which the perpendicular from the origin is of length 2 units and this perpendicular makes an angle of 240° with the x-axis.

Solution. Here $\quad p = 2$ and $\theta = 240°$

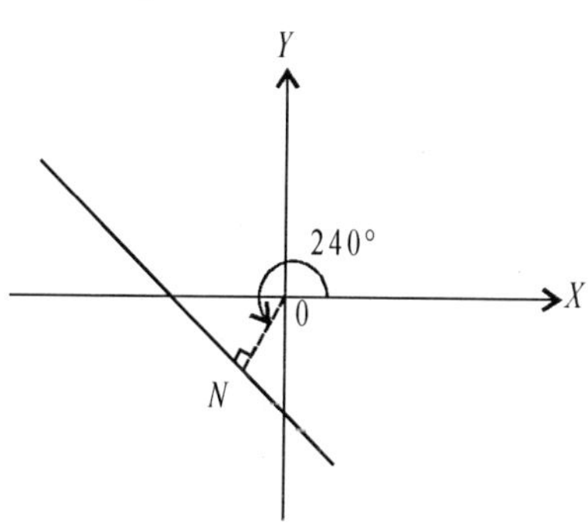

The equation of the line is

$$x \cos 240° + y \sin 240° = 2$$

$$x \times -\frac{1}{2} + y \times \frac{-\sqrt{3}}{2} = 2$$

$$\Rightarrow \quad x + \sqrt{3}y + 4 = 0$$

Example 11. A straight line is drawn through the point A (2, 1) making an angle of 30° with the positive direction of x-axis. Find the coordinates of two points on it at a distance of 2 units from A on either side of it.

Solution. Here $\theta = 30°$ and the line passes through A (2, 1) using distance form, the equation of the line is

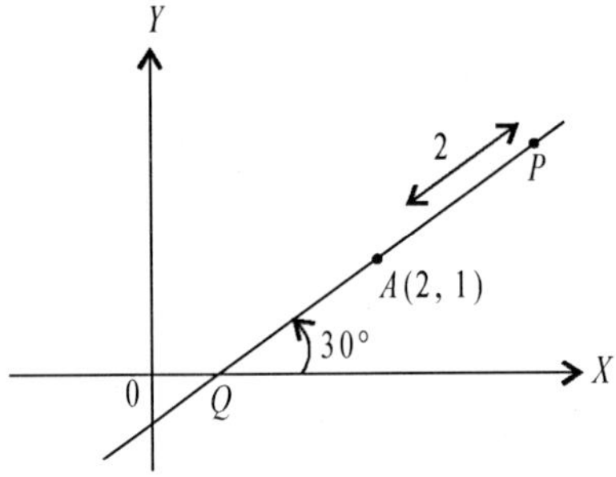

$$\frac{x-2}{\cos 30°} = \frac{y-1}{\sin 30°} = r \Rightarrow \frac{x-2}{\frac{\sqrt{3}}{2}} = \frac{y-1}{1/2} = r$$

Taking $r = 2, -2$, the two points on the line are given by

$$x - 2 = \frac{\sqrt{3}}{2}.2, y - 1 = \frac{1}{2}.2$$

and

$$x - 2 = \frac{\sqrt{3}}{2}(-2), y - 1 = \frac{1}{2}(-2)$$

$$\Rightarrow \quad x = 2 + \sqrt{3},\ y = 2 \text{ and } x = 2 - \sqrt{3},\ y = 0$$

Hence, the required points are $(2 + \sqrt{3}, 2)$ and $(2 - \sqrt{3}, 0)$

Example 12. Find the distance of the point A (4, 1) from the line $4x - y = 0$ measured along the line making an angle of 135° with the positive direction of x-axis.

Solution. The equation of the line through A (4, 1) and inclined at angle of 135° with x-axis is

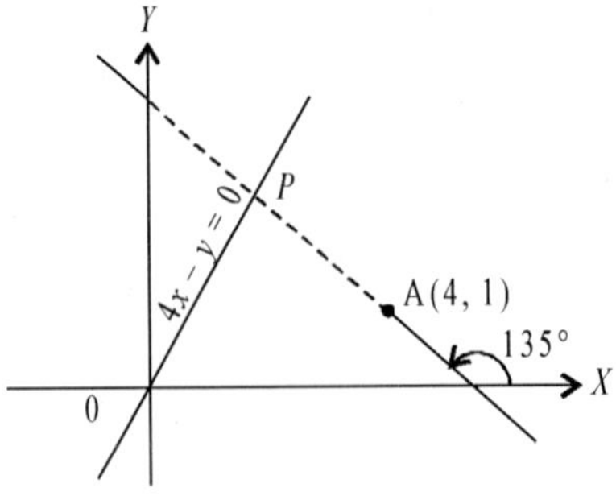

$$\frac{x-4}{\cos 135°} = \frac{y-1}{\sin 135°} = \text{r} \qquad \text{(Distance form)}$$

$$\Rightarrow \qquad \frac{x-4}{-\frac{1}{\sqrt{2}}} = \frac{y-1}{\frac{1}{\sqrt{2}}} = r$$

$\therefore$ Any point on line is $p\left(4 - \frac{r}{\sqrt{2}}, 1 + \frac{r}{\sqrt{2}}\right)$, where $|r|$ is the distance of the point p from the point A.

The points p lies on the line $4x - y = 0$ if

$$4\left(4 - \frac{r}{\sqrt{2}}\right) - \left(1 + \frac{r}{\sqrt{2}}\right) = 0$$

i.e. if $\qquad 15 = 5 \cdot \frac{r}{\sqrt{2}}$

i.e. if $\qquad r = 3\sqrt{2}$

Hence, the required distance $= |r| = |3\sqrt{2}| = 3\sqrt{2}$.

Example 13. Prove that the locus of the point which is equi distant from the points $(-3, 7)$ and $(2, -5)$ is a straight line.

Solution. Let $A(-3, 7)$ and $B(2, -5)$be the given points and $p(x, y)$ be any point on the locus, then $|AP| = |BP|$(given)

$$\Rightarrow \qquad \sqrt{(x+3)^2 + (y-7)^2} = \sqrt{(x-2)^2 + (y+5)^2}$$

$$\Rightarrow \quad x^2 + 6x + 9 + y^2 - 14y + 49 = x^2 - 4x + 4 + y^2 + 10y + 25$$

$$\Rightarrow \qquad 10x - 24y + 29 = 0$$

which is a first degree equation in x and y and so it represents a straight line. Hence the locus is a straight line.

EXERCISE 3.12

1. Find the equation of a straight line parallel to x-axis at a distance (*i*) 3 units above it, (*ii*) 3 units below it.
2. Find the equation of a straight line parallel to y-axis at a distance (*i*) 2 units to the right (*ii*) 2 units to the left of it.
3. Find the equation of a straight line parallel to the x-axis and having intercept -2 on the y-axis.
4. Find the equation of a vertical line passing through $(-7, 3)$.
5. Find the equation of a horizontal line passing through $(5, -2)$.
6. Find the equation of the line passing through the point $(2, -5)$ and making an intercept of -3 on the y-axis.
7. Find the equation of the straight line joining the points :
   (*i*) $(2, 3), (2, -4)$ (*ii*) $(2, 3), (4, 1)$

8. Find the equation of that straight line which cuts an intercept 2 from y-axis and makes an angle $\tan^{-1}\frac{5}{4}$ with positive direction of x-axis.

9. If the length of perpendicular left fall from the origin on the line $\frac{x}{a}+\frac{y}{b}=1$ is p, prove that

$$\frac{1}{p^2}=\frac{1}{a^2}+\frac{1}{b^2}$$

10. Find the equation of the line which cuts off intercepts 3 and 4 from the axes.

11. If the straight line $\frac{x}{a}+\frac{y}{b}=1$ passes through the points (12, – 15) and (8, – 9), find the values of a and b.

12. The slope of a straight line through the point A (3, 2) is $\frac{3}{4}$. Find the equation of the line in symmetrical form. Also find the coordinates of the points on the line that are 5 units away from A.

[**Hint.** Let θ be the inclination of the line then $\tan\theta=\frac{3}{4} \Rightarrow \theta$ is a acute and $\sin\theta=\frac{3}{5}$, $\cos\theta=\frac{4}{5}$.

13. A straight line passing through the point A (1, 2) makes an angle of 60° with x-axis and intercepts the line $x+y=6$ at p. Find the length of AP.

ANSWERS

1. (*i*) $y-3=0$ (*ii*) $y+3=0$

2. (*i*) $x-2=0$ (*ii*) $x+2=0$

3. $y+2=0$

4. $x+7=0$

5. $y+2=0$

6. $x+y+3=0$

7. (*i*) $x-2=0$ (*ii*) $x+y-5=0$

8. $5x-4y+8=0$

9. $4x+3y=12$

10. $a=2, b=3$

11. $\frac{x-3}{\frac{4}{5}}=\frac{y-4}{\frac{3}{5}}=r$; (7, 5) (– 1, – 1)

13. $AP=3(\sqrt{3}-1)$

3.22. REDUCTION OF GENERAL EQUATION TO STANDARD FORMS

1. To reduce the general equation $Ax + By + C = 0$ to the form $y = mx + c$

The given general equation of the line is

$$Ax+By+C=0 \quad ...(1)$$

$$BY=-Ax-c$$

$$y=-\frac{A}{B}x-\frac{C}{B} \quad ...(2)$$

Which is in the required slope intercept form $y = mx + c$, where $m = -\frac{A}{B}$

and $$c = -\frac{C}{B}$$

Rule to reduce the general equation to the slope intercept form
Solve the given equation for y in terms of x.

2. To reduce the general equation $Ax + By + C = 0$ to the intercept form

The given general equation of the line is $Ax + By + C = 0$ ——(1). It can be written as $Ax + By = -C$

Dividing both sides by $-C$, we get

$$\frac{Ax}{-C} + \frac{By}{-C} = 1$$

$$\frac{x}{-\frac{C}{A}} + \frac{y}{-\frac{C}{B}} = 1 \qquad ...(2)$$

Which is the required intercept form $\frac{x}{a} + \frac{y}{b} = 1$ where $a = -\frac{C}{A}$ and $b = -\frac{C}{B}$

Rule to reduce the general equation to the intercept form :

(*i*) Transpose the constant term to R. H. S.

(*ii*) Divide both sides by the constant term on R. H. S.

(*iii*) Take the coefficients of x and y to the denominator.

Rule to find the intercepts made by a line on the coordinate axes :

Since a line cuts the x-axis at a point whose ordinate is zero and it cuts the y-axis at a point whose abscissa is zero, therefore.

(*i*) To obtain x-intercepts, put $y = 0$ in the equation of the line and find the value of x.

(*ii*) To obtain y-intercepts, put $x = 0$ in the equation of the line and find the value of y.

3. To reduce the general equation $Ax + By + C = 0$ to the normal form

The given general equation of the line is $Ax + By + C = 0$... (1) we note that the normal form

$$x \cos\theta + y \sin\theta = p \qquad ... (2)$$

has the following characteristics :

1. (coeff. of x)^2 + (coeff. of y)^2 = 1 $\qquad (\because \cos^2\theta + \sin^2\theta = 1)$
2. The constant term occurs on R. H. S. and is positive. Multiplying both sides of equation (1) by k ($\neq 0$), we get

$$k\,Ax + k\,By + k\,C = 0 \Rightarrow k\,Ax + k\,By = -k\,C \qquad ...(3)$$

The equation (3) is in the normal form if $(k\,A)^2 + (+k\,B)^2 = 1$

$$\Rightarrow \quad k^2(A^2 + B^2) = 1$$

$$\Rightarrow \quad k = \frac{1}{\pm\sqrt{A^2 + B^2}}$$

Putting this value of k in equation (3), we get

$$\frac{A}{\pm\sqrt{A^2+B^2}}x+\frac{B}{\pm\sqrt{A^2+B^2}}y=-\frac{C}{\pm\sqrt{A^2+B^2}} \quad \text{... (4)}$$

since p is always positive the sign of $\pm\sqrt{A^2+B^2}$ is to be so chosen that the expressions on R. H. S is + *ve*.

Two cases arise :

Case I: When C is + *ve*.

Here R. H. S. of (4) will be + *ve* if we take – *ve* sign with the radical.

∴ The required normal form is

$$-\frac{A}{\sqrt{A^2+B^2}}x-\frac{B}{\sqrt{A^2+B^2}}y=\frac{C}{\sqrt{A^2+B^2}}$$

where $\cos\theta = -\frac{A}{\sqrt{A^2+B^2}}$, $\sin\theta = -\frac{B}{\sqrt{A^2+B^2}}$ and

$$p=\frac{C}{\sqrt{A^2+B^2}}$$

Case II: When C is – *ve*.

Here R.H.S of (4) will be + *ve* if we take + *ve* sign with the radical.

∴ The required normal form is

$$\frac{A}{\sqrt{A^2+B^2}}x+\frac{B}{\sqrt{A^2+B^2}}y=-\frac{C}{\sqrt{A^2+B^2}}$$

where $\cos\theta = \frac{A}{\sqrt{A^2+B^2}}$, $\sin\theta = \frac{B}{\sqrt{A^2+B^2}}$ and $p = -\frac{C}{\sqrt{A^2+B^2}}$

Rule to reduce the general equation to the normal form

(*i*) Transpose the constant term to R. H. S.

(*ii*) After transposition if the constant term on R.H.S. is – *ve*, make it + *ve* by changing the sign throughout.

(*iii*) Divide both sides by $\sqrt{(\text{coeff. of } x)^2+(\text{coeff. of } y)^2}$

ILLUSTRATIVE EXAMPLES

Example 1. Reduce the equation $2x - 3y - 5 = 0$ to slope-intercept form and find the intercept it makes on y-axis.

Solution. The given equation is $2x - 3y - 5 = 0$ ... (1)

It can be written as $-3y = -2x + 5$

$$y=\frac{2}{3}x-\frac{5}{3} \quad \text{... (2)}$$

which is in the required slope-intercept form.

Comparing (2) with $y = mx + c$, we get $c = -\frac{5}{3}$

Hence, the intercept made by the given line on y-axis $= -\frac{5}{3}$.

Example 2. Find the value of k so that the line $2x + ky - 9 = 0$ may be

(*i*) parallel to $3x - 4y + 7 = 0$

(*ii*) perpendicular to $3y + 2x - 1 = 0$

Solution. Given $\quad 2x + ky - 9 = 0 \quad$ ...(1)

$$\Rightarrow \quad ky = -2x + 9$$

$$\Rightarrow \quad y = \frac{-2}{k}x + \frac{9}{k}$$

$\therefore$ Slope of line (1) is $-\frac{2}{k}$

(*i*) $\quad 3x - 4y + 7 = 0 \quad$ ... (2)

$$4y = 3x + 7$$

$$y = \frac{3}{4}x + \frac{7}{4}$$

$\therefore$ Slope of line (2) is $\frac{3}{4}$

As the lines (1) and (2) are parallel

$$\therefore \quad -\frac{2}{k} = \frac{3}{4} \qquad [\because m_1 = m_2]$$

$$\Rightarrow \quad k = -\frac{8}{3}$$

(*ii*) Given line is $\quad 3y + 2x - 1 = 0$

$$2x + 3y - 1 = 0 \qquad ...(3)$$

Slope of line (3) is $-\frac{2}{3}$

As the lines (1) and (3) are perpendicular

$$\left(-\frac{2}{k}\right)\left(-\frac{2}{3}\right) = -1 \qquad [\because m_1 m_2 = -1]$$

$$\Rightarrow \quad 4 = -3k$$

$$k = -\frac{4}{3}$$

Example 3. Find the equation of the line through the point (2, – 5) and

(*i*) parallel to $2x - 3y = 7$

(*ii*) perpendicular to $3x + 7y = 11$

Solution. (*i*) Given line is $2x - 3y - 7 = 0 \quad$ ...(1)

$$\text{Its slope} = \frac{2}{3}$$

$\therefore$ The slope of a line parallel to (1) is $\frac{2}{3} \qquad [m_1 = m_2]$

The equation of the line through (2, – 5) and parallel to (1) is

$$\Rightarrow \quad y-(-5) = \frac{2}{3}(x-2) \qquad \text{(point slope form)}$$

$$\Rightarrow \quad 3y+15 = 2x-4$$

$$\Rightarrow \quad 2x-3y-19 = 0$$

(*ii*) Given line is $3x + 7y - 11 = 0$ ... (2)

$$\text{Its slope} = -\frac{3}{7}$$

$\therefore$ Slope of a line perpendicular to (2) $= \frac{7}{3}$

$$\left[m_2 = -\frac{1}{m_1}\right]$$

$\therefore$ The equation of the line through (2, – 5) and perpendicular to (2) is

$$y-(-5) = \frac{7}{3}(x-2) \qquad \text{(point slope form)}$$

$$\Rightarrow \quad 3y+15 = 7x-14$$

$$\Rightarrow \quad 7x-3y-29 = 0$$

Example 4. Reduce the equation $3x - 2y + 4 = 0$ to intercept form.

Solution. The given equation is $3x - 2y + 4 = 0$

$$\Rightarrow \quad 3x-2y = -4$$

Dividing both sides by – 4, we get

$$\frac{3x}{-4} - \frac{2y}{-4} = 1$$

$$\frac{x}{-\frac{4}{3}} + \frac{y}{2} = 1, \text{ which is the required intercept form.}$$

$$\text{Here a (}x\text{-intercept)} = -\frac{4}{3} \text{ and b (}y\text{-intercept)} = 2$$

Example 5. Find the equation of the line whose y-intercept is – 3 and that is perpendicular to $3x = 2y - 5$

Solution. The given line is $3x = 2y - 5$

$$3x-2y+5 = 0 \qquad \text{... (1)}$$

$$\text{Its slope} = \frac{3}{2}$$

$\therefore$ Slope of a line perpendicular to (1) is $-\frac{2}{3}$

$\therefore$ The equation of the line with y-intercept – 3 and perpendicular to (1) is

$$y = -\frac{2}{3}x + (-3)$$

$$\Rightarrow \quad 3y = -2x-9$$

$$\Rightarrow \quad 2x+3y+9 = 0$$

EXERCISE 3.13

1. Reduce the following lines to slope intercept form also find their slope and y-intercept :
   (*i*) $3x + 5y - 7 = 0$ (*ii*) $5x - 7y + 35 = 0$
2. If $6x + 5y - 11 = 0$ and $2k\,x + 5y + 3 = 0$ are parallel lines, find the value of k.
3. If the lines $\frac{x}{3}+\frac{y}{4}-5 = 0$ and $3x + ky = 9$ are perpendicular to each other, then find the value of k.
4. Reduce the following equations to normal form :
   (*i*) $4x + 3y - 9 = 0$ (*ii*) $2x - 3y + 7 = 0$
   (*iii*) $x + y + \sqrt{2} = 0$ (*iv*) $x + 3 = 0$
5. Find the equation of a line parallel to $x - 2y + 8 = 0$ and passing through the point (1, 2).
6. Find the equation of a straight line perpendicular to $3x + 8y = 12$ and passing through the point (– 1, – 2).
7. Find the equation of a straight line perpendicular to $2x = 3y + 5$ and with x-intercept– 4.
   [**Hint.** x-intercept = – 4, the line passing through (– 4, 0)].

ANSWERS

1. (*i*) $y = -\frac{3}{5}x+\frac{7}{5}; m = -\frac{3}{5}, c = \frac{7}{5}$

(*ii*) $y = \frac{5}{7}x+5, m = \frac{5}{7}, c = 5$

2. 3 **3.** – 4

4. (*i*) $\frac{4}{5}x+\frac{3}{5}y = \frac{9}{5}$

(*ii*) $-\frac{2}{\sqrt{13}}x+\frac{3}{\sqrt{13}}y = \frac{7}{\sqrt{13}}$

(*iii*) $x \cos 225° + y \sin 225° = 1$

(*iv*) $x \cos 180° + y \sin 180° = 3$

5. $x - 2y + 3 = 0$

6. $8x - 3y + 2 = 0$

7. $3x + 2y + 12 = 0$

3.23. CIRCLE

A circle is the focus of a point which moves in a plane so that its distance from a fixed point, called the centre, is equal to a given distance called radius.

3.24. EQUATION OF A CIRCLE WHOSE CENTRE (0, 0) AND RADIUS *R*

Let $P(x, y)$ be any point on the circle whose centre is O (0, 0) and radius is r. Join OP, then

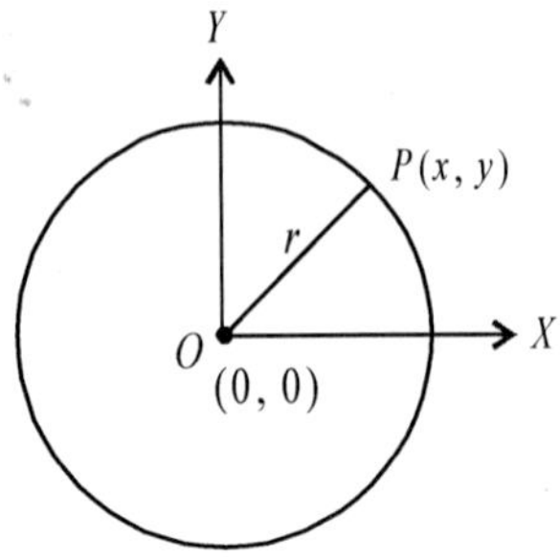

$$OP = r$$

$$\Rightarrow \quad \sqrt{(x-0)^2 + (y-0)^2} = r$$

$$\Rightarrow \quad \sqrt{(x^2 + y^2} = r$$

$$\Rightarrow \quad x^2 + y^2 = r^2$$

3.25. EQUATION OF A CIRCLE WHOSE CENTRE (*h*, *k*) AND RADIUS *R*

Let $P(x, y)$ be any point on the circle whose centre is $C(h, k)$ and radius is r. Join CP, then

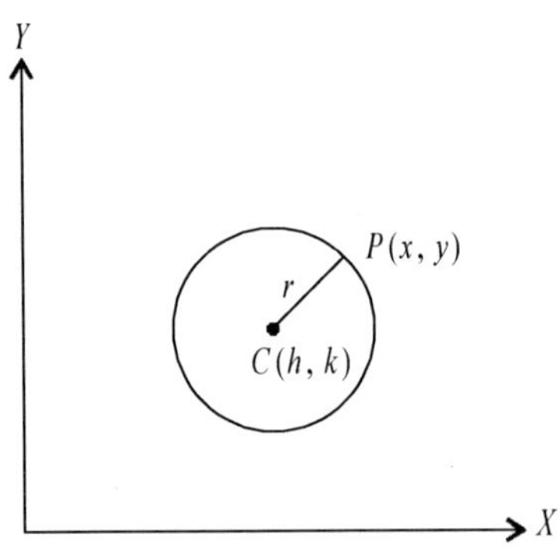

$$CP = r$$

$$\Rightarrow \quad \sqrt{(x-h)^2 + (y-k)^2} = r$$

$$\Rightarrow \quad (x-h)^2 + (y-k)^2 = r^2$$

3.26. GENERAL EQUATION OF CIRCLE

A second degree equation in x and y of the form $x^2 + y^2 + 2gx + 2fy + c = 0$ is called the general equation of circle with centre $(-g, -f)$ and radius $\sqrt{g^2 + f^2 - c}$. In the general equation of circle, the coefficient of x^2 is always equal to the coefficient of y^2. If $g^2 + f^2 > c$, the radius of the circle is real.

If $g^2 + f^2 = c$, the radius vanishes *i.e.* the circle becomes a point coinciding with the point (–g, – f). Such a circle is called a point circle .
If $g^2 + f^2 < c$, the radius of the circle is imaginary. In this case the equation does not represent any geometrical locus or it is a circle with the real centre and an imaginary radius.

3.27. DIFFERENT FORMS OF EQUATION OF A CIRCLE

1. If the centre of the circle (h, k) lies on x-axis, then $k = 0$ and the equation of circle reduces to
$(x - h)^2 + y^2 = r^2$

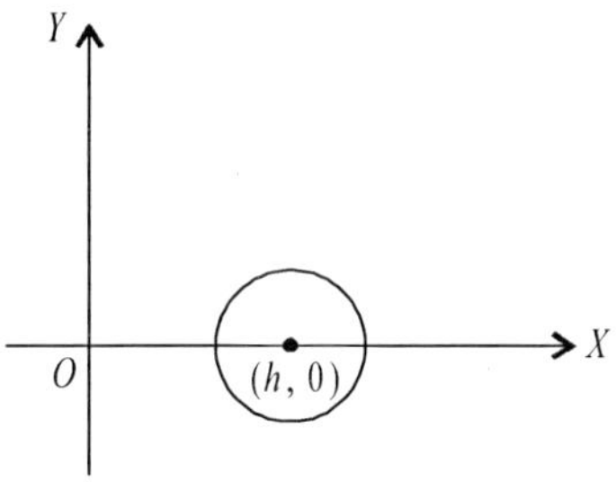

2. If the centre of the circle (h, k) lies on y-axis, then $h = 0$ and the equation of circle reduces to
$x^2 + (y - k)^2 = r^2$

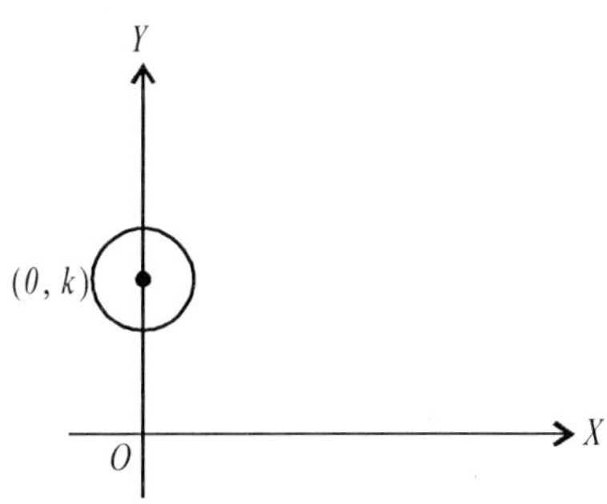

3. If the circle passes through the origin (0, 0), then we have $h^2 + k^2 = r^2$, and the equation reduced to
$x^2 + y^2 - 2hx - 2ky = 0$

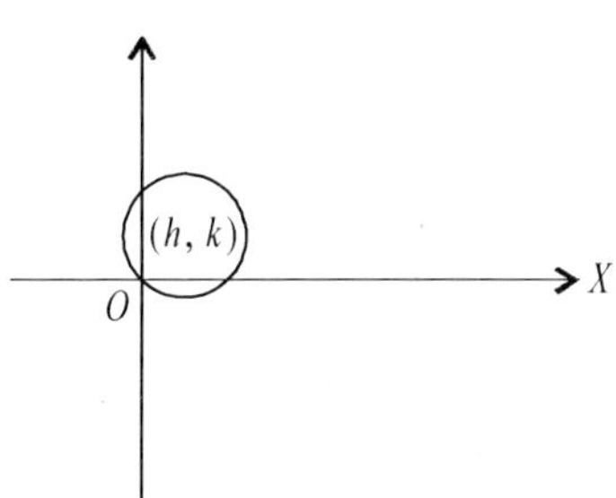

4. If the centre lies on x-axis and circle passes through origin, then $r = h$, $k = 0$ and equation of circle reduces to
$x^2 + y^2 - 2hx = 0$

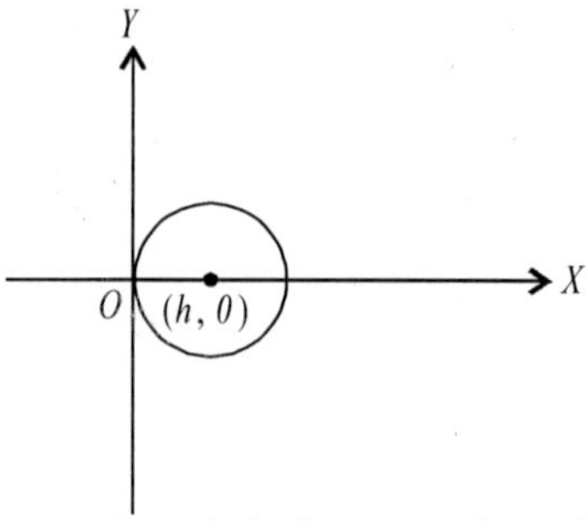

5. If the centre lies on y-axis and circle passes through origin, then $r = k$, $h = 0$ and equation of circle reduces to
$x^2 + y^2 - 2ky = 0$

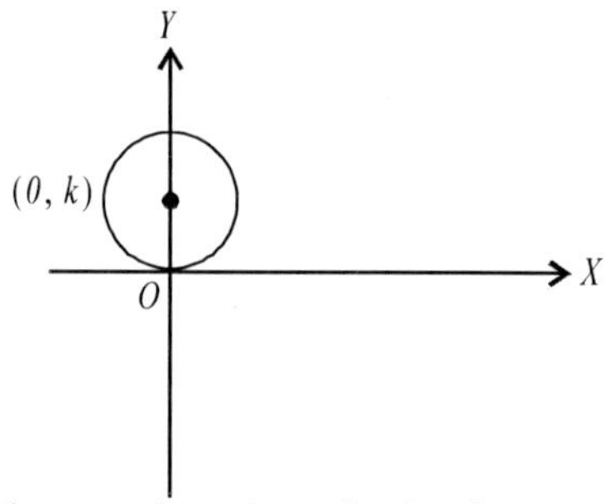

6. If the circle touches x-axis and y-axis simultaneously in Ist quadrant, then $h = k = r$ and the equation of circle reduces to $(x - r)^2 + (y - r)^2 = r^2$

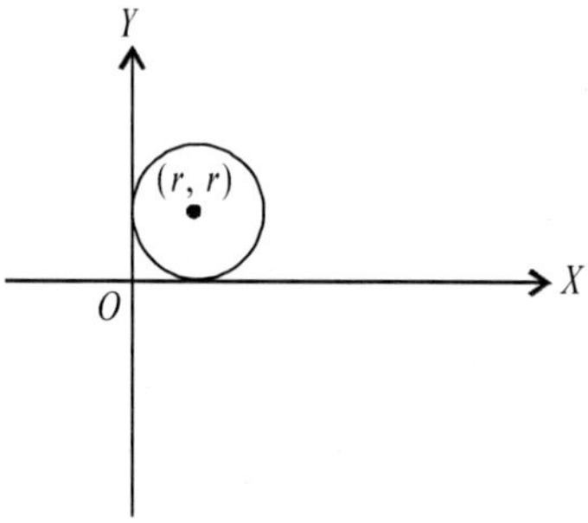

3.28. EQUATION OF THE CIRCLE DESCRIBED ON THE LINE JOINING THE POINTS (x_1, y_1) AND (x_2, y_2) AS DIAMETER

Let A and B be the points (x_1, y_1) and (x_2, y_2) respectively, and the coordinates of any point p on the circle be (x, y).
Join AP and BP.

Slope of AP is $\quad m_1 = \dfrac{y - y_1}{x - x_1}$

and the slope of BP is $\quad m_2 = \dfrac{y - y_2}{x - x_2}$

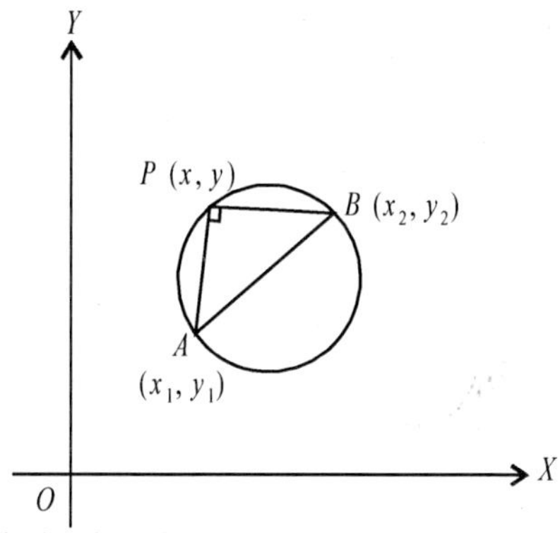

since APB is a semi-circle, therefore $\angle APB = 90°$ and so the lines AP and BP are at right angles.

$$\therefore \quad m_1 m_2 = -1$$

$$\Rightarrow \quad \frac{y-y_1}{x-x_1} \times \frac{y-y_2}{x-x_2} = -1$$

$$\Rightarrow \quad (x-x_1)(x-x_2)+(y-y_1)(y-y_2) = 0$$

ILLUSTRATIVE EXAMPLES

Example 1. Find the equation of the circle whose centre is (2, 3) and radius is 3.

Solution. Here $(h, k) = (2, 3)$ and $r = 3$

$\therefore$ Equation of circle is

$$(x-2)^2+(y-3)^2 = 3^2$$
$$x^2+4-4x+y^2+9-6y = 9$$
$$x^2+y^2-4x-6y+4 = 0$$

Example 2. Find the equation of the circle whose centre is (3, – 2) and radius is 5.

Solution. Here $(h, k) = (3, -2)$ and $r = 5$

$\therefore$ Equation of circle is

$$(x-3)^2+(y+2)^2 = 5^2$$
$$x^2+9-6x+y^2+4+4y = 25$$
$$x^2+y^2-6x+4y-12 = 0$$

Example 3. Find the equation of the circle whose centre is (2, – 1) and passes through the point (3, 6).

Solution. Here $(h, k) = (2, -1)$, point (3, 6)

Radius of the circle is r = distance between the points (2, – 1) and (3, 6)

$$r = \sqrt{(3-2)^2+(6+1)^2}$$
$$r = \sqrt{1+49} = \sqrt{50}$$

$\therefore$ The required equation of the circle is

$$(x-h)^2+(y-k)^2 = r^2$$
$$(x-2)^2+(y+1)^2 = 50$$
$$x^2-4x+4+y^2+2y+1 = 50$$
$$x^2+y^2-4x+2y-45 = 0$$

Example 4. Find the centres and the radius of the circle whose equations are :

(*i*) $x^2 + y^2 + 12y = 0$

(*ii*) $7x^2 + 7y^2 - 4x - y = 3$

(*iii*) $(x - 2)(x - 4) + (y - 1)(y - 3) = 0$

Solution. (*i*) The equation of the circle is $x^2 + y^2 + 12y = 0$

Here $g = 0, f = 6, c = 0$

The centre is $(-g, -f)$, *i.e.* $(0, -6)$

$$\text{radius} = \sqrt{g^2 + f^2 - c} = \sqrt{0 + 6^2 - 0} = 6$$

Hence, centre is $(0, -6)$ and radius $= 6$

(*ii*) The equation of the circle is

$$7x^2 + 7y^2 - 4x - y = 3$$

$$x^2 + y^2 - \frac{4}{7}x - \frac{1}{7}y - \frac{3}{7} = 0$$

Here $\qquad g = -\frac{2}{7}, f = -\frac{1}{14}, c = -\frac{3}{7}$

The centre is $(-g, -f)$ *i.e.* $\left(\frac{2}{7}, \frac{1}{14}\right)$

$$\text{radius} = \sqrt{g^2 + f^2 - c}$$

$$r = \sqrt{\left(-\frac{2}{7}\right)^2 + \left(-\frac{1}{14}\right)^2 - \left(-\frac{3}{7}\right)}$$

$$= \sqrt{\frac{4}{49} + \frac{1}{14\times14} + \frac{3}{7}}$$

$$= \frac{1}{7}\sqrt{4 + \frac{1}{2\times2} + 3\times7}$$

$$r = \frac{1}{14}\sqrt{16+1+84} = \frac{1}{14}\sqrt{101}$$

Hence, the centre is $\left(\frac{2}{7}, \frac{1}{14}\right)$ and radius is $\frac{1}{14}\sqrt{101}$.

(*iii*) The equation of the circle is

$$(x - 2)(x - 4) + (y - 1)(y - 3) = 0$$

$$x^2 - 6x + 8 + y^2 - 4y + 3 = 0$$

$$x^2 + y^2 - 6x - 4y + 11 = 0$$

Hence $\qquad g = -3, f = -2, c = 11$

The centre is $(-g, -f)$ *i.e.* $(3, 2)$

$$\text{radius} = \sqrt{g^2 + f^2 - c}$$

$$r = \sqrt{9 + 4 - 11} = \sqrt{2}$$

Hence, the centre is $(3, 2)$ and radius $r = \sqrt{2}$

Example 5. Find the equation of the circle which passes through the point (1, 0) (0, – 6), and (3, 4)

Solution. Let the equation of the circle be

$$x^2 + y^2 + 2gx + 2fy + c = 0 \qquad \text{... (1)}$$

It passes through the points (1, 0), (0, – 6) and (3, 4)

$$1 + 2g + c = 0 \qquad \text{... (2)}$$

$$36 - 12f + c = 0 \qquad \text{... (3)}$$

$$25 + 6g + 8f + c = 0 \qquad \text{... (4)}$$

Subtracting (3) from (2) and (4) respectively, we have

$$2g + 12f - 35 = 0$$

$$6g + 20f - 11 = 0$$

Solving these equations, we have

$$\frac{g}{-132+700} = \frac{f}{-210+22} = \frac{1}{40-72}$$

$$\frac{g}{568} = \frac{f}{-188} = \frac{1}{-32}$$

$$g = -\frac{568}{32} = -\frac{71}{4}$$

$$f = \frac{188}{32} = \frac{47}{8}$$

Putting the values of g in (2), we get

$$1 - \frac{71}{2} + c = 0$$

$$c = \frac{71}{2} - 1 = \frac{69}{2}$$

Putting the values of g, f and c in (1), we get

$$x^2 + y^2 - 2 \times \frac{71}{4} x + 2 \times \frac{47}{8} y + \frac{69}{2} = 0$$

$$4x^2 + 4y^2 - 142x + 47y + 138 = 0$$

Example 6. Find the equation of the circle which passes through the points (1, – 2) and (4, – 3) and has its centre on the line $3x + 4y - 5$

Solution. Let the equation of the circle be

$$x^2 + y^2 + 2gx + 2fy + c = 0 \qquad \text{...(1)}$$

It passes through the points (1, – 2) and (4, – 3)

$$1 + 4 + 2g - 4f + c = 0$$

$$2g - 4f + c + 5 = 0 \qquad \text{...(2)}$$

and

$$16 + 9 + 8g - 6f + c = 0$$

$$8g - 6f + c + 25 = 0 \qquad \text{...(3)}$$

The centre of the circle $(-g, -f)$ lies on the line $3x + 4y = 5$

$$\therefore \quad -3g - 4f = 5$$

$$\Rightarrow \quad 3g + 4f + 5 = 0 \qquad \text{...(4)}$$

Subtracting the equation (2) from (3) we have

$$6g - 2f + 20 = 0$$

$$3g - f + 10 = 0 \qquad \text{...(5)}$$

Subtracting (5) from (4), we have

$$5f - 5 = 0 \Rightarrow f = 1$$

Putting the value of f in (5), we have

$$3g - 1 + 10 = 0 \Rightarrow g = -3$$

∴ Putting the values of g and f in (2), we get

$$2(-3) - 4 \times 1 + c + 5 = 0$$

$$c = 6 + 4 - 5 = 5$$

Hence putting the values of g, f and c in (1), we get

$$x^2 + y^2 - 6x + 2y + 5 = 0$$

Example 7. Find the equation of circle concentric with the circle $x^2 + y^2 + 8x + 10y - 40 = 0$ and which passes through the point (1, 6) *(I.U. 2007, 08)*

Solution. $x^2 + y^2 + 8x + 10y - 40 = 0$

$$\text{Centre} = (-4, -5)$$

Circles with the same centre are called concentric. Equation of circle with centre (– 4, – 5)

$$(x + 4)^2 + (y + 5)^2 = r^2 \quad \text{... (1)}$$

It passes through (1, 6)

$$(1 + 4)^2 + (6 + 5)^2 = r^2$$

$$25 + 121 = r^2$$

$$146 = r^2$$

$$\therefore \quad (x + 4)^2 + (y + 5)^2 = 146$$

$$x^2 + 16 + 8x + y^2 + 25 + 10y = 146$$

$$x^2 + y^2 + 8x + 10y - 105 = 0$$

Example 8. Find the value of p for which the line $x \cos\alpha + y \sin\alpha = p$ may touch the circle $x^2 + y^2 - 2ax \cos\alpha - 2ay \sin\alpha = 0$ *(I.U. 2007, 08)*

Solution. $\quad x \cos\alpha + y \sin\alpha = p \quad$ ...(1)

$$x^2 + y^2 - 2ax \cos\alpha - 2ay \sin\alpha = 0 \quad \text{...(2)}$$

∴ Centre $(a \cos\alpha, a \sin\alpha)$, $r = a$

The line (1) will be a tangent to the circle (2) iff length of ⊥ form centre $(a \cos\alpha, a \sin\alpha)$ of the circle (2) to the line (1) = Radius of circle (2)

$$\left|\frac{a\cos\alpha \times \cos\alpha + a\sin\alpha \times \sin\alpha - p}{\sqrt{\cos^2\alpha + \sin^2\alpha}}\right| = a$$

$$\left|\frac{a\cos^2\alpha + a\sin^2 - p}{1}\right| = a$$

$$|a - p| = a$$

$$a - p = a$$

$$\Rightarrow \quad \boxed{p = 0}$$

EXERCISE 3.14

1. Find the equations of the circles whose
   (*i*) Centre is $(3, -2)$ and radius is 5.
   (*ii*) Centre is $(a, -b)$ and radius is $a + b$.
2. Find the coordinates of the centres and radii of the circles whose equations are:
   (*i*) $x^2 + y^2 + 4x - 4y - 1 = 0$
   (*ii*) $3x^2 + 3y^2 + 6x - 9y + 1 = 0$
   (*iii*) $2x^2 + 2y^2 = 18$
3. Find the equation of the circle concentric with the circle $(x - 2)^2 + y^2 = 4$ and whose radius is 10.
4. Find the equation of the circle whose centre is at the point $(2, -3)$ and which passes through the point $(0, 1)$.
5. Find the equation of the circle concentric with the circle $x^2 + y^2 - 8x + 6y - 5 = 0$ and passing through the point $(-2, -7)$
6. Find the equation of the circle that passes through the points $(3, -2)$ and $(-2, 0)$ and having centre on the straight line $2x - y = 3$.
7. Find the equations of the circles which pass through the points :
   (*i*) $(1, 0)$, $(0, -6)$ and $(3, 4)$
   (*ii*) $(-3, 4)$, $(-2, 0)$ and $(1, 5)$
8. Find the equations of the circle which passes through the points $(7, 10)$ and $(-7, -4)$ and has radius equal to 10.

ANSWERS

1. (*i*) $x^2 + y^2 - 6x + 4y - 12 = 0$
   (*ii*) $x^2 + y^2 - 2ax + 2by = 2ab$
2. (*i*) $(-2, 2)$; 3
   (*ii*) $\left(-1, \frac{3}{2}\right); \frac{1}{2}\sqrt{\frac{35}{3}}$
   (*iii*) $(0, 0)\,; 3$
3. $x^2 + y^2 - 4x - 96 = 0$
4. $x^2 + y^2 - 4x + 6y - 7 = 0$
5. $x^2 + y^2 - 8x + 6y - 27 = 0$
6. $x^2 + y^2 + 3x + 12y + 2 = 0$
7. (*i*) $4x^2 + 4y^2 - 142x + 47y + 138 = 0$
   (*ii*) $x^2 + y^2 + x - 5y - 2 = 0$
8. $x^2 + y^2 - 2x - 4y - 95 = 0$; $x^2 + y^2 + 2x - 8y - 83 = 0$

3.29. PARABOLA

Parabola is the locus of a point which moves so that its distance from a fixed point called focus is always equal to its perpendicular distance from a fixed straight line called directrix.

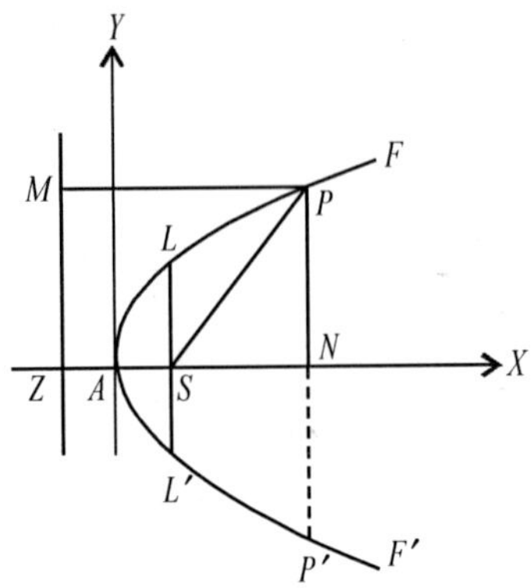

The straight line passing through the focus and perpendicular to the directrix is called the axis of parabola.
The ratio of the distance of the moving point from the fixed point (focus) and from the directrix is called the eccentricity and is denoted by e.
The point where the parabola meets its axis is called its vertex. In the adjoining figure I, $FPAP'F'$ is a parabola.
S is focus
MZ is directrix
AX is the axis
A is the vertex

and $$\frac{PS}{PM} = e = 1$$

3.30. EQUATION OF A PARABOLA

Let S be the fixed point, ZM the directrix and $P(x, y)$ be a variable point on the parabola. Draw SZ perpendicular to the directrix and bisect SZ at A. Produce ZA to X, then AX is the axis of the parabola.
The point A is clearly a point on the parabola because $AS = AZ$, and is the vertex of the parabola.
Take A as origin, AX as axis of x and AY, perpendicular to AX as the axis of y, if ZA or $AS = a$, then the coordinates of S are $(a, 0)$
Join SP and draw perpendicular PN and PM to AX and ZM respectively. Then from the definition of the parabola,

$$PS = PM$$
$$PS = NZ = NA + AZ$$
$$\Rightarrow \quad \sqrt{(x-a)^2 + (y-0)^2} = x + a$$
$$\Rightarrow \quad (x - a)^2 + y^2 = (x + a)^2$$
$$x^2 - 2ax + a^2 + y^2 = x^2 + 2ax + a^2$$
$$\Rightarrow \quad y^2 = 4ax$$

This is the standard form of the equation of a parabola whose focus is $(a, 0)$ vertex is $(0, 0)$, directrix is $x + a = 0$ and axis is x-axis or $y = 0$

Other standard forms of parabola

(*a*) $y^2 = -4ax$

(*b*) $x^2 = 4ay$

(*c*) $x^2 = -4ay$

where $a > 0$

(*a*) The parabola $y^2 = -4ax$

Vertex is A (0, 0)

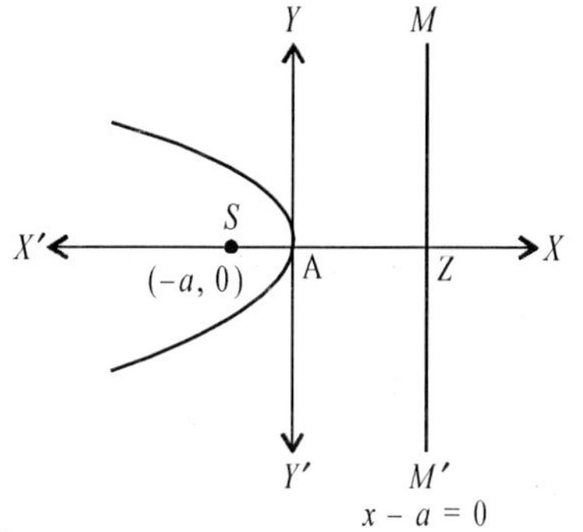

Focus is S $(-a, 0)$

Directrix ZM is $x - a = 0$

Axis is $y = 0$

(*b*) The parabola $x^2 = 4ay$

Vertex is A (0, 0)

Focus is S $(0, a)$

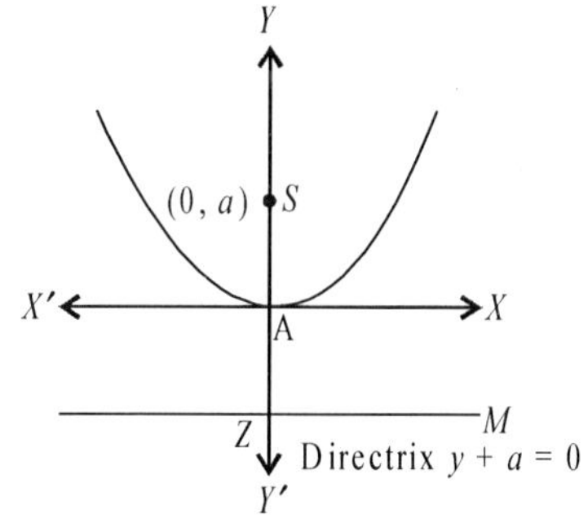

Directrix is $y + a = 0$

Axis is $x = 0$

(*c*) The parabola $x^2 = -4ay$, $a > 0$

Vertex is A (0, 0)

Focus is S $(0, -a)$

Directrix is $y - a = 0$

Axis is $x = 0$

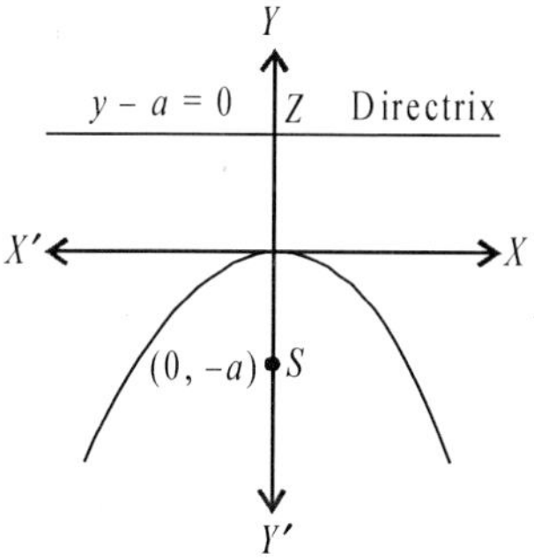

3.31. LATUS RECTUM AND FOCAL CHORD

Any chord passing through the focus of a parabola is called the focal chord. It joins two points on the parabola. The focal chord of the parabola which is perpendicular to the axis of the parabola is called the latus rectum.
In figure I, LSL' is the latus rectum.
We have, SL = the distance of L from the directrix

$$SL = SZ = 2a$$
$$LSL' = 2SZ = 4a$$

Hence, the length of the latus rectum is $4a$.
The distance of any point p on the parabola from its focus is called the focal distance. Thus PS is the focal distance of the point P.
The coordinates of the ends L and L' of the latus rectum are $(a, 2a)$ and $(a, -2a)$.
Note : For all the parabolas in the standard forms which are $y^2 = 4ax$, $y^2 = -4ax$, $x^2 = 4ay$ and $x^2 = -4ay$, the latus rectum are all equal to $4a$.

3.32. PARAMETRIC EQUATIONS OF THE PARABOLA $y^2 = 4ax$

The equations $x = at^2$ and $y = 2at$ are called parametric equations of the parabola $y^2 = 4ax$.
Useful information about parabola in standard forms

Form	$y^2 = 4ax$	$y^2 = -4ax$	$x^2 = 4ay$	$x^2 = -4ay$
Vertex	(0, 0)	(0, 0)	(0, 0)	(0, 0)
Focus	$(a, 0)$	$(-a, 0)$	$(0, a)$	$(0, -a)$
Axis	$y = 0$	$y = 0$	$x = 0$	$x = 0$
Directrix	$x = -a$	$x = a$	$y = -a$	$y = a$
Latus rectum	$4a$	$4a$	$4a$	$4a$
Tangent at vertex	$x = 0$	$x = 0$	$y = 0$	$y = 0$

ILLUSTRATIVE EXAMPLES

Example 1. Find the focus, directrix and the coordinates of the ends of the latus rectum of the parabolas :
(*i*) $y^2 = 15x$ (*ii*) $x^2 = 6y$

Solution. (*i*) Equation of the parabola is $y^2 = 15x$

Here $4a = 15$

$$\Rightarrow \quad a = \frac{15}{4}$$

Focus is $(a, 0)$ *i.e.* $\left(\frac{15}{4}, 0\right)$

Directrix is given by $x = -a$

$$\Rightarrow \quad x = -\frac{15}{4}$$

$$\Rightarrow \quad 4x + 15 = 0$$

The coordinates of the ends of latus rectum are

$(a, 2a)$ and $(a, -2a)$

$$\Rightarrow \quad \left(\frac{15}{4}, 2\times\frac{15}{4}\right) \text{ and } \left(\frac{15}{4}, -2\times\frac{15}{4}\right)$$

$$\Rightarrow \quad \left(\frac{15}{4}, \frac{15}{2}\right) \text{ and } \left(\frac{15}{4}, -\frac{15}{2}\right)$$

(*ii*) Equation of the parabola is $x^2 = 6y$

Comparing this equation with the standard equation

$x^2 = 4ay$, we find that

$$6 = 4a \quad \Rightarrow \quad a = \frac{3}{2}$$

$\therefore$ Focus is $(0, a)$ *i.e.* $\left(0, \frac{3}{2}\right)$

Directrix is given by $\quad y = -a$

$$\Rightarrow \quad y = -\frac{3}{2}$$

$$\Rightarrow \quad 2y + 3 = 0$$

The coordinates of the ends of latus rectum are

$(2a, a)$ and $(-2a, a)$

$$\Rightarrow \quad \left(2\times\frac{3}{2}, \frac{3}{2}\right) \text{ and } \left(-2\times\frac{3}{2}, \frac{3}{2}\right)$$

$$\Rightarrow \quad \left(3, \frac{3}{2}\right) \text{ and } \left(-3, \frac{3}{2}\right)$$

Example 2. Find the vertex, focus, axis, latus rectum and directrix of the parabolas :

(*i*) $y^2 + 4x - 6y + 13 = 0$

(*ii*) $x^2 - 4x - 3y + 10 = 0$

Solution. (*i*) Equation of the parabola is

$$y^2 + 4x - 6y + 13 = 0$$

$$\Rightarrow \quad y^2 - 6y + 9 = -4x - 4$$

$$\Rightarrow \quad (y-3)^2 = -4(x+1)$$

To transfer the origin at the point $(-1, 3)$, we put

$$x + 1 = X \text{ and } y - 3 - Y$$

$\therefore$ The equation of the given parabola takes the form $Y^2 = -4X$

Here $\quad 4a = 4 \quad \Rightarrow \quad a = 1$

For vertex $\quad X = 0, Y = 0$

$$\Rightarrow \quad x + 1 = 0, y - 3 = 0$$

$$\Rightarrow \quad x = -1, y = 3$$

$\therefore$ Vertex is $(-1, 3)$

For focus $\quad X = -a, Y = 0$

$$x + 1 = -1, y - 3 = 0$$

$$x = -2, y = 3$$

$\therefore$ Focus is $(-2, 3)$

Axis is $\quad Y = 0 \Rightarrow y - 3 = 0$

$\Rightarrow \quad y = 3$

Latus rectum $= 4a = 4 \times 1 = 4$

Directrix is $\quad X = 1 \Rightarrow x + 1 = 1$

$\Rightarrow \quad x = 0$

(*ii*) Equation of the given parabola is

$$x^2 - 4x - 3y + 10 = 0$$

$$\Rightarrow \quad x^2 - 4x = 3y - 10$$

$$\Rightarrow \quad x^2 - 4x + 4 = 3y - 10 + 4$$

$$\Rightarrow \quad (x-2)^2 = 3y - 6$$

$$\Rightarrow \quad (x-2)^2 = 3(y-2)$$

To transfer the origin to the point (2, 2) we put

$$x - 2 = X \text{ and } Y = y - 2$$

Then the equation of the given parabola becomes

$$X^2 = 3Y$$

Here $\quad 4a = 3 \Rightarrow a = \dfrac{3}{4}$

For vertex $\quad X = 0, Y = 0$

$\Rightarrow \quad x - 2 = 0, y - 2 = 0$

$\Rightarrow \quad x = 2, y = 2$

$\therefore$ Vertex is (2, 2)

For focus $\quad X = 0, Y = a$

$\Rightarrow \quad x - 2 = 0, \; y - 2 = \dfrac{3}{4}$

$\Rightarrow \quad x = 2, \; y = \dfrac{11}{4}$

$\therefore \quad$ focus is $\left(2, \dfrac{11}{4}\right)$

Axes is $\quad X = 0 \Rightarrow x - 2 = 0 \Rightarrow x = 2$

Latus rectum $\quad = 4a = 4 \times \dfrac{3}{4} = 3$

Directrix is $\quad Y = -a$

$\Rightarrow \quad y - 2 = -\dfrac{3}{4} \Rightarrow y = 2 - \dfrac{3}{4}$

$\Rightarrow \quad 4y = 5$

Example 3. Find the point on the parabola $y^2 = 12x$, whose ordinate is three times its abscissa.

Solution. Let the abscissa of the point be x_1. Then its ordinate is $3x_1$. Therefore the point is $(x_1, 3x_1)$.

$\therefore$ The point $(x_1, 3x_1)$ lies on the parabola $y^2 = 12x$.

$$\Rightarrow \quad (3x_1)^2 = 12\,x_1$$

$$\Rightarrow \quad 9x_1^{\,2} = 12x_1$$

$$\Rightarrow \quad 3x_1\,(3x_1 - 4) = 0$$

$$\Rightarrow \quad x_1 = 0, \; x_1 = \frac{4}{3}$$

$\therefore$ The points are (0, 0) and $\left(\frac{4}{3}, 4\right)$

Since the point (0, 0) does not satisfy the given condition, therefore the required point is $\left(\frac{4}{3}, 4\right)$.

Example 4. Find the equation of the parabola if its focus is (0, 0) and vertex is (0, 2).

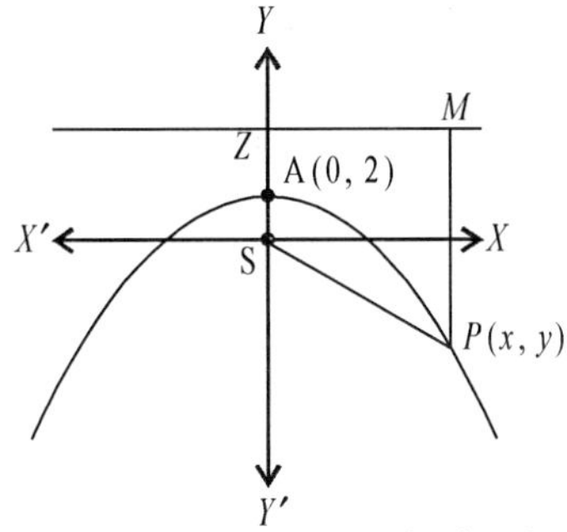

Solution. Let A (0, 2) and S (0, 0) be respectively, the vertex and focus of the parabola. Then its axis is AS. Produce SA to Z such that

$$AZ = SA = 2$$

and draw $ZM \perp SZ$. Then $SZ = SA + AZ = 2 + 2$

and thus, the equation of the directrix ZM is $y = 4$.

Take any point P on the parabola and draw PM perpendicular to the directrix. Then from the definition of the parabola

$$PS = PM$$

$$\Rightarrow \quad \sqrt{(x-0)^2+(y-0)^2} = \frac{y-4}{\sqrt{1^2+0^2}}$$

$$\Rightarrow \quad \sqrt{x^2+y^2} = y-4$$

$$\Rightarrow \quad x^2 + y^2 = (y-4)^2$$

$$\Rightarrow \quad x^2 + y^2 = y^2 - 8y + 16$$

$$\Rightarrow \quad x^2 + 8y - 16 = 0$$

Example 5. Find the equation of the parabola whose focus is (– 1, 1) and directrix in $x + y + 1 = 0$.

Solution. Let $P(x, y)$ be any point on the parabola and $S(-1, 1)$ be its focus. Then

$$PS = \sqrt{(x+1)^2+(y-1)^2}$$

If PM be the perpendicular to the directrix $x + y + 1 = 0$, then

$$PM = \frac{x+y+1}{\sqrt{1^2+1^2}} = \frac{x+y+1}{\sqrt{2}}$$

$\therefore$ From the definition of the parabola,

$$PS = PM$$

$$\Rightarrow \qquad \sqrt{(x+1)^2+(y-1)^2} = \frac{x+y+1}{\sqrt{2}}$$

$$\Rightarrow \qquad (x+1)^2+(y-1)^2 = \frac{(x+y+1)^2}{2}$$

$$\Rightarrow \qquad x^2-2xy+y^2+2x-6y+3 = 0$$

EXERCISE 3.15

1. Find the equation of the parabola with
   (*i*) focus at (5, 0) and directrix $x = -5$
   (*ii*) focus at (– 3, 0) and directrix $x - 3 = 0$
2. Find the focus, directrix and eccentricity of the conic represented by the equation $3y^2 = 8x$.
3. Find the length of latus-rectum, coordinates of focus and the equation of directrix for the following parabolas :
   (*i*) $y^2 = -4x$ (*ii*) $x^2 = -12y$ (*iii*) $3x^2 = 4y$
4. Find the equation of the parabola with vertex at origin and directrix the line $y + 3 = 0$. Also find its focus.
5. Find the vertex, focus, axis, latus rectum and directrix of the parabolas
   (*i*) $(y+3)^2 = 2(x+2)$ (*ii*) $x^2 - 6x - 4y - 11 = 0$
6. The focal distance of a point on the parabola $y^2 = 8ax$ is 4. Find its coordinates.

ANSWERS

1. (*i*) $y^2 = 20x$; 20 (*ii*) $y^2 = -12x$; 12
2. focus $= \left(\frac{2}{3}, 0\right)$, $x + \frac{2}{3} = 0$, $e = 1$
3. (*i*) 4 ; (– 1, 0) ; $x - 1 = 0$ (*ii*) 12 ; (0, – 3) ; $y - 3 = 0$
   (*iii*) $\frac{4}{3}$; $\left(0, \frac{1}{3}\right)$; $3y + 1 = 0$
4. $x^2 = 12y$; (0, 3)
5. (*i*) (– 2, – 3) ; $\left(-\frac{3}{2}, -3\right)$; $y + 3 = 0$, 2 ; $2x + 5 = 0$
   (*ii*) (3, – 5) ; (3, – 4), $x = 3$, 4 ; $y + 6 = 0$
6. (2, 4) , (2, – 4)

3.33. ELLIPSE

An ellipse is a locus of a point which moves in a plane so that its distance from a fixed point *S*, called the focus bears to its perpendicular distance from fixed straight line *ZK*, called the directrix, a ratio always less than unity.

The ratio is called the eccentricity and is denoted by *e*. Thus if *P* is any point on the ellipse then

$$PS = e\,PM, \text{ where } e < 1$$

and $PM \perp ZK$ (the directrix)

3.34. EQUATION OF AN ELLIPSE

Let S be the focus, ZK the directrix and e the eccentricity. Draw $SZ \perp ZK$.
Let the points A and A' divide the line segment SZ in the ratio $e : 1$ internally and externally respectively.

Then
$$\frac{SA}{SZ} = e \text{ and } \frac{SA'}{A'Z} = e \qquad \text{... (1)}$$

By definition the points A and A' lie on the ellipse. Let C be the middle point of AA' and $AA' = 2a$.
From equation (1), we have
$$SA = e\,AZ \qquad \text{...(2)}$$
and
$$SA' = e\,A'Z \qquad \text{...(3)}$$

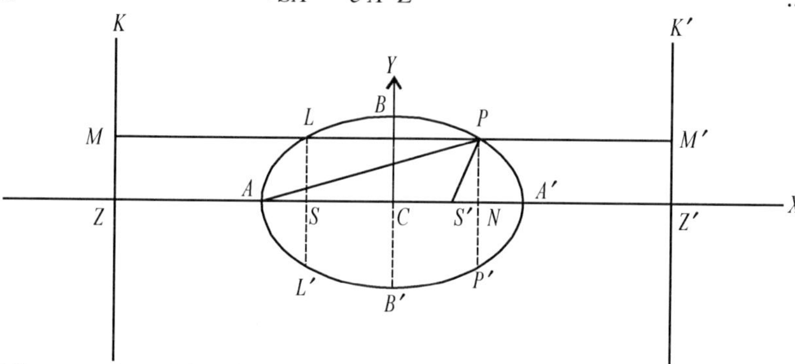

Adding equations (2) and (3)
$$SA + SA' = e(AZ + A'Z)$$
$$\Rightarrow \quad AA' = e\,[(CZ - CA) + (CZ + CA')]$$
$$\Rightarrow \quad 2a = e\,[2cz] \qquad [\because CA' = CA]$$
$$\Rightarrow \quad CZ = \frac{a}{e}$$

Again subtracting equation (2) from (3), we have
$$SA' - SA = e\,(A'Z - AZ)$$
$$\Rightarrow \quad (CA' + CS) - (CA - CS) = e\,AA'$$
$$\Rightarrow \quad 2CS = e.\,2a \qquad [\because CA' = CA]$$
$$\Rightarrow \quad CS = ae$$

Now let C be the origin, CA' the axis of x and a line through C perpendicular to AA' as the axis of y. Then the coordinates of S are $(-ae, 0)$ and the equation of ZM is

$$x = -\frac{a}{e}.$$

Let $P\,(x, y)$ be any point on the ellipse. Join SP and draw PM perpendicular upon the directrix and PN perpendicular upon AA'. Then
$$SP = e\,PM$$
$$\Rightarrow \quad SP = e\,NZ = e\,(CN + CZ)$$
$$\Rightarrow \quad \sqrt{(x+ae)^2 + y^2} = e\left(x + \frac{a}{e}\right) = (ex + a)$$

$\Rightarrow \qquad (x + ae)^2 + y^2 = (ex + a)^2$

$\Rightarrow \qquad x^2 + 2aex + a^2e^2 + y^2 = e^2x^2 + 2aex + a^2$

$\Rightarrow \qquad x^2 + y^2 + a^2e^2 = e^2x^2 + a^2$

$\Rightarrow \qquad x^2 (1 - e^2) + y^2 = a^2 (1 - e^2)$

$$\Rightarrow \qquad \frac{x^2}{a^2} + \frac{y^2}{a^2(1-e^2)} = 1$$

Which is the equation of the ellipse.

Putting $b^2 = a^2 (1 - e^2)$, where $a > b$, the above equation takes the form.

$$\frac{x^2}{a^2} + \frac{y^2}{b^2} = 1$$

Which is the equation of the ellipse in standard form.

Note : Clearly $b = a\sqrt{1-e^2}$ is less than a. It is called semiminor axis because if we put $x = 0$ in the equation

$$\frac{x^2}{a^2} + \frac{y^2}{a^2(1-e^2)} = 1$$

we get $$y = \pm a\sqrt{1-e^2}$$

Therefore, the point $(0, a\sqrt{1-e^2})$ and $(0, -a\sqrt{1-e^2})$ which lies on y-axis, also lie on the ellipse. Let these points be B and B'. Then

$$CB = a\sqrt{1-e^2} \text{ and } CB' = CB$$

The length BB' is called the minor axis and the length CB or CB' is called the semi-minor axis both of which are $a\sqrt{1-e^2}$ or b.

The length AA' which is $2a$ is called the major axis.

Special Feature of the Ellipse

The ellipse represented by the equation

$$\frac{x^2}{a^2} + \frac{y^2}{b^2} = 1, \quad \text{where } a > b$$

(*i*) Has its vertices $(a, 0)$, $(-a, 0)$, $(0, b)$ and $(0, -b)$.

(*ii*) The line segment joining the two vertices on x-axis is the major axis and is of length $2a$.

(*iii*) The line segment joining the two vertices on the y-axis is the major axis which is of length $2b$.

(*iv*) The point $(0, 0)$ is the centre of the ellipse.

(*v*) The eccentricity e is given by

$$b^2 = a^2 (1 - e^2) \text{ or } e = \sqrt{1-\frac{b^2}{a^2}}$$

3.35. ANOTHER FORM OF ELLIPSE

The ellipse $\frac{x^2}{a^2}+\frac{y^2}{b^2}=1$ where $b > a$ is another from of the ellipse in which the major axis lies along the y-axis and the minor axis lies along the x-axis. The slope of the ellipse is shown in the adjoining figure. Here,

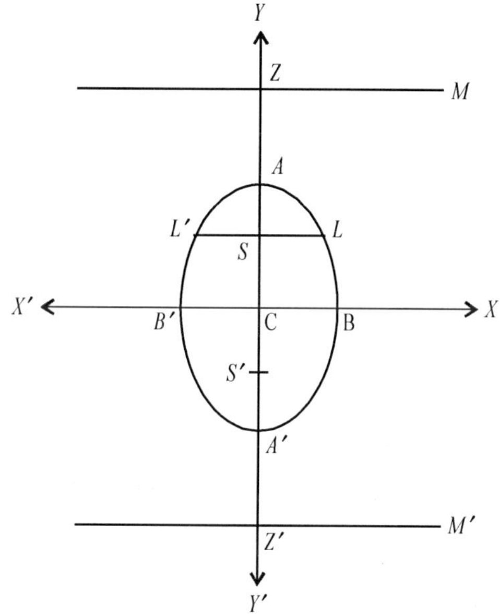

(*i*) AA' is major axis of the ellipse which is along the axis of y. Its length is $2b$ and equation is $x = 0$.

(*ii*) BB' is minor axis. Its length is $2a$ and equation is $y = 0$.

(*iii*) Vertices are A $(0, b)$ and A' $(0, -b)$.

(*iv*) Foci are S $(0, be)$ and S' $(0, -be)$

(*v*) Centre C is $(0, 0)$

(*vi*) For eccentricity $a^2 = b^2(1-e^2)$

(*vii*) The equation of directrices ZM and $Z'M'$ are $y=\frac{b}{e}$ and $y=-\frac{b}{e}$ respectively.

3.36. THE LENGTH OF LATUS RECTUM OF THE ELLIPSE

$\frac{x^2}{a^2}+\frac{y^2}{b^2}=1$ **where $a > b$.**

The length of latus rectum $= \frac{2b^2}{a}$

3.37. USEFUL INFORMATION ABOUT ELLIPSE IN STANDARD FORMS

The standard equation of an ellipse is $\frac{x^2}{a^2}+\frac{y^2}{b^2}=1$ where $a > b$ or $b > a$.

S.No.	Form	$\frac{x^2}{a^2}+\frac{y^2}{b^2}=1$ **where** $a > b$	$\frac{x^2}{a^2}+\frac{y^2}{b^2}=1$ $b > a$
1	Centre	(0, 0)	(0 , 0)
2	Length of major axis	$2a$	$2b$
3	Length of minor axis	$2b$	$2a$
4	Vertices	$(\pm a, 0)$	$(0, \pm b)$
5	Foci	$(\pm ae, 0)$	$(0, \pm be)$
6	e is given by	$b^2 = a^2(1-e^2)$	$a^2 = b^2(1-e^2)$
7	Length of latus rectum	$\frac{2b^2}{a}$	$\frac{2a^2}{b}$
8	Equation of directrices	$x = \pm ae$	$y = \pm be$

3.38. AUXILIARY CIRCLE

The circle which is described on the major axis of an ellipse as diameter is called the auxiliary circle of the ellipse.

Let the ellipse be $\frac{x^2}{a^2}+\frac{y^2}{b^2}=1$, $a > b$

Then the major axis $= 2a$

$\therefore$ Radius of the auxiliary circle $= a$

Centre of the auxiliary circle = centre of the ellipse *i.e.* (0, 0)

Hence the equation of the auxiliary circle is

$$x^2 + y^2 = a^2$$

3.39. PARAMETRIC EQUATION OF THE ELLIPSE $\frac{x^2}{a^2}+\frac{y^2}{b^2}=1$

The equations $x = a\cos\phi$, $y = b\sin\phi$ are parametric equations of the ellipse and ϕ is the parameter.

ILLUSTRATIVE EXAMPLES

Example 1. Find the eccentricity, the coordinates of the foci, length of latus rectum and equation of directrices of the ellipse $25x^2 + 9y^2 = 225$

Solution. The equation of the given ellipse is

$$25x^2 + 9y^2 = 225$$

$$\Rightarrow \quad \frac{x^2}{9}+\frac{y^2}{225} = 1$$

Here $a^2 = 9$ and $b^2 = 225$. Thus $a^2 < b^2$, therefore the major axis lies along the y-axis and minor axis along the x-axis.

The eccentricity e is given by

$$a^2 = b^2 (1 - e^2)$$

$$9 = 25 (1 - e^2)$$

$$\frac{9}{25} = 1 - e^2$$

$$e = \frac{4}{5}$$

The coordinates of the foci are $(0, \pm be)$ *i.e.*

$$\left(0, \pm 5 \times \frac{4}{5}\right) \text{ i.e. } (0, \pm 4)$$

$$\text{Latus rectum} = \frac{2a^2}{b} = \frac{2 \times 9}{5} = \frac{18}{5}$$

Equations of directrices are $y = \pm \frac{b}{e}$

$$y = \pm \frac{5}{\frac{4}{5}}$$

$$y = \pm \frac{25}{4}$$

Example 2. Find the length of the axes, the coordinates of the foci and vertices, eccentricity, length of latus rectum and the equations of directrices of the conic represented by $x^2 + 4y^2 = 100$

Solution. The equation of the given conic is

$$x^2 + 4y^2 = 100$$

i.e.

$$\frac{x^2}{100} + \frac{y^2}{25} = 1 \quad \dots (1)$$

Here $a^2 = 100, b^2 = 25 \Rightarrow a = 10, b = 5$

Length of major axis $= 2a = 2 \times 10 = 20$

Length of minor axis $= 2b = 2 \times 5 = 10$

The coordinates of the vertices are $(a, 0)$ and $(-a, 0)$ *i.e.*

$(10, 0)$ and $(-10, 0)$

$$\text{Length of latus rectum} = \frac{2b^2}{a} = 2 \times \frac{25}{10} = 5$$

Let e be the eccentricity of the ellipse them

$$b^2 = a^2 (1 - e^2)$$

$$25 = 100 (1 - e^2)$$

$$\Rightarrow \qquad e^2 = \frac{3}{4} \quad \Rightarrow \quad e = \frac{\sqrt{3}}{2} \qquad (\because e > 0)$$

$$\therefore \qquad ae = 10 \times \frac{\sqrt{3}}{2} = 5\sqrt{3}$$

$$\frac{a}{e} = \frac{10}{\frac{\sqrt{3}}{2}} = \frac{20}{\sqrt{3}}$$

The coordinates of the foci are $(ae, 0)$ and $(-ae, 0)$ *i.e.* $(5\sqrt{3}, 0)$ and $(-5\sqrt{3}, 0)$

The equations of the directrices are $x - \frac{a}{e} = 0$ and $x + \frac{a}{e} = 0$

$$i.e. \qquad x - \frac{20}{\sqrt{3}} = 0 \text{ and } x + \frac{20}{\sqrt{3}} = 0$$

Example 3. Find the length of the axes, coordinates of the foci, the eccentricity, length of latus rectum of the conic $16x^2 + 9y^2 = 144$

Solution. The equation of the given conic is $16x^2 + 9y^2 = 144$

$$i.e. \qquad \frac{x^2}{9} + \frac{y^2}{16} = 1$$

Here $\quad a^2 = 9, b^2 = 16$

The major axis lies along the y-axis and minor axis along the x-axis

Length of major axis $= 2b = 2 \times 4 = 8$

Length of minor axis $= 2a = 2 \times 3 = 6$

The eccentricity e is given by

$$a^2 = b^2 (1 - e^2)$$

$$9 = 16 (1 - e^2)$$

$$\frac{9}{16} = 1 - e^2$$

$$\Rightarrow \qquad e^2 = 1 - \frac{9}{16} = \frac{7}{16} \Rightarrow e = \frac{\sqrt{7}}{4}$$

The coordinates of the foci are

$$(0, \pm be) \; i.e. \left(0, \pm 4 \times \frac{\sqrt{7}}{4}\right) i.e. \; (0, \pm\sqrt{7})$$

The length of latus rectum $= \dfrac{2a^2}{b} = \dfrac{2 \times 9}{4} = \dfrac{18}{4} = \dfrac{9}{2}$

Example 4. Find the eccentricity of the ellipse

$$4x^2 + 9y^2 - 8x - 36y + 4 = 0$$

Solution. The equation of the given ellipse is

$$4x^2 + 9y^2 - 8x - 36y + 4 = 0$$

$$4(x^2 - 2x) + 9(y^2 - 4y) = -4$$

$$4(x^2-2x+1)+9(y^2-4y+4) = 4+36-4$$
$$4(x-1)^2+9(y-2)^2 = 36$$
$$\frac{(x-1)^2}{9}+\frac{(y-2)^2}{4} = 1$$

Here $a^2 = 9$ and $b^2 = 4$. Clearly $a^2 > b^2$

If e is the eccentricity of the ellipse then

$$b^2 = a^2(1-e^2)$$
$$4 = 9\ (1-e^2)$$
$$\frac{4}{9} = 1-e^2$$

$$\Rightarrow \qquad e^2 = 1-\frac{4}{9} = \frac{5}{9}$$

$$\Rightarrow \qquad e = \frac{\sqrt{5}}{3}$$

Example 5. Find the equation of the ellipse whose centre is origin latus rectum is 5 and eccentricity is $\frac{2}{3}$.

Solution Let the equation of the ellipse be

$$\frac{x^2}{a^2}+\frac{y^2}{b^2} = 1 \qquad \text{... (1)}$$

Where $a > b$, then its latus rectum is $\frac{2b^2}{a}$

$\therefore$ According to equation

$$\frac{2b^2}{a} = 5 \qquad \text{... (2)}$$
$$2b^2 = 5a$$
$$2a^2(1-e^2) = 5a$$
$$2a\left(1-\frac{4}{9}\right) = 5$$
$$2a \times \frac{5}{9} = 5$$
$$a = \frac{9}{2}$$

$\therefore$ From equation (2), we have

$$2b^2 = 5 \times \frac{9}{2}$$
$$b^2 = \frac{45}{4}$$

Putting these values of a and b^2 in equation (1), we get

$$\frac{x^2}{\frac{81}{4}}+\frac{y^2}{\frac{45}{4}} = 1$$

$$\frac{4x^2}{81}+\frac{4y^2}{45} = 1$$

$\Rightarrow$ $$20x^2 + 36y^2 = 405$$

Example 6. Find the equation of the ellipse if its foci are (±2, 0) and the length of latus rectum is $\frac{10}{3}$.

Solution. Let the equation of the ellipse be $\frac{x^2}{a^2}+\frac{y^2}{b^2}=1$ where $a > b$

them its latus rectum is $\frac{2b^2}{a}$

$\therefore$ According to question

$$\frac{2b^2}{a} = \frac{10}{3} \qquad \text{... (1)}$$

$$b^2 = \frac{5}{3}a$$

$\Rightarrow$ $$a^2(1-e^2) = \frac{5}{3}a$$

$\Rightarrow$ $$a^2 - a^2e^2 = \frac{5}{3}a$$

Here $$ae = 2$$

$$a^2-(2)^2 = \frac{5}{3}a$$

$$3a^2 - 5a - 12 = 0$$

$$(a-3)(3a+4) = 0$$

$$a = 3, -\frac{4}{3}$$

$\therefore$ $$a = 3 \qquad (\because a > 0)$$

Now from equation (1)

$$\frac{2b^2}{a} = \frac{10}{3}$$

$$2b^2 = \frac{10\times 3}{3} = 10$$

$$b^2 = 5$$

Putting these values of a and b^2 in the ellipse we get

$$\frac{x^2}{9}+\frac{y^2}{5} = 1$$

$$\Rightarrow \quad 5x^2 + 9y^2 = 45$$

EXERCISE 3.16

1. Find the lengths of axes, coordinates of foci and vertices, eccentricity, length of latus rectum and equation of directrix of the following ellipses :
   (*i*) $x^2 + 16y^2 = 16$ (*ii*) $9x^2 + 16y^2 = 144$
   (*iii*) $\frac{x^2}{169}+\frac{y^2}{25}=1$ (*iv*) $4x^2 + 3y^2 = 36$
2. Find the equation of the ellipse whose eccentricity is $\frac{1}{2}$ and whose foci are $(\pm 2, 0)$
3. Find the equations of ellipse satisfying the following conditions :
   (*i*) vertices at $(\pm 5, 0)$, foci at $(\pm 4, 0)$
   (*ii*) vertices at $(0, \pm 10)$, $e = \frac{4}{5}$
4. If the eccentricity of an ellipse is zero prove that the ellipse is a circle
5. Find the eccentricity of the ellipse $9x^2 + 16y^2 = 144$
6. Find the length of latus rectum of the ellipse $5x^2 + 4y^2 = 1$
7. Find the equation of the ellipse whose foci are (4, 0) and (– 4, 0) and eccentricity is $\frac{1}{3}$

ANSWERS

1. (*i*) 8, 2 ; $(\pm\sqrt{15}, 0)$, $(\pm 4, 0)$; $\frac{\sqrt{15}}{4}$; $\frac{1}{2}$; $x - +\frac{16}{\sqrt{15}}$
   (*ii*) 8, 6 ; $(+\sqrt{7}, 0), (\pm 4, 0)$; $\frac{\sqrt{7}}{4}$;$\frac{9}{2}$; $x = \pm\frac{16}{\sqrt{7}}$
   (*iii*) 26, 10 ; $(\pm 12, 0), (\pm 13, 0)$; $\frac{12}{13}$; $\frac{50}{13}$; $x = \pm\frac{169}{12}$
   (*iv*) $4\sqrt{3}, 6$; $(0, \pm\sqrt{3})$, $(0, \pm 2\sqrt{3})$; $\frac{1}{2}$; $3\sqrt{3}$; $y = \pm\sqrt{3}$
2. $3x^2 + 4y^2 = 48$
3. (*i*) $9x^2 + 25y^2 = 225$ (*ii*) $25x^2 + 9y^2 = 900$
5. $\frac{\sqrt{7}}{4}$ 6. $\frac{4}{5}$ 7. $8x^2 + 9y^2 - 1152$

3.40. HYPERBOLA

A hyperbola is a locus of a point which moves in a plane such that its distance from a fixed point S, called the focus, bears to its perpendicular distance from a fixed straight line ZK, called the directrix, a ratio always greater than unity. The ratio is called the eccentricity and is denoted by e.

Thus, if P is any point on the hyperbola, then

$$PS = e\,PM$$

Where $e > 1$, and $PM \perp ZK$ (directrix).

3.41. EQUATION OF A HYPERBOLA IN STANDARD FORM

Let S be the focus and ZK be the directrix. Draw SZ perpendicular to the directrix.

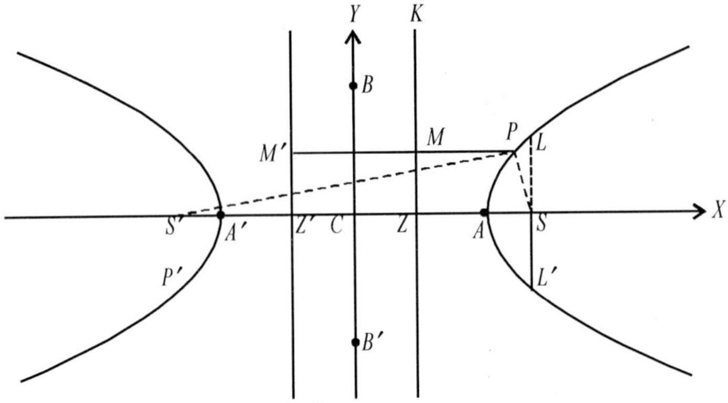

There will be a point A on SZ such that

$$SA = e\,AZ \qquad ...(1)$$

Since $e > 1$ there will be another point A', on SZ produced such that

$$SA' = e\,A'Z \qquad ...(2)$$

Let $AA' = 2a$ and C be the middle point of AA'. Then

$$CA = CA' = a$$

Subtracting equation (1) from (2), we have

$$SA' - SA = e(A'Z - AZ)$$

$$\Rightarrow \quad AA' = e(A'Z - AZ)$$

$$2a = e\,[CA' + CZ] - e\,[CA - CZ]$$

$$2a = e\,.\,2CZ$$

$$\Rightarrow \quad CZ = \frac{a}{e}$$

Adding equation (1) and (2), we have

$$SA + SA' = e(AZ + A'Z)$$

$$\Rightarrow \quad (CS - CA) + (CS + CA') = e\,AA'$$

$$\Rightarrow \quad 2CS = e\,.\,2a \qquad (\because CA' = CA)$$

$$\Rightarrow \quad CS = ae$$

Let C be the origin, CAX as axis of x and a line through C perpendicular to CAX as axis of Y.

Let P (x, y) be any point on the hyperbola. Join PS and draw PM, perpendicular upon the directrix ZK and PN perpendicular to AA'.
The focus S is $(ae, 0)$. The equation of the directrix ZK is

$$x = \frac{a}{e}$$

Now, form the definition of the hyperbola
$PS = e\,PM = e\,NZ = e(CN - CZ)$

$$\Rightarrow \qquad \sqrt{(x-ae^2)+y^2} = e\left(x - \frac{a}{e}\right)$$

$$\Rightarrow \qquad (x - ae)^2 + y^2 = (ex - a)^2$$

$$\Rightarrow \qquad x^2 - 2\,aex + a^2\,e^2 + y^2 = e^2x^2 - 2aex + a^2$$

$$\Rightarrow \qquad x^2 + a^2e^2 + y^2 = e^2x^2 + a^2$$

$$\Rightarrow \qquad x^2\,(e^2 - 1) - y^2 = a^2\,(e^2 - 1)$$

$$\Rightarrow \qquad \frac{x^2}{a^2} - \frac{y^2}{a^2(e^2-1)} = 1$$

Putting $b^2 = a^2\,(e^2 - 1)$ above equation takes the form

$$\frac{x^2}{a^2} - \frac{y^2}{b^2} = 1$$

Which is the required equation.
Note 1. The curve meets the axis of y *i.e.* $x = 0$ at points given by

$$\frac{0}{a^2} - \frac{y^2}{b^2} = 1$$

$$\Rightarrow \qquad y = \pm\sqrt{-b^2} = \text{imaginary}$$

Therefore the hyperbola does not meet the axis of y in real points.
2. If $b = a$ then the standard equation of the hyperbola

$$\frac{x^2}{a^2} - \frac{y^2}{b^2} = 1$$

becomes $x^2 - y^2 = a^2$
This is called a rectangular or equilateral hyperbola. In rectangular hyperbola both the axes are of equal length.

3.42. SOME IMPORTANT TERMS OF HYPERBOLA

(*i*) **Vertices :** The point A and A' are called the vertices of the hyperbola and their coordinates are $(a, 0)$ and $(-a, 0)$ respectively.

(*ii*) **Centre :** The point C (0, 0) is called the centre of the hyperbola.

(*iii*) **Transverse Axis :** AA' is the transverse axis of the hyperbola. Its length is $2a$ and equation is $y = 0$.

(*iv*) **Conjugate axis :** BB' is the conjugate axis of the hyperbola, where B and B' are on y-axis such that $BC = CB' = b$. Thus the length of the conjugate axis is $2b$ and its equation is $x = 0$.

(*v*) **Foci :** S and S' are foci of the hyperbola and their coordinates are $(ae, 0)$ and $(-ae, 0)$ respectively.

(*vi*) **Directrices :** ZM and $Z'M'$ are directrices and their equations are $x = \frac{a}{e}$ and $x = -\frac{a}{e}$ respectively

(*vii*) **Eccentricity :** For the hyperbola $\frac{x^2}{a^2} - \frac{y^2}{b^2} = 1$, the eccentricity is given by

$$b^2 = a^2(e^2 - 1)$$

$$\Rightarrow \qquad \frac{b^2}{a^2} = e^2 - 1$$

$$\Rightarrow \qquad e^2 = \frac{b^2}{a^2} + 1 = \frac{b^2 + a^2}{a^2}$$

$$\Rightarrow \qquad e = \frac{\sqrt{a^2 + b^2}}{a}$$

For the rectangular hyperbola for which $a = b$, we have

$$e = \frac{\sqrt{a^2 + a^2}}{a} = \sqrt{2}$$

The length of latus-rectum of the hyperbola $\frac{x^2}{a^2} - \frac{y^2}{b^2} = 1$

The length of latus rectum $= LSL'$

$$= 2\,LS = 2\,l$$

$$= \frac{2b^2}{a}$$

Note : The coordinates of L and L' are $\left(ae, \frac{b^2}{a}\right)$ and $\left(ae, \frac{-b^2}{a}\right)$ respectively.

The difference of focal distances of any point on a hyperbola

Let the equation of the hyperbola be

$$\frac{x^2}{a^2} - \frac{y^2}{b^2} = 1$$

Then its transverse axis $= 2a$

The difference of focal distances of any point on the hyperbola is constant and is equal to the length of the transverse axis.

3.43. THE PARABOLIC EQUATION OF THE HYPERBOLA

Let the equation of the hyperbola be

$$\frac{x^2}{a^2} - \frac{y^2}{b^2} = 1$$

It is satisfied by $x = a \sec \phi$, $y = b \tan \phi$ for all values of ϕ. Therefore ϕ is a parameter and the parametric equation of the hyperbola are $x = a \sec \phi$ and $y = b \tan \phi$.

ILLUSTRATIVE EXAMPLES

Example 1. Find the length of the axes, eccentricity, the coordinates of foci and latus rectum of the hyperbola $4x^2 - 9y^2 = 36$.

Solution. The given equation can be written as

$$\frac{x^2}{9} - \frac{y^2}{4} = 1$$

Here $a^2 = 9$ and $b^2 = 4$

Therefore $a = 3$ and $b = 2$

The transverse axis $= 2a = 6.$

and the conjugate axis $= 2b = 4$

The eccentricity e is given by $b^2 = a^2 (e^2 - 1)$

$$\Rightarrow \quad e^2 = \frac{a^2 + b^2}{a^2} = \frac{9+4}{3}$$

$$\therefore \quad e = \frac{\sqrt{13}}{3}$$

For foci, we have

$$x = \pm ae, y = 0$$

$$\Rightarrow \quad x = \pm 3 \times \frac{\sqrt{13}}{3}, y = 0$$

$$\Rightarrow \quad x = \pm\sqrt{13}, y = 0$$

$\therefore$ The foci are $(\pm\sqrt{13}, 0)$

$$\text{Latus rectum} = \frac{2b^2}{a} = \frac{2 \times 4}{3} = \frac{8}{3}$$

Example 2. Find the eccentricity of the hyperbola $4x^2 - 16y^2 = 9$ and also its length of latus rectum.

Solution. The given equation is

$$4x^2 - 16y^2 = 9$$

$$\frac{4x^2}{9} - \frac{16y^2}{9} = 1$$

$$\frac{x^2}{\frac{9}{4}} - \frac{y^2}{\frac{9}{16}} = 1$$

Here $a^2 = \frac{9}{4}$ and $b^2 = \frac{9}{16}$

The eccentricity e is given by

$$b^2 = a^2 (e^2 - 1)$$

$$\frac{9}{16} = \frac{9}{4}(e^2 - 1)$$

$$\frac{1}{4} = e^2 - 1$$

$$e^2 = 1 + \frac{1}{4} = \frac{5}{4}$$

$$e = \frac{\sqrt{5}}{2}$$

The length of latus rectum $= \dfrac{2b^2}{a} = \dfrac{2 \times \frac{9}{16}}{\frac{3}{2}}$

$$= \frac{3}{4}$$

Example 3. Find the equation of the hyperbola whose conjugate axis is 5 and the distance between the foci is 13.

Solution. Let the equation of the hyperbola be

$$\frac{x^2}{a^2} - \frac{y^2}{b^2} = 1 \qquad \dots (1)$$

Then we have $2b = 5 \Rightarrow b = \dfrac{5}{2}$

and $2ae = 13 \Rightarrow ae = \dfrac{13}{2}$

$\therefore$ $b^2 = a^2e^2 - a^2$

$\Rightarrow$ $\left(\dfrac{5}{2}\right)^2 = \left(\dfrac{13}{2}\right)^2 - a^2$

$\Rightarrow$ $a^2 = \dfrac{169}{4} - \dfrac{25}{4} = \dfrac{144}{4}$

$\Rightarrow$ $a^2 = 36$

Putting the value of a^2 and b^2 in (1) we get

$$\frac{x^2}{36} - \frac{y^2}{\frac{24}{4}} = 1$$

$\Rightarrow$ $\dfrac{x^2}{36} - \dfrac{4y^2}{25} = 1$

$\Rightarrow$ $25x^2 - 144y^2 = 900$

Example 4. Find the equation of the hyperbola whose directrix is $2x+y=1$, focus is (1, 1) and eccentricity is $\sqrt{3}$.

Solution. Let $P(x, y)$ be any point on the hyperbola, $S(1, 1)$ its focus and PM be perpendicular to the directrix

$$2x+y = 1$$

Then, we have $PS = e\, PM$

$$\Rightarrow \quad \sqrt{(x-1)^2+(y-1)^2} = \sqrt{3}\left(\frac{2x+y-1}{\sqrt{4+1}}\right)$$

$$\Rightarrow \quad \sqrt{5}\left[\sqrt{(x-1)^2+(y-1)^2}\right] = \sqrt{3}(2x+y-1)$$

$$\Rightarrow \quad 5x^2+5y^2-10x-10y+10 = 3\,(4x^2+y^2+4xy-4x-2y+1)$$

$$\Rightarrow 7x^2-2y^2+12xy-2x+4y-7=0$$

Example 5. If the eccentricities of the hyperbola $\frac{x^2}{a^2}-\frac{y^2}{b^2}=1$ and $\frac{x^2}{a^2}-\frac{y^2}{b^2}=-1$ are e_1 and e_2 respectively, prove that $\frac{1}{e_1^2}+\frac{1}{e_2^2}=1$

Solution. For the hyperbola $\frac{x^2}{a^2}-\frac{y^2}{b^2}=1$, we have

$$b^2 = a^2\,(e_1^2-1)$$

$$\Rightarrow \quad \frac{b^2}{a^2} = e_1^2-1$$

$$\Rightarrow \quad e_1^2 = \frac{b^2}{a^2}+1$$

$$\Rightarrow \quad e_1^2 = \frac{b^2+a^2}{a^2}$$

$$\Rightarrow \quad \frac{1}{e_1^2} = \frac{a^2}{a^2+b^2} \qquad \ldots (1)$$

Again for the hyperbola $\frac{x^2}{a^2}-\frac{y^2}{b^2}=-1$

$$\Rightarrow \quad \frac{y^2}{b^2}-\frac{x^2}{a^2} = 1,$$

we have

$$a^2 = b^2\,(e_2^2-1)$$

$$\Rightarrow \quad \frac{a^2}{b^2}+1 = e_2^2$$

$$\Rightarrow \qquad e_2^2 = \frac{a^2+b^2}{b^2}$$

$$\Rightarrow \qquad \frac{1}{e_2^2} = \frac{b^2}{a^2+b^2} \qquad ...(2)$$

Adding equations (1) and (2), we get

$$\frac{1}{e_1^2}+\frac{1}{e_2^2} = \frac{a^2}{a^2+b^2}+\frac{b^2}{a^2+b^2} = 1$$

$$\Rightarrow \qquad \frac{1}{e_1^2}+\frac{1}{e_2^2} = 1$$

EXERCISE 3.17

1. Find the lengths of axis, coordinates of vertices and foci, eccentricity, length of latus rectum and the equation of the directrices of the following hyperbolas :
   (*i*) $4x^2 - 25y^2 = 100$ (*ii*) $25x^2 - 9y^2 = 225$
2. Find the equation of the hyperbola whose eccentricity is $\frac{3}{2}$ and whose foci are $(\pm 2, 0)$.
3. Find the equation of the hyperbola with vertices at $(0, \pm 6)$ and $e = \frac{5}{3}$. Also find its foci.
4. Find the eccentricity, equation of directrices and latus rectum of the hyperbola $3x^2 - y^2 = 4$
5. Find the equation of the hyperbola whose transverse axis is 8 and foci are $(\pm 5, 0)$
6. Find the equation of the hyperbola whose directrix is $x + 2y + 1 = 0$, coordinates of whose focus are (2, 3) and eccentricity is $\sqrt{2}$.
7. Find the equation of the hyperbola whose foci are (6, 4) and (– 4, 4) and eccentricity is 2.

ANSWERS

1. (*i*) 10, 4 ; $(\pm 5, 0)$, $(\pm\sqrt{29}, 0)$; $\frac{\sqrt{29}}{5}$; $\frac{8}{5}$; $x = \pm\frac{25}{\sqrt{29}}$

   (*ii*) 6, 10 ; $(\pm 3, 0)$, $(\pm\sqrt{34}, 0)$; $\frac{\sqrt{34}}{3}$; $\frac{50}{3}$; $x = \pm\frac{9}{\sqrt{34}}$
2. $45x^2 - 36y^2 = 80$
3. $16x^2 - 9y^2 = 576$; $(0, \pm 10)$
4. 2 : $x\sqrt{3} \pm 1 = 0$; $4\sqrt{3}$
5. $9x^2 - 16y^2 = 144$
6. $3x^2 - 8xy - 3y^2 - 24x - 38y + 63 = 0$
7. $12x^2 - 4y^2 - 24x + 32y - 127 = 0$

UNIT – IV

Probability and Distributions

4.1 DEFINITIONS AND NOTATIONS

Experiment and Trial

Any operation that results in two or more outcomes is called an experiment and performing of an experiment is called trial.

Random Experiment

A random experiment is defined as an experiment in which all possible outcomes are known and which can be repeated under identical conditions but it is not possible to predict the outcome of any particular trial in advance. *e.g.*, Tossing a coin or throwing a die.

Sample Space

The possible outcomes of a random experiment are called the basic outcomes, and the set of all basic outcomes is called the sample space. The sample space is generally denoted by the letter S.

e.g., In throwing a fair die, sample space is $S=\{1,\ 2,\ 3,\ 4,\ 5,\ 6\}$. In tossing of two unbiased coins, sample space is $S=\{HH,\ HT,\ TH,\ TT\}$.

Event

Any subset of the sample space is defined as an event. An event which is sure to occur is called a certain event. Also an event is called an impossible event if it can never occur.

For example, In throwing a fair die, the event getting a number less than 7 is a certain event and the event getting a number 7 is an impossible event.

Complement of an Event

Let A be an event in the sample space S. The set of basic outcomes of a random experiment belonging to S but not to A is called the complement of A and is denoted by $\bar{A}$. The probability that event A will not occur is denoted by $P(\bar{A})$.

Union

The union of two events A and B is the event that consists of all the elements (outcomes) contained in A, in B, or in both. The union of A and B is denoted in set theory notation as $A \cup B$.

The union of three events A and B and C would be denoted as $A \cup B \cup C$.

Note that the symbol $\cup$ is read as "OR".

Intersection

The intersection of two events A and B is the event that consists of all the elements (outcomes) contained in both A and B. The intersection of A and B is denoted symbolically as $A \cap B$.

Note that the symbol $\cap$ is read as "AND".

Exhaustive Events

The total numbers of all possible outcomes in any trial are known as exhaustive events. *e.g.*, In tossing of a coin, there are two exhaustive events, head and tail.

Dependent and Independent Events

If the occurrence of event A changes the probability of event B, then events A and B are dependent. On the other hand, if the occurrence of event A does not change the probability of Event B, then events A and B are independent. *e.g.*, In case a card is drawn from a pack of well shuffled cards and is not replaced, then the second draw of the card is dependent on the first draw. However, if the first card drawn is replaced before drawing the second card, the result of the second draw is independent of the first draw.

Mutually Exclusive or Disjoint Events

Two events are mutually exclusive or disjoint if they cannot occur at the same time. *e.g.*, In tossing of a coin, the events head and tail are mutually exclusive, since if the outcome is head, the possibility of getting tail in the same trial is ruled out

4.2 DEFINITION OF PROBABILITY

The probability of event A is the number of ways event A can occur divided by the total number of possible outcomes, *i.e.*,

$$P(A) = \frac{\text{The number of ways event } A \text{ can occur}}{\text{The total number of possible outcomes}}$$

Probability Postulates

1. For any event A in the sample space S, $0 \le P(A) \le 1$.
2. Let A denote an event and let O_i denote the basic outcomes. Then $P(A) = \sum_A P(O_i)$.
3. $P(S) = 1$ for any sample space S.

4.3 CONDITIONAL PROBABILITY

The conditional probability of an event, such as *A*, occurring given that another event, such as *B*, has occurred is called the conditional probability. The conditional probability of event *A*, given event *B*, is denoted by *P*(*A*/*B*). Conditional probabilities are based on knowledge of one of the variables. The conditional probability of an event, such as *A*, occurring given that another event, such as *B*, has occurred is expressed as:

$$P(A/B) = \frac{P(A \cap B)}{P(B)}, \quad P(B) \neq 0$$

Similarly, the conditional probability of an event, such as *B*, occurring given that another event, such as *A*, has occurred is expressed as:

$$P(B/A) = \frac{P(A \cap B)}{P(A)}, \quad P(A) \neq 0$$

Note: The probability that events *A* and *B* both occur is the probability of the intersection of *A* and *B*. The probability of the intersection of events *A* and *B* is denoted by $P(A \cap B)$. If events *A* and *B* are mutually exclusive, then $P(A \cap B) = 0$.

The probability that events *A* or *B* occur is the probability of the union of *A* and *B*. The probability of the union of events *A* and *B* is denoted by $P(A \cup B)$.

4.4 PROBABILITY RULES

Complement Rule

Let *A* be an event and $\bar{A}$ its complement, then

$$P(\bar{A}) = 1 - P(A)$$

Addition Rule of Probability

If *A* and *B* be two events associated with a random experiment, then:

$$P(A \cup B) = P(A) + P(B) - P(A \cap B)$$

or

$$P(A \text{ or } B) = P(A) + P(B) - P(A \text{ and } B)$$

Notice that, the equation, $P(A \text{ or } B) = P(A) + P(B) - P(A \text{ and } B)$ contains special events: An event (*A* and *B*) which is the intersection of set/events *A* and *B*, and another event (*A* or *B*) which is the union (*i.e.*, either/or) of sets *A* and *B*.

The above rule is known also as the Inclusion-Exclusion Formula. It can be extended to more than two events. If *A*, *B* and *C* are three events associated with a random experiment, then

$$P(A \cup B \cup C) = P(A) + P(B) + P(C) - P(A \cap B) - P(B \cap C) - P(A \cap C) + P(A \cap B \cap C)$$

Special Case of Addition Rule

1. If *A* and *B* are mutually exclusive events, then $P(A \cap B) = 0$, therefore,

$$P(A \cup B) = P(A) + P(B)$$

2. If A, B and C are mutually exclusive events, then,

$$P(A \cap B) = P(B \cap C) = P(A \cap C) = P(A \cap B \cap C) = 0$$

therefore, $P(A \cup B \cup C) = P(A) + P(B) + P(C)$

Multiplication Rule of Probability

If A and B be two events associated with a random experiment, then

$$P(A \cap B) = P(A)\ P(B/A) \text{ if } P(A) \neq 0$$

or

$$P(A \cap B) = P(B)\ P(A/B) \text{ if } P(B) \neq 0$$

where $P(A/B)$ and $P(B/A)$ are conditional probabilities.

Special Case of Multiplicative Rule

1. If A and B be two independent events associated with a random experiment, then

$$P(A \cap B) = P(A) \times P(B)$$

2. If $A_1, A_2, \ldots\ldots. A_n$ are independent events associated with a random experiment, then

$$P(A_1 \cap A_2 \ldots \cap A_n) = P(A_1) \times P(A_2) \ldots \times P(A_n)$$

ILLUSTRATIVE EXAMPLES

Example 1. Find the probability of getting a head in throw of a coin.

Solution. Here, $S = \{H, T\}$

Event of getting tail $E = \{H\}$

Clearly, $n(E) = 1$ and $n(S) = 2$

Probability of getting a tail is given by

$$P(E) = \frac{n(E)}{n(S)} = \frac{1}{2}$$

Example 2. Three coins are tossed, find the probability of getting at least two heads.

Solution. Here, $S = \{HHH, HHT, HTH, THH, THT, TTH, HTT, TTT\}$

If E is the required event, then

$$E = \{HHH, HHT, HTH, THH\}$$

Clearly, $n(E) = 4$ and $n(S) = 8$

$$P(E) = \frac{n(E)}{n(S)} = \frac{4}{8} = \frac{1}{2}$$

Example 3. If there are two children in a family, find the probability that there is atleast one boy in the family.

Solution. Here, $S = \{BB, BG, GB, GG\}$ where B and G stands for 'Boy' and 'Girl' respectively.

If E is the required event, then $E = \{BB, BG, GB\}$

Clearly, $n(E) = 3$ and $n(S) = 4$

$$P(E) = \frac{n(E)}{n(S)} = \frac{3}{4}$$

Example 4. What is the probability of getting an odd number in the throw of an unbiased die?

Solution. Here, $S = \{1, 2, 3, 4, 5, 6\}$

If E is the required event, then $E = \{1, 2, 3\}$

Clearly, $n(E) = 3$ and $n(S) = 6$

$$P(E) = \frac{n(E)}{n(S)} = \frac{3}{6} = \frac{1}{2}$$

Example 5. What is the chance that leap-year, selected at random, will contain 53 Fridays?

Solution. There are 366 days in a leap year and we can write $366 = (7 \times 52) + 2$. It means that the leap year will contain at least 52 Fridays. The possible combinations for the remaining two days can be as follows:

(*i*) Sunday, Monday
(*ii*) Monday, Tuesday
(*iii*) Tuesday, Wednesday
(*iv*) Wednesday, Thursday
(*v*) Thursday, Friday
(*vi*) Friday, Saturday
(*vii*) Saturday, Sunday

Of these seven likely cases only (*v*) and (*vi*) are favourable.

Hence, the required probability $= \frac{2}{7}$

Example 6. A bag contains 7 red, 12 white and 4 green balls. What is the probability that

(*i*) 3 balls drawn are all white
(*ii*) 3 balls drawn are one of each colour?

Solution. Here, the total balls are $= 7 + 12 + 4 = 23$

3 balls out of these 23 balls can be drawn in ${}^{23}C_3 = \frac{23 \times 22 \times 21}{3 \times 2 \times 1} = 1771$ ways.

The sample space for this experiment contains 1771 sample points, *i.e.*, $n(S) = 1771$.

(*i*) Let E_1 be the event that the 3 balls drawn are all white. Now 3 white balls can be drawn from 12 white balls in $^{12}C_3 = \dfrac{12\times11\times10}{3\times2\times1} = 220$ ways.

So,
$$n(E_1) = 220$$
$$P(E_1) = \frac{n(E_1)}{n(S)} = \frac{220}{1771}$$

(*ii*) Let E_2 be the event that the 3 balls drawn are of each colour. Now 1 red ball can be drawn from 7 red balls in $^7C_1 = 7$ ways, 1 white ball can be drawn from 12 white balls in $^{12}C_1 = 12$ ways and 1 green ball can be drawn from 4 green balls in $^4C_1 = 4$ ways.

3 balls one of each colour can be drawn in $7\times12\times4 = 336$ ways.

So,
$$n(E_2) = 336$$
$$P(E_2) = \frac{n(E_2)}{n(S)} = \frac{336}{1771}$$

Example 7. *A* and *B* are mutually exclusive events of an experiment. If $P(\text{'not } A\text{'}) = 0.65$, $P(A\cup B) = 0.65$ and $P(B) = p$, find the value of *p*.

Solution. By addition theorem for mutually exclusive events, we have

$$P(A\cup B) = P(A) + P(B)$$
$$P(A\cup B) = 1 - P(\text{'not } A\text{'}) + P(B) \qquad [\because P(A) = 1 - P(\bar{A})]$$
$$0.65 = 1 - 0.65 + p$$
$$p = 0.30$$

Example 8. The probability that at least one of the events *A* and *B* occurs is 0.6. If *A* and *B* occur simultaneously with probability 0.2, then find $P(\bar{A}) + P(\bar{B})$.

Solution. Here, $P(A\cup B) = 0.6$ and $P(A\cap B) = 0.2$

Now,
$$P(A\cup B) = P(A) + P(B) - P(A\cap B)$$
$$0.6 = P(A) + P(B) - 0.2$$
$$0.6 = 1 - P(\bar{A}) + 1 - P(\bar{B}) - 0.2 = 1.8 - \left[P(\bar{A}) + P(\bar{B})\right]$$
$$P(\bar{A}) + P(\bar{B}) = 1.8 - 0.6 = 1.2$$

Example 9. *A*, *B* and *C* are three mutually exclusive and exhaustive events associated with a random experiment. Find *P*(*A*), it is given that $P(B) = \dfrac{3}{2}P(A)$ and $P(C) = \dfrac{1}{2}P(B)$.

Solution. Let $P(A) = p$, then $P(B) = \frac{3}{2}P(A)$ or $P(B) = \frac{3}{2}p$

and $P(C) = \frac{1}{2}P(B)$ or $P(C) = \frac{3}{4}p$

Since *A*, *B* and *C* are mutually exclusive and exhaustive events associated with a random experiment, therefore, $A \cup B \cup C = S$.

Now, $P(A \cup B \cup C) = P(S)$ or $P(A \cup B \cup C) = 1$ $[\because P(S) = 1]$

$$P(A) + P(B) + P(C) = 1$$

$$p + \frac{3}{2}p + \frac{3}{4}p = 1 \text{ or } p = \frac{4}{13}.$$

Example 10. *A* card is drawn from a pack of 52 cards. Find the probability of getting a king or heart or a red card.

Solution. Consider the following events:

A = getting a king, B= getting a heart and C = getting a red card.

We have $P(A) = \frac{^4C_1}{^{52}C_1} = \frac{4}{52}$, $P(B) = \frac{^{13}C_1}{^{52}C_1} = \frac{13}{52}$ and $P(C) = \frac{^{26}C_1}{^{52}C_1} = \frac{26}{52}$

$$P(A \cap B) = P(\text{getting a king of heart}) = \frac{1}{52},$$

$$P(B \cap C) = P(\text{getting a heart}) = \frac{13}{52}$$

$$P(C \cap A) = P(\text{getting a red king}) = \frac{2}{52}$$

$$P(A \cap B \cap C) = P(\text{getting a king of heart}) = \frac{1}{52}$$

$$P(\text{getting a king or heart or red}) = P(A \cup B \cup C)$$

$$= P(A) + P(B) + P(C) - P(A \cap B) - P(B \cap C) - P(A \cap C) + P(A \cap B \cap C)$$

$$= \frac{4}{52} + \frac{13}{52} + \frac{26}{52} - \frac{1}{52} - \frac{13}{52} - \frac{2}{52} + \frac{1}{52} = \frac{28}{52} = \frac{7}{13}$$

Example 11. If *A* and *B* are two events such that $P(A) = 0.5$, $P(B) = 0.6$ and $P(A \cup B) = 0.8$, find $P(A/B)$ and $P(B/A)$.

Solution. We have,

$$P(A \cup B) = P(A) + P(B) - P(A \cap B)$$

$$P(A \cap B) = P(A) + P(B) - P(A \cup B) = 0.5 + 0.6 - 0.8 = 0.3$$

Now, $$P(A/B) = \frac{P(A \cap B)}{P(B)} = \frac{0.3}{0.6} = 0.5$$

and $$P(B/A) = \frac{P(A \cap B)}{P(A)} = \frac{0.3}{0.5} = 0.6$$

Example 12. *A* jar contains black and white marbles. Two marbles are chosen without replacement. The probability of selecting a black marble and then a white marble is 0.34, and the probability of selecting a black marble on the first draw is 0.47. What is the probability of selecting a white marble on the second draw, given that the first marble drawn was black ?

Solution. Consider the following events :

$$A = \text{Selecting a black marble}$$
$$B = \text{Selecting a white marble}$$

Here, $P(A) = 0.47$ and $P(A \cap B) = 0.34$

$$P(B/A) = \frac{P(A \cap B)}{P(A)} = \frac{0.34}{0.47} = 0.72$$

Example 13. Consider an experiment throwing a pair of dice. Let *A* and *B* be the events given by *A* = the sum of points is 8, *B* = there is an even number on first die. Find $P(A/B)$ and $P(B/A)$.

Solution. We have, $A = \{(2,6),(3,5),(4,4),(5,3),(6,2)\}$

and $B = \{(2,1),\ldots,(2,6),(4,1),\ldots,(4,6),(6,1),\ldots,(6,6)\}$

$$P(A) = \frac{5}{36} \text{ and } P(B) = \frac{18}{36}$$

Now $P(A/B)$ = Probability of occurrence of *A* when *B* has occurred

= Probability of getting 8 as the sum when there is an even number on the first die

$$= \frac{n(A \cap B)}{n(B)} = \frac{3}{18} = \frac{1}{6}$$

and $P(B/A)$ = Probability of occurrence of *B* when *A* has occurred

= Probability of getting an even number on the first die when the sum of the numbers on two dice is 8

$$= \frac{n(A \cap B)}{n(A)} = \frac{3}{5}$$

Example 14. *A* bag contains 10 white and 15 black balls. Two balls are drawn in succession without replacement. What is the probability that first is white and second is black?

Solution. Consider the following events:

$$A = \text{Getting a white ball in first draw}$$
$$B = \text{Getting a black ball in second draw}$$

Required probability = Probability of getting a white ball in first draw and black ball in second draw

$$= P(A \text{ and } B) = P(A \cap B) = P(A)P(B/A)$$

Now, $P(A) = \frac{{}^{10}C_1}{{}^{25}C_1} = \frac{10}{25} = \frac{2}{5}$

and $P(B/A)$ = Probability of getting a black ball in second draw when a white ball has already been selected in first draw

$$= \frac{{}^{10}C_1}{{}^{24}C_1} = \frac{10}{24} = \frac{5}{8}$$

Required probability $= P(A \cap B) = P(A)P(B/A)$

$$= \frac{2}{5} \times \frac{5}{8} = \frac{1}{4}$$

Example 15. A coin is tossed twice and the four possible outcomes are assumed to be equally likely. If A is the event, 'both head and tail have appeared', and B be the event, 'at most one tail is observed', find $P(A)$, $P(B)$, $P(A/B)$ and $P(B/A)$.

Solution. Here $S = \{HH, HT, TH, TT\}$, $A = \{HT, TH\}$ and $B = \{HH, HT, TH\}$

$$A \cap B = \{HT, TH\}$$

Now, $P(A) = \frac{n(A)}{n(S)} = \frac{2}{4} = \frac{1}{2}$

$$P(B) = \frac{n(B)}{n(S)} = \frac{3}{4}$$

and $P(A \cap B) = \frac{n(A \cap B)}{n(S)} = \frac{2}{4} = \frac{1}{2}$

$$P(A/B) = \frac{P(A \cap B)}{P(B)} = \frac{2}{3}$$

$$P(B/A) = \frac{P(A \cap B)}{P(A)} = 1$$

Example 16. Events A and B are independent. Find $P(B)$ if $P(A) = 0.35$ and $P(A \cup B) = 0.6$.

Solution. We have

$$P(A \cup B) = P(A) + P(B) - P(A \cap B)$$

$$P(A \cup B) = P(A) + P(B) - P(A)P(B)$$

($\because$ A and B are independent)

$$= P(A) + P(B)\ [1 - P(A)]$$

$$0.6 = 0.35 + P(B)\ [1 - 0.35]$$

$$P(B) = \frac{0.25}{0.65} = \frac{5}{13}$$

Example 17. *X* can solve 90% of the problems given in a book and *Y* can solve 70%. What is the probability that at least one of them will solve the problem, selected at random from the book?

Solution. Consider the following events:

$$A = \text{X solves the problem}$$

$$B = \text{Y solves the problem}$$

Here, *A* and *B* are independent events such that

$$P(A) = \frac{90}{100} = \frac{9}{10}$$

and
$$P(B) = \frac{70}{100} = \frac{7}{10}$$

$$\text{Now required probability } = P(A \cup B)$$

$$= 1 - P(\overline{A})P(\overline{B}) \qquad (\because A \text{ and } B \text{ are independent})$$

$$= 1 - \left(1 - \frac{9}{10}\right)\left(1 - \frac{7}{10}\right)$$

$$= 1 - \frac{1}{10} \times \frac{3}{10} = 0.97$$

4.5 BAYE'S THEOREM

Let $E_1, E_2, ..., E_k$ be mutually exclusive and collectively exhaustive events and let *A* be some other event. The conditional probability of E_i given *A* can be expressed as Baye's theorem.

$$P(E_i/A) = \frac{P(A/E_i)P(E_i)}{P(A/E_1)P(E_1) + P(A/E_2)P(E_2) + ... + P(A/E_k)P(E_k)}$$

ILLUSTRATIVE EXAMPLES

Example 1. An urn I contains 3 white and 4 red balls and urn II contains 5 white and 6 red balls. One ball is drawn at random from one of the urns and is found to be white. Find the probability that it was drawn from urn I. *(I.U. 2007, 08)*

Solution. Let E_1 and E_2 denote the events that the urn I and II are chosen respectively and let *A* be the event that the one ball drawn from the selected urn is white. Then

$$P(E_1) = P(E_2) = \frac{1}{2}$$

$$P(A/E_1) = \frac{^3C_1}{^7C_1} = \frac{3}{7}$$

and
$$P(A/E_2) = \frac{^5C_1}{^{11}C_1} = \frac{5}{11}$$

$$P(E_1/A) = \text{The probability that one white ball drawn at random came from urn I}$$

$$P(E_1/A) = \frac{P(E_1)P(A/E_1)}{P(E_1)P(A/E_1)+P(E_2)P(A/E_2)}$$

$$= \frac{\frac{1}{2}\times\frac{3}{7}}{\frac{1}{2}\times\frac{3}{7}+\frac{1}{2}\times\frac{5}{11}} = \frac{3}{7}\times\frac{77}{68} = \frac{33}{68}$$

Example 2. The contents of urns I, II and III are as follows:

1 white, 2 black and 3 red balls,

2 white, 1 black and 1 red balls, and

4 white, 5 black and 3 red balls.

One urn is chosen at random and two balls drawn, they happen to be white and red. What are the probabilities that they come from urns I, II and III?

Solution. Let E_1, E_2 and E_3 denote the events that the urn I, II and III are chosen respectively and let A be the event that the two balls taken from the selected urn are white and red. Then

$$P(E_1) = P(E_2) = P(E_3) = \frac{1}{3}$$

$$P(A/E_1) = \frac{1\times 3}{{}^6C_2} = \frac{1}{5},\ P(A/E_2) = \frac{2\times 1}{{}^4C_2} = \frac{1}{3}$$

and

$$P(A/E_3) = \frac{4\times 3}{{}^{12}C_2} = \frac{2}{11}$$

$$P(E_1/A) = \text{The probability that two balls drawn at random came from urn I}$$

$$P(E_1/A) = \frac{P(E_1)P(A/E_1)}{P(E_1)P(A/E_1)+P(E_2)P(A/E_2)+P(E_3)P(A/E_3)}$$

$$= \frac{\frac{1}{3}\times\frac{1}{5}}{\frac{1}{3}\times\frac{1}{5}+\frac{1}{3}\times\frac{1}{3}+\frac{1}{3}\times\frac{2}{11}} = \frac{1}{5}\times\frac{165}{118} = \frac{33}{118}$$

$$P(E_2/A) = \text{The probability that two balls drawn at random came from urn II}$$

$$P(E_2/A) = \frac{P(E_2)P(A/E_2)}{P(E_1)P(A/E_1)+P(E_2)P(A/E_2)+P(E_3)P(A/E_3)}$$

$$= \frac{\frac{1}{3}\times\frac{1}{3}}{\frac{1}{3}\times\frac{1}{5}+\frac{1}{3}\times\frac{1}{3}+\frac{1}{3}\times\frac{2}{11}} = \frac{1}{3}\times\frac{165}{118} = \frac{55}{118}$$

$$P(E_3/A) = \text{The probability that two balls drawn at random came from urn III}$$

$$P(E_3/A) = 1-[P(E_1/A)+P(E_2/A)]$$

$$= 1-\left(\frac{33}{118}+\frac{55}{118}\right) = 1-\frac{88}{118} = \frac{30}{118}$$

Example 3. In a bolt factory machines *X*, *Y*, *Z* manufacture respectively 25%, 35% and 40% of the total. Of their output 5, 4 and 2 per cent are defective bolts. A bolt is drawn at random from the product and is found to be defective. What are the probabilities that it was manufactured by machine *X*, *Y* and *Z*?

Solution. Let E_1, E_2 and E_3 denote the events that a bolt selected at random is manufactured by the machines *X*, *Y* and *Z* respectively and let *A* denote the event of its being defective. Then

$$P(E_1) = \frac{25}{100} = 0.25,\ P(E_2) = \frac{35}{100} = 0.35,$$

$$P(E_3) = \frac{40}{100} = 0.40$$

The probability of drawing a defective bolt manufactured machine *X* is

$$P(A/E_1) = \frac{5}{100} = 0.05$$

Similarly, $$P(A/E_2) = \frac{4}{100} = 0.04$$

and $$P(A/E_3) = \frac{2}{100} = 0.02$$

$$P(E_1/A) = \text{The probability that a defective bolt selected at random is manufactured by machine } X$$

$$P(E_1/A) = \frac{P(E_1)P(A/E_1)}{P(E_1)P(A/E_1)+P(E_2)P(A/E_2)+P(E_3)P(A/E_3)}$$

$$= \frac{0.25\times0.05}{0.25\times0.05+0.35\times0.04+0.40\times0.02} = \frac{125}{345} = \frac{25}{69}$$

$$P(E_2/A) = \text{The probability that a defective bolt selected at random is manufactured by machine } Y$$

$$P(E_2/A) = \frac{P(E_2)P(A/E_2)}{P(E_1)P(A/E_1)+P(E_2)P(A/E_2)+P(E_3)P(A/E_3)}$$

$$= \frac{0.35\times0.04}{0.25\times0.05+0.35\times0.04+0.40\times0.02} = \frac{140}{345} = \frac{28}{69}$$

$$P(E_3/A) = \text{The probability that a defective bolt selected at random is manufactured by machine } Z$$

$$P(E_3/A) = 1-[P(E_1/A)+P(E_2/A)]$$

$$= 1-\left(\frac{25}{69}+\frac{28}{69}\right) = 1-\frac{53}{69} = \frac{16}{69}$$

Example 4. The chance that doctor X will diagnose disease Y correctly is 60%. The chance that a patient will die by his treatment after correct diagnosis is 40% and the chance of death by wrong diagnosis is 70%. A patient of doctor X, who had disease Y, died. What is the chance that his disease was correctly diagnosed?

Solution. Let E_1 and E_2 denote the events that the diagnosis was correct and incorrect respectively. Let A be the event that the patient dies. Then

$$P(E_1) = \frac{60}{100} = 0.6 \text{ and } P(E_2) = 1 - P(E_1) = 1 - 0.6 = 0.4$$

$$P(A/E_1) = \frac{40}{100} = 0.4 \text{ and } P(A/E_2) = \frac{70}{100} = 0.7$$

$P(E_1/A)$ = The probability that a patient was correctly diagnosed, given that he had died

$$P(E_1/A) = \frac{P(E_1)P(A/E_1)}{P(E_1)P(A/E_1) + P(E_2)P(A/E_2)}$$

$$= \frac{0.6 \times 0.4}{0.6 \times 0.4 + 0.4 \times 0.7} = \frac{0.24}{0.52} = 0.4615$$

EXERCISE 4.1

1. From a pack of 52 cards 3 are drawn at random. Find the chance that they are a king, a queen and a knave.
2. Three unbiased coins are tossed, find the probability of getting
   (*i*) all heads (*ii*) two heads (*iii*) one head
   (*iv*) at least one head (*v*) at least two heads.
3. A bag contains 7 white, 6 red and 5 black balls. Two balls are drawn at random. Find the probability that they will both be white?
4. Four cards are drawn from a pack of 52 cards. Find the probability that
   (*i*) all are diamonds (*ii*) there is one red card of each suit
   (*iii*) there are two spades and two hearts.
5. Two dice are thrown simultaneously. Find the probability of getting
   (*i*) an even number as the sum
   (*ii*) the sum as a prime number
   (*iii*) a total of at least 10
   (*iv*) a doublet of even number.
6. Tickets numbered from 1 to 20 are mixed up together and then a ticket is drawn at random. What is the probability that the ticket has a number which is multiple of 3 or 7?
7. A bag contains 50 tickets numbered 1, 2, 3, ..., 50 of which five are drawn at random and arranged in ascending order of magnitude $(x_1 < x_2 < x_3 < x_4 < x_5)$. Find the probability that $x_3 = 30$.

8. If A, B, C are mutually and exhaustive events, find $P(B)$, if $\frac{1}{3}P(C)=\frac{1}{2}P(A)=P(B)$.

9. If $P(A)=a$ and $P(B)=b$, then show that $P(A/B)\geq(a+b-1)/b$.

10. Two cards are drawn from a pack of 52 cards. What is the probability that either both are red or both are king?

11. Given two mutually exclusive events A and B such that $P(A)=\frac{1}{2}$ and $P(B)=\frac{1}{3}$. Find $P(A \text{ or } B)$.

12. A die is thrown twice and the sum of numbers appearing is observed to be 6. What is the conditional probability that the number 4 has appeared at least once?

13. A bag contains 19 tickets, numbered from 1 to 19. A ticket is drawn and then another ticket is drawn with out replacement. Find the probability that both tickets will show even numbers.

14. If A and B are two events such that $P(A)=0.3, P(B)=0.6$ and $P(B/A)=0.5$, find $P(A/B)$ and $P(A\cup B)$.

15. A bag contains 3 red and 4 black balls and another bag has 4 red and 2 black balls. One bag is selected at random and from the selected bag a ball is drawn. Let A be the event that the first bag is selected, B be the event that second bag is selected and C be the event that the ball drawn is red. Find $P(A), P(B), P(C/A)$ and $P(C/B)$.

16. If $P(A)=0.4, P(B)=p, P(A\cup B)=0.6$ and A and B are given to be independent events, find the value of p.

17. A bag contains 5 white, 7 red and 4 black balls. If 4 balls are drawn one by one with replacement, what is the probability that none is white?

18. Two dice are thrown. Find the probability of getting an odd number on the first die and a multiple of 3 on the other.

19. A problem of Statistics is given to 3 students whose chances of solving it are $\frac{1}{2},\frac{1}{3},\frac{1}{4}$. What is the probability that the problem is solved?

20. Two boxes contain respectively 4 white and 2 black and 1 white and 3 black balls. One ball is transferred from the first box into the second and then one ball is drawn from the second. It turns out to be black. What is the probability that the transferred ball was white?

21. In a class of 75 students, 15 were considered to be very intelligent, 45 as medium and the rest below average. The probability that a very intelligent student fail in a *viva-voce* examination is 0.005, the medium student failing has a probability 0.05, and the corresponding probability for a below average student is 0.15. If a student is known to have passed the *viva-voce* examination, what is the probability that he is below average?

22. A factory has two machines. The empirical evidence has established that machines I and II produce 30% and 70% of the output respectively. It has also been established that 5% and 1% of the output produced by these machines was defective respectively. A defective item is drawn at random. What is the probability that the defective item was produced by machine II?

23. Suppose that there is chance for a newly constructed flyover to collapse whether the design is faulty or not. The chance that the design is faulty is 5%. The chance that the flyover collapses if the design is faulty is 95%, otherwise it is 30%. A flyover collapsed. What is the probability that it is collapsed because of faulty design?

ANSWERS

1. $\frac{16}{5525}$ **2.** (*i*) $\frac{1}{8}$ (*ii*) $\frac{3}{8}$ (*iii*) $\frac{3}{8}$ (*iv*) $\frac{7}{8}$ (*v*) $\frac{1}{2}$ **3.** $\frac{7}{51}$

4. (*i*) $\frac{11}{4165}$ (*ii*) $\frac{2197}{20825}$ (*iii*) $\frac{468}{20825}$ **5.** (*i*) $\frac{1}{2}$ (*ii*) $\frac{5}{12}$ (*iii*) $\frac{1}{6}$ (*iv*) $\frac{1}{12}$

6. $\frac{2}{5}$ **7.** $\frac{551}{15134}$ **8.** $\frac{1}{6}$ **10.** $\frac{55}{221}$

11. $\frac{5}{6}$ **12.** $\frac{2}{5}$ **13.** $\frac{4}{19}$ **14.** $\frac{1}{4}$; 0.75

15. $\frac{1}{2}$; $\frac{1}{2}$; $\frac{3}{7}$; $\frac{2}{3}$ **16.** $\frac{1}{3}$ **17.** $\left(\frac{11}{16}\right)^4$ **18.** $\frac{1}{6}$

19. $\frac{3}{4}$ **20.** 0.6 **21.** 0.18 **22.** 0.318

23. 0.1428

4.6 PROBABILITY DISTRIBUTIONS

There are two types of probability distributions:

I. Discrete Distributions

A discrete probability function is a function that can take a discrete number of values (not necessarily finite). This is most often the non-negative integers or some subset of the non-negative integers. The mathematical definition of a discrete probability function, $p(x)$, is a function that satisfies the following properties:

1. The probability that x can take a specific value is $p(x)$. That is
   $P(X = x) = p(x)$
2. $p(x)$ is non-negative for all real x.
3. The sum of $p(x)$ over all possible values of x is 1, that is $\Sigma p_j = 1$

Where j represents all possible values that x can have and p_j is the probability at x_j.

II. Continuous Distributions

Since continuous probability functions are defined for an infinite number of points over a continuous interval, the probability at a single point is always zero. Probabilities are measured over intervals, not single points. The property that the integral must equal one is equivalent to the property for discrete distributions that the sum of all the probabilities must equal one.

The mathematical definition of a continuous probability function, $f(x)$, is a function that satisfies the following properties:

1. The probability that x is between two points a and b is

$$p(a \le x \le b) = \int_a^b f(x)\ dx$$

2. It is non-negative for all real x.

3. The integral of the probability function is one, that is $\int_{-\infty}^{+\infty} f(x)\ dx = 1$

Note: Discrete probability functions are referred to as probability mass functions and continuous probability functions are referred to as probability density functions. The term probability functions cover both discrete and continuous distributions. When we are referring to probability functions in generic terms, we may use the term probability density functions to mean both discrete and continuous probability functions.

4.7 BINOMIAL DISTRIBUTION

Binomial distribution was discovered by James Bernoulli in the year 1700. It is a discrete probability distribution. Binomial distribution is used when there are exactly two mutually exclusive outcomes of a trial. These outcomes are appropriately called "success" and "failure". Let a random variable X denote the number of successes in these n trials. The binomial distribution is used to obtain the probability of observing r successes in n trials, with the probability of success on a single trial denoted by p. Each single trial is called a Bernoulli trial satisfying the following conditions:

1. Each trial results in one of two possible, mutually exclusive, outcomes. One of the possible outcome is denoted (arbitrarily) as a success, and the other is denoted a failure.
2. The probability of a success, denoted by p, remains constant from trial to trial. The probability of a failure, 1-p, is denoted by q.
3. The trials are independent; that is, the outcome of any particular trial is not affected by the outcome of any other trial.

The probability of r successes in n trials in a specified order (say) SSFFFS...FSSF (where S stands for success and F for failure) is given by

$$\begin{aligned} P(SSFFFS...FSSF) &= P(S)P(S)P(F)P(F)...P(S)P(F) \\ &= ppqqq...qppq \\ &= \underbrace{ppp...p}_{r\ factors}\quad \underbrace{qqq...q}_{(n-r)\ factors} \\ &= p^r q^{n-r} \end{aligned}$$

But r successes in n trials can occur in ${}^nC_r$ ways and the probability for each of these ways is $p^r q^{n-r}$. Hence, the probability of r successes in n trials is given by:

$$P(X = r) = {}^nC_r \ p^r \ q^{n-r}$$

Where $p+q=1$ and $r = 0,1,2,...,n$

The probability distribution of the number of successes so obtained is called the Binomial probability distribution and X is called the Binomial Variate.

Note:

1. The number of observations n is fixed.
2. Each observation is independent.
3. Each observation represents one of two outcomes 'success' or 'failure'.
4. The probability of 'success' p is the same for each outcome.
5. n and p are called the parameters of the distribution.
6. $P(X = r)$ is usually written as $P(r)$.
7. If n independent trials repeated N times then the expected frequency of r successes is given by $N.P(r)$.

4.8 APPLICATIONS OF BINOMIAL DISTRIBUTION

Binomial distribution is mainly applied in the following problems

1. Number of defectives in a sample from production line.
2. Estimation of reliability of system
3. Radar detection
4. Number of rounds fired from a gun hitting a target.

4.9 RECURRENCE FORMULA FOR BINOMIAL DISTRIBUTION

We have $$P(r) = {}^nC_r \ p^r \ q^{n-r}$$

and $$P(r+1) = {}^nC_{r+1} \ p^{r+1} \ q^{n-(r+1)}$$

Now, $$\frac{P(r+1)}{P(r)} = \frac{{}^nC_{r+1} \ p^{r+1} \ q^{n-(r+1)}}{{}^nC_r \ p^r \ q^{n-r}}$$

$$= \frac{n!}{(r+1)!(n-r-1)!} \times \frac{r!(n-r)!}{n!} \times \frac{p^{r+1}q^{n-r-1}}{p^r q^{n-r}}$$

$$= \frac{r! \ (n-r) \ (n-r-1!}{(r+1) \ r! \ (n-r-1)!} \times \frac{p}{q} = \frac{n-r}{r+1} \times \frac{p}{q}$$

or $$P(r+1) = \frac{n-r}{r+1}.\frac{p}{q}.P(r)$$

which is the required recurrence formula. Using this formula successively, we can find $P(1)$, $P(2)$,... if $P(0)$ is known.

4.10 MEAN AND VARIANCE OF BINOMIAL DISTRIBUTION

The mean and the variance of the binomial distribution is given by np and npq respectively.

ILLUSTRATIVE EXAMPLES

Example 1. The items produced by a company contain 5% defective items. What is the probability of getting 3 defective items in a sample of 10 items?

Solution. Here,
$$p = \frac{5}{100} = \frac{1}{20}, \; n = 10, r = 3$$
$$q = 1 - p = 1 - \frac{1}{20} = \frac{19}{20}$$

We know that
$$P(r) = {}^nC_r p^r q^{n-r}$$
$$P(3 \text{ defective items}) = P(r = 3)$$
$$= {}^{10}C_3 \left(\frac{1}{20}\right)^3 \left(\frac{19}{20}\right)^{10-3}$$
$$= \frac{10\times9\times8}{3\times2} \cdot \frac{(19)^7}{(20)^{10}} = \frac{120\times(19)^7}{(20)^{10}}$$

Example 2. A die is thrown 6 times. If getting an even number is a success, what is the probability of:

(*i*) no success (*ii*) exactly 5 successes
(*iii*) at least 5 successes (*iv*) at most 5 successes

Solution. Here, $S = \{1, 2, 3, 4, 5, 6\}$. Let A denote 'getting an even number'.
$A = \{2, 4, 6\}$

$$p = \frac{n(A)}{n(S)} = \frac{3}{6} = \frac{1}{2}, \; n = 6$$
$$q = 1 - p = 1 - \frac{1}{2} = \frac{1}{2}$$

We know that
$$P(r) = {}^nC_r \, p^r \, q^{n-r}$$

(*i*)
$$P(\text{no success}) = P(r = 0)$$
$$= {}^6C_0 \left(\frac{1}{2}\right)^0 \left(\frac{1}{2}\right)^{6-0} = \left(\frac{1}{2}\right)^6 = \frac{1}{64}$$

(*ii*)
$$P(\text{exactly 5 successes}) = P(r = 5)$$
$$= {}^6C_5 \left(\frac{1}{2}\right)^5 \left(\frac{1}{2}\right)^{6-5} = 6\times\left(\frac{1}{2}\right)^6 = \frac{3}{32}$$

(*iii*)
$$P(\text{at least 5 successes}) = P(r \geq 5)$$
$$= P(5 \text{ or } 6 \text{ successes})$$
$$= P(5) + P(6)$$

$$= {}^6C_5\left(\frac{1}{2}\right)^5\left(\frac{1}{2}\right)^{6-5} + {}^6C_6\left(\frac{1}{2}\right)^6\left(\frac{1}{2}\right)^{6-6}$$

$$= \frac{3}{32} + \frac{1}{64} = \frac{7}{64}$$

(*iv*) $P(at\ most\ 5\ successes) = P(r \le 5)$

$$= 1 - P(r > 5) \quad = 1 - P(r = 6)$$

$$= 1 - \frac{1}{64} = \frac{63}{64}$$

Example 3. A pair of dice thrown 10 times. If getting a doublet (same number on both) is considered a success, find the probability of

(*i*) no success (*ii*) 3 successes.

Solution. A doublet can be obtained when a pair of dice is thrown in

(1,1), (2, 2), (3, 3), (4, 4), (5, 5), (6, 6) *i.e.*, 6 ways.

The two dice can be thrown in $6^2 = 36$ ways.

$$p = P(getting\ doublet) = \frac{6}{36} = \frac{1}{6},\ n = 10$$

$$q = 1 - p = 1 - \frac{1}{6} = \frac{5}{6}$$

We know that $P(r) = {}^nC_r p^r q^{n-r}$

(*i*) $P(no\ success) = P(r = 0)$

$$= {}^{10}C_0\left(\frac{1}{6}\right)^0\left(\frac{5}{6}\right)^{10-0} = \left(\frac{5}{6}\right)^{10} = \frac{9765625}{60466176}$$

(*ii*) $P(3\ successes) = P(r = 3)$

$$= {}^{10}C_3\left(\frac{1}{6}\right)^3\left(\frac{5}{6}\right)^{10-3} = \frac{5}{9} \times \left(\frac{5}{6}\right)^7 = \frac{390625}{2519424}$$

Example 4. Five cards are drawn successively with replacement from a well-shuffled pack of 52 cards. What is the probability that:

(*i*) none is spade (*ii*) only 3 cards are spades?

Solution. Here, $p = P(spade\ card) = \frac{13}{52} = \frac{1}{4},\ n = 5$

$$q = 1 - p = 1 - \frac{1}{4} = \frac{3}{4}$$

We know that $P(r) = {}^nC_r p^r q^{n-r}$

(i) $P(none\ is\ spade) = P(r=0)$

$$= {}^5C_0 \left(\frac{1}{4}\right)^0 \left(\frac{3}{4}\right)^{5-0} = \left(\frac{3}{4}\right)^5 = \frac{243}{1024}$$

(ii) $P(only\ 3\ cards\ are\ spade) = P(r=3)$

$$= {}^5C_3 \left(\frac{1}{4}\right)^3 \left(\frac{3}{4}\right)^{5-3} = \frac{20}{2}\times\left(\frac{1}{4}\right)^3\times\left(\frac{3}{4}\right)^2 = \frac{45}{512}$$

Example 5. Find the parameters of the binomial distribution for which mean = 4 and variance = 3.

Solution. We know that for a Binomial distribution,

$$\text{Mean} = np \text{ and variance} = npq$$

Here, $np = 4$ and $npq = 3$

Now, $$\frac{npq}{np} = \frac{3}{4} \Rightarrow q = \frac{3}{4}$$

$$p = 1 - q = 1 - \frac{3}{4} = \frac{1}{4}$$

and $$n\times\frac{1}{4} = 4 \Rightarrow n = 16.$$

Example 6. Comment on the following statement. The mean of a Binomial distribution is 3 and standard deviation is $\sqrt{5}$.

Solution. We know that for a Binomial distribution,

$$\text{Mean} = np \text{ and standard deviation } = \sqrt{npq}$$

Here, $np = 3$ and $\sqrt{npq} = \sqrt{5} \Rightarrow npq = 5$

Now, $$\frac{npq}{np} = \frac{5}{3} \Rightarrow q = \frac{5}{3} > 1$$ (which is not possible)

Thus, the given statement is wrong.

Example 7. The probability of a man hitting a target is $\frac{1}{4}$. How many times must he fire so that the probability of his hitting the target at least once is greater than $\frac{2}{3}$?

Solution. Let the man hits the target n times.

Here $$p = \frac{1}{4} \text{ and } q = 1 - p = 1 - \frac{1}{4} = \frac{3}{4}$$

$P(hitting\ the\ target\ at\ least\ once) = P(r \geq 1)$

$$= 1 - P(r < 1) = 1 - P(r = 0)$$

According to the given condition

$$[1-P(r=0)] > \frac{2}{3}$$

$$1-{}^{n}C_0\left(\frac{1}{4}\right)^0\left(\frac{3}{4}\right)^{n-0} > \frac{2}{3}$$

$$1-(0.75)^n > \frac{2}{3} \text{ or } (0.75)^n < \frac{1}{3} \text{ or } (0.75)^n < 0.3333$$

Taking log on both the sides, we have

$$n\log(0.75) < \log(0.3333)$$

$$n(-0.1249) < (-0.47712)$$

$$n(0.1249) > (0.47712)$$

$$n > \frac{(0.47712)}{(0.1249)} \quad or \quad n > 3.82$$

$$n = 4.$$

Example 8. Six dice are thrown 729 times. How many times do you expect at least three dice to show a five or six? (*I.U. 2007, 08*)

Solution. Here, p = the probability of getting 5 or 6 with one die

$$= \frac{2}{6} = \frac{1}{3}$$

$$q = 1-p = 1-\frac{1}{3} = \frac{2}{3}, \quad n=6, \quad N=729$$

The expected numbers of times at least three dice showing five or six $= N \cdot P(r \geq 3)$

$$= 729 \times [P(r=3)+P(r=4)+P(r=5)+P(r=6)]$$

$$= 729 \times \left[{}^6C_3\left(\frac{2}{3}\right)^3\left(\frac{1}{3}\right)^3 + {}^6C_4\left(\frac{2}{3}\right)^4\left(\frac{1}{3}\right)^2 + {}^6C_5\left(\frac{2}{3}\right)^5\left(\frac{1}{3}\right)^1 + {}^6C_6\left(\frac{2}{3}\right)^0\left(\frac{1}{3}\right)^6\right]$$

$$= \frac{729}{3^6}[160+60+12+1] = 233$$

Example 9. Out of 800 families with 5 children each, how many families would be expected to have

(*i*) 3 boys (*ii*) 5 girls (*iii*) either 2 or 3 boys?

Assume equal probabilities for boy and girl.

Solution. Here, $n=5, \; N=800$

$$p = P(a\ boy) = \frac{1}{2},$$

$$q = P(a\ girl) = 1-p = 1-\frac{1}{2} = \frac{1}{2}$$

(*i*) The probability of having 3 boys out of 5 children

$$= {}^5C_3\left(\frac{1}{2}\right)^3\left(\frac{1}{2}\right)^{5-3} = \frac{20}{2}\times\left(\frac{1}{2}\right)^5 = \frac{10}{32} = 0.3125$$

The expected number of families $= N.P(r=3) = 800\times 0.3125 = 250$.

(*ii*) The probability of having 5 girls out of 5 children $= P(r=0)$

$$= {}^5C_0\left(\frac{1}{2}\right)^0\left(\frac{1}{2}\right)^{5-0} = \left(\frac{1}{2}\right)^5 = \frac{1}{32} = 0.03125$$

The expected number of families $= N.P(r=0) = 800\times 0.03125 = 25$.

(*iii*) The probability of having 2 or 3 boys out of 5 children $= P(r=2 \text{ or } r=3)$

$$= p(r=2) + p(r=3)$$

$$= {}^5C_2\left(\frac{1}{2}\right)^2\left(\frac{1}{2}\right)^{5-2} + {}^5C_3\left(\frac{1}{2}\right)^3\left(\frac{1}{2}\right)^{5-3}$$

$$= 10\times\left(\frac{1}{2}\right)^5 + 10\times\left(\frac{1}{2}\right)^5$$

$$= 20\times\left(\frac{1}{2}\right)^5 = \frac{20}{32} = 0.625$$

The expected number of families $= N.P(r=2 \text{ or } r=3) = 800\times 0.625 = 500$.

Example 10. Four coins are tossed 160 times. The number of times *r* heads occur is given below :

r	0	1	2	3	4
Number of times	8	34	69	43	6

Fit a binomial distribution to this data on the hypothesis that coins are unbiased.

Solution. The coins are unbiased so the probability of getting head is $= \frac{1}{2}$.

So $$p = \frac{1}{2},\ q = 1-p = 1-\frac{1}{2} = \frac{1}{2}$$

Now, $$P(r=0) = {}^4C_0\left(\frac{1}{2}\right)^0\left(\frac{1}{2}\right)^{4-0} = \left(\frac{1}{2}\right)^4 = \frac{1}{16}$$

Using recurrence relation, we have

$$P(r+1) = \frac{n-r}{r+1}.\frac{p}{q}.p(r)$$

$$P(1) = \frac{4-0}{0+1}\cdot P(0) = 4\times P(0) = 4\times\frac{1}{16} = \frac{1}{4}$$

$$P(2) = \frac{4-1}{1+1} \cdot P(1) = \frac{3}{2} \times P(1) = \frac{3}{2} \times \frac{1}{4} = \frac{3}{8}$$

$$P(3) = \frac{4-2}{2+1} \cdot P(2) = \frac{2}{3} \times P(2) = \frac{2}{3} \times \frac{3}{8} = \frac{1}{4}$$

$$P(4) = \frac{4-3}{3+1} \cdot P(3) = \frac{1}{4} \times P(3) = \frac{1}{4} \times \frac{1}{4} = \frac{1}{16}$$

r	$P(r)$	*Theoretical frequencies* = $N.\ p(r)$
0	$\frac{1}{16}$	$160 \times \frac{1}{16} = 10$
1	$\frac{1}{4}$	$160 \times \frac{1}{4} = 40$
2	$\frac{3}{8}$	$160 \times \frac{3}{8} = 60$
3	$\frac{1}{4}$	$160 \times \frac{1}{4} = 40$
4	$\frac{1}{6}$	$160 \times \frac{1}{16} = 10$

EXERCISE 4.2

1. A pair of dice is thrown 6 times. If getting a total of 9 is considered a success. What is the probability of at least 5 successes?
2. A die is thrown 6 times. If getting an odd number is a success. Find the probability of :

   (*i*) no successes (*ii*) 5 successes

   (*iii*) at least 5 successes (*iv*) at most 5 successes.
3. A coin is tossed 5 times. What is the probability of getting at least 3 heads?
4. Find the probability distribution of the number of heads observed when a coin is tossed 3 times.
5. If on an average one ship in every ten is wrecked, find the probability that out of 5 ships expected to arrive, 4 at least will arrive safely.
6. A pair of dice is thrown 4 times. What is the probability of getting doublets at least twice?
7. "The mean and variance of a binomial distribution are respectively 6 and 9". Is this statement correct?
8. A student is given a true-false examination with 8 questions. If he gets 6 or more correct answers, he passes the examination. Given that he guesses the answer to each question, find the probability he passes the examination.

9. In a box containing 60 bulbs, 6 are defective. What is the probability that out of a sample of 5 bulbs

(*i*) none is defective (*ii*) exactly 2 are defective?

10. The sum of mean and variance of a binomial distribution is 15 and the sum of their squares is 117. Determine the distribution.

11. Out of 2000 families with 4 children each, how many would you expect to have

(*i*) at least 1 boy (*ii*) 2 boys

(*iii*) 1 or 2 girls (*iv*) no girls?.

Assume equal probabilities for boys and girls.

12. In a sampling a large number of parts manufactured by a machine, the mean number of defectives in a sample of 20 is 2. Out of 1000 such samples, how many would be expected to contain at least 3 defective.

13. Assuming that 20% of the population of a city is literate, so that the chance of an individual being literate is 0.2 and assuming that 100 investigators each take 10 individuals to see whether they are literate. How many investigators would you expect to report 3 or less were literate?

ANSWERS

1. $\dfrac{49}{96}$ **2.** (*i*) $\dfrac{1}{64}$ (*ii*) $\dfrac{3}{32}$ (*iii*) $\dfrac{7}{64}$ (*iv*) $\dfrac{63}{64}$ **3.** $\dfrac{1}{2}$

4.

r	0	1	2	3
$p(r)$	1/8	3/8	3/8	1/8

5. $\dfrac{7}{5}\left(\dfrac{9}{10}\right)^4$ **6.** $\dfrac{19}{44}$

7. No **8.** $\dfrac{37}{256}$ **9.** (*i*) $\left(\dfrac{9}{10}\right)^5$ (*ii*) $\dfrac{729}{10000}$

10. ${}^{27}C_r\left(\dfrac{1}{3}\right)^r\left(\dfrac{2}{3}\right)^{27-r}$, $r = 0, 1, 2, \ldots, 27$

11. (*i*) 1875 (*ii*) 750 (*iii*) 1250 (*iv*) 125 **12.** 323

13. 88 (approx.)

4.11 POISSON DISTRIBUTION

Poisson distribution was discovered by Simeon Denis Poisson in 1837. It is a discrete probability distribution.

In a random experiment, let p be the probability of the occurrence of an event and let n trials be made in such a way that

1. p is very small, *i.e.*, $p \to 0$
2. n is very large, *i.e.*, $n \to \infty$
3. $np = \lambda$ (say) is finite.

Then the probability of occurrence of this event r times is given by the

$$P(X = r) = e^{-\lambda}\frac{\lambda^r}{r!}, \text{ where } r = 0, 1, 2, \ldots$$

Note:

1. The events must be random and independent of each other.
2. Events must be rare events.
3. λ is called the parameter of the distribution.
4. $P(X = r)$ is usually written as $P(r)$.
5. If n independent trials repeated N times then the expected frequency of r successes is given by $N.P(r)$.

4.12 APPLICATIONS OF POISSON DISTRIBUTION

Poisson distribution is mainly applied in the following problems:

1. The demand for a product.
2. The number of phone calls at a call centre per minute.
3. The number of times a web server is accessed per minute.
4. The number of soldiers killed by horse-kicks each year.
5. The number of deaths in a area by rare disease.
6. The number of printing mistakes at each page of the book.

4.13 RECURRENCE FORMULA FOR POISSON DISTRIBUTION

We have $$P(r) = e^{-\lambda}\frac{\lambda^r}{r!} \text{ and } P(r+1) = e^{-\lambda}\frac{\lambda^{r+1}}{(r+1)!}$$

Now, $$\frac{P(r+1)}{P(r)} = \frac{e^{-\lambda}\dfrac{\lambda^{r+1}}{(r+1)!}}{e^{-\lambda}\dfrac{\lambda^r}{r!}} = \frac{\lambda^{r+1}}{(r+1)!}\times\frac{r!}{\lambda^r}$$

$$= \frac{\lambda}{(r+1)\ r!}\times r! = \frac{\lambda}{(r+1)}$$

$$p(r+1) = \frac{\lambda}{(r+1)}p(r)$$

which is the required recurrence formula. Using this formula successively, we can find $P(1)$, $P(2)$,... if $P(0)$ is known.

4.14 MEAN AND VARIANCE OF POISSON DISTRIBUTION

The mean and the variance of the Poisson distribution is same and given by λ.

ILLUSTRATIVE EXAMPLES

Example 1. Using Poisson distribution, find the probability that the aces of spades will be drawn from a pack of well-shuffled cards at least once in 104 consecutive trials. (Given $e^{-2} = 0.1353$)

Solution. Here, $$p = \frac{1}{52},\ n = 104$$

$$\lambda = np = 104\times\frac{1}{52} = 2$$

We know that $P(r) = e^{-\lambda}\dfrac{\lambda^r}{r!}$

$$P(at\ least\ once) = P(r \geq 1) = 1 - P(r < 1)$$
$$= 1 - P(r=0) = 1 - e^{-2}\frac{2^0}{0!}$$
$$= 1 - e^{-2} = 1 - 0.1353 = 0.8647$$

Example 2. Suppose a book of 580 pages contain 40 typographical mistakes. If these mistakes are randomly distributed throughout the book. What is the probability that 10 pages, selected at random will be free of mistakes? (Given $e^{-0.69} = 0.50157$)

Solution. Here, $p = \dfrac{40}{580} = 0.069,\ n = 10$

$$\lambda = np = 10 \times 0.069 = 0.69$$

We know that $P(r) = e^{-\lambda}\dfrac{\lambda^r}{r!}$

Required probability $= P(r = 0)$

$$= e^{-0.69}\frac{(0.69)^0}{0!}$$
$$= e^{-0.69} = 0.50157$$

Example 3. A car hire firm has two cars, which it hires out day-by-day. The number of demands for a car on each day is distributed as a Poisson distribution with mean 1.5. Calculate the proportion of days on which neither car is used, the proportion of days on which some demand is refused. (Given $e^{-1.5} = 0.2231$)

Solution. Here, the mean of the Poisson distribution is λ.

$$\lambda = 1.5$$

We know that $P(r) = e^{-\lambda}\dfrac{\lambda^r}{r!}$

The proportion of days on which neither car is used

= Probability of there being no demand for the car

$$= P(r = 0)$$
$$= e^{-1.5}\frac{(1.5)^0}{0!}$$
$$= e^{-1.5} = 0.2231$$

The proportion of days on which some demand is refused

= Probability for the number of demands to be more than two

$$= p(r > 2) = 1 - p(r \leq 1)$$
$$= 1 - [p(r = 0) + p(r = 1) + p(r = 2)]$$

$$= 1 - \left[e^{-1.5} + e^{-1.5} \frac{(1.5)}{1!} + e^{-1.5} \frac{(1.5)^2}{2!} \right]$$

$$= 1 - [1 + 1.5 + 1.125]e^{-1.5}$$

$$= 1 - 3.625 \times 0.2231 = 0.1913$$

Example 4. It is given that 2% of the electric bulbs manufactured by a company are defective. Using Poisson distribution, find the probability that a sample of 200 bulbs will contain

(*i*) no defective bulb

(*ii*) 2 defective bulbs

(*iii*) at most 3 defective bulb

(*iv*) at least 3 defective bulbs (Given $e^{-4} = 0.0183$)

Solution. Here,

$$p = \frac{2}{100} = 0.02, \; n = 200$$

$$\lambda = np = 200 \times 0.02 = 4$$

We know that

$$P(r) = e^{-\lambda} \frac{\lambda^r}{r!}$$

(*i*) Probability of no defective bulb $= P(r = 0)$

$$= e^{-4} \frac{(4)^0}{0!}$$

$$= e^{-4} = 0.0183$$

(*ii*) Probability of 2 defective bulbs $= P(r = 2)$

$$= e^{-4} \frac{(4)^2}{2!}$$

$$= 0.0183 \times \frac{16}{2} = 0.1464$$

(*iii*) Probability of at most 3 defective bulbs $= P(r \le 3)$

$$= P(r = 0) + P(r = 1) + P(r - 2) + P(r = 3)$$

$$= e^{-4} + e^{-4} \frac{(4)^1}{1!} + e^{-4} \frac{(4)^2}{2!} + e^{-4} \frac{(4)^3}{3!}$$

$$= e^{-4}(1 + 4 + 8 + \frac{32}{3})$$

$$= 0.0183 \times \frac{71}{3} = 0.4331$$

(*iv*) Probability of at least 3 defective bulbs $= P(r \geq 3)$

$$= 1 - P(r < 3)$$

$$= 1 - [P(r=0) + P(r=1) + P(r=2)]$$

$$= 1 - [e^{-4} + 4e^{-4} + 8e^{-4}]$$

$$= 1 - (e^{-4} \times 13) = 1 - (0.0183 \times 13)$$

$$= 1 - 0.2379 = 0.7621$$

Example 5. A manufacturer knows that the razor blades he makes contain on the average 0.5% defectives. He packs them in packets of 5. What is the probability that a packet picked at random contains 3 or more defective blades?

(Given $e^{-0.025} = 0.9753$)

Solution. Here, $p = \dfrac{0.5}{100} = 0.005,\ n = 5$

$$\lambda = np = 5 \times 0.005 = 0.025$$

We know that $P(r) = e^{-\lambda} \dfrac{\lambda^r}{r!}$

Probability of 3 or more defective blades $= P(r \geq 3)$

$$= 1 - P(r < 3)$$

$$= 1 - [P(r=0) + P(r=1) + P(r=2)]$$

$$= 1 - \left[e^{-0.025} + e^{-0.025} \frac{(0.025)^1}{1!} + e^{-0.025} \frac{(0.025)^2}{2!} \right]$$

$$= 1 - e^{-0.025} [1 + 0.025 + 0.0003125]$$

$$= 1 - 0.9753 \times 1.0253 = 1 - 0.999975$$

$$= 0.00002491$$

Example 6. If the variance of the Poisson distribution is 2, find probabilities for $r = 1, 2, 3, 4$ using recurrence relation. Also find $P(r \geq 4)$. (Given $e^{-2} = 0.1353$)

Solution. Here, variance $= 2 \Rightarrow \lambda = 2$

We know that $P(r) = e^{-\lambda} \dfrac{\lambda^r}{r!}$

$$P(0) = e^{-2} \frac{(2)^0}{0!} = e^{-2} = 0.1353$$

We know that the recurrence relation for Poisson distribution is

$$P(r+1) = \frac{\lambda}{(r+1)} P(r)$$

Now, putting $r = 0, 1, 2, 3$ in the recurrence relation, we have

$$P(1) = \frac{\lambda}{1} P(0) = 2 \times 0.1353 = 0.2706$$

$$P(2) = \frac{\lambda}{2} P(1) = \frac{2}{2} \times 0.2706 = 0.2706$$

$$P(3) = \frac{\lambda}{3} P(2) = \frac{2}{3} \times 0.2706 = 0.1804$$

$$P(4) = \frac{\lambda}{4} P(3) = \frac{2}{4} \times 0.1804 = 0.0902$$

and

$$\begin{aligned} P(r \geq 4) &= 1 - P(r < 4) \\ &= 1 - [P(r=0) + P(r=1) + P(r=2) + P(r=3)] \\ &= 1 - [P(0) + P(1) + P(2) + P(3)] \\ &= 1 - (0.1353 + 0.2706 + 0.2706 + 0.1804) \\ &= 1 - 0.8569 = 0.1431 \end{aligned}$$

Example 7. A manufacturer who produces medicine bottles finds that 0.1% of the bottles are defective. The bottles are packed in boxes containing 500 bottles. A drug manufacturer buys 1000 boxes from the producer of the bottles. Using Poisson distribution, find how many boxes will contain:

(*i*) no defective bottle (*ii*) at least two defective bottles. (Given $e^{-0.5} = 0.6065$)

Solution. Here,

$$p = \frac{0.1}{100} = 0.001, \ n = 500$$

$$\lambda = np = 500 \times 0.001 = 0.5, \ N = 1000$$

We know that

$$P(r) = e^{-\lambda} \frac{\lambda^r}{r!}$$

Number of boxes containing no defective bottle $= N.P(r = 0)$

$$\begin{aligned} &= 1000 \times e^{0.5} \frac{(0.5)^0}{0!} \\ &- 1000 \times 0.6065 \\ &= 606.50 = 606 \text{ (approx.)} \end{aligned}$$

Number of boxes containing at least two defective bottles $= N.P(r \geq 2)$

$$\begin{aligned} &= N.[1 - P(r < 2)] \\ &= N.[1 - \{P(r=0) + P(r=1)\}] \\ &= 1000 \times \left[1 - \left(e^{-0.5} + e^{-0.5} \frac{(0.5)^1}{1!}\right)\right] \\ &= 1000 \times [1 - (0.6065 \times 1.5)] \\ &= 1000 \times 0.09025 \\ &= 90.25 = 90 \text{ (approx.)} \end{aligned}$$

Example 8. For a Poisson variate X, calculate $P(X > 0)$, if it is given that

$$4P(X=4) = 5P(X=5).$$

Solution. Given,

$$4P(X=4) = 5P(X=5)$$

$$4.e^{-\lambda}\frac{\lambda^4}{4!} = 5.e^{-\lambda}\frac{\lambda^5}{5!}$$

$$4.\frac{\lambda^4}{4!} = 5.\frac{\lambda^5}{5\times 4!}$$

$$4.\lambda^4 = \lambda^5 \Rightarrow \lambda = 4$$

Now,

$$P(X>0) = P(r>0)$$

$$= 1 - P(r \le 0) = 1 - P(r=0)$$

$$= 1 - e^{-4}\frac{(4)^0}{0!} = 1 - e^{-4}$$

$$= 1 - 0.0183 = 0.9817$$

Example 9. A typist kept a record of mistakes per day during 300 working days:

Mistakes per day	0	1	2	3	4
Number of days	143	90	44	14	9

Fit a Poisson distribution for the above data and calculate theoretical frequencies.

Solution. Here, $\lambda = \text{mean} = \dfrac{\Sigma fx}{\Sigma f}$

$$= \frac{0+90+88+42+36}{300} = \frac{256}{300} = 0.853,\ N = 300$$

$$P(r) = e^{-\lambda}\frac{\lambda^r}{r!} = e^{-0.853}\frac{(0.853)^r}{r!} = (0.426)\cdot\frac{(0.853)^r}{r!}$$

$$P(0) = (0.426)\cdot\frac{(0.853)^0}{0!} = 0.426$$

$$P(1) = (0.426)\cdot\frac{(0.853)^1}{1!} = 0.426\times 0.853 = 0.363$$

$$P(2) = (0.426)\cdot\frac{(0.853)^2}{2!} = 0.426\times 0.3638 = 0.155$$

$$P(3) = (0.426)\cdot\frac{(0.853)^3}{3!} = 0.426\times 0.1034 = 0.044$$

$$P(4) = (0.426)\cdot\frac{(0.853)^4}{4!} = 0.426\times 0.0221 = 0.009$$

r	$N.P(r)$	*Theoretical frequencies*
0	127.8	128
1	108.9	109
2	46.5	47
3	13.2	13
4	2.7	3

Example 10. Fit a Poisson distribution to the following and calculate theoretical frequencies:

Deaths	0	1	2	3	4
Frequencies	122	60	15	2	1

(*I.U. 2007, 08*)

Solution. Here, $\lambda = \text{mean} = \dfrac{\Sigma fx}{\Sigma f}$

$$= \frac{0+60+30+6+4}{200} = \frac{100}{200} = 0.5,\ N = 200$$

$$P(r) = e^{-\lambda}\frac{\lambda^r}{r!} = e^{-0.5}\frac{(0.5)^r}{r!} = (0.61)\cdot\frac{(0.5)^r}{r!}$$

$$P(0) = (0.61)\cdot\frac{(0.5)^0}{0!} = 0.61$$

$$P(1) = (0.61)\cdot\frac{(0.5)^1}{1!} = 0.61\times 0.5 = 0.305$$

$$P(2) = (0.61)\cdot\frac{(0.5)^2}{2!} = 0.61\times 0.125 = 0.07625$$

$$P(3) = (0.61)\cdot\frac{(0.5)^3}{3!} = 0.61\times 0.0208 = 0.012708$$

$$P(4) = (0.61)\cdot\frac{(0.5)^4}{4!} = 0.61\times 0.002604 = 0.001588$$

r	$N.P(r)$	*Theoretical frequencies*
0	122	122
1	61	61
2	15.25	15
3	2.54	3
4	0.3176	0

EXERCISE 4.3

1. Suppose a book of 600 pages contain 40 printing mistakes. If the mistakes are randomly distributed throughout the book, what is the probability that 10 pages, selected at random, will be free of mistakes? (Given $e^{-0.67} = 0.51$)

2. Suppose 300 misprints are distributed randomly throughout a book of 500 pages. Find the probability that a given page contains:

(*i*) exactly 2 misprints (*ii*) 2 or more misprints.

3. Suppose 2% of the items made by a factory are defective. Find the probability that there are 3 defective items in a sample of 100 items. (Given $e^{-2} = 0.135$)

4. If the probability that an individual suffers a bad reaction from a certain injection is 0.001, find the probability that out of 2000 individuals

(*i*) exactly 3 individuals

(*ii*) more than 2 individuals

(*iii*) none

(*iv*) more than 1 individual, will suffer a bad reaction.

5. An insurance company finds that 0.005% of the population dies from a certain kind of accident each year. What is the probability that the company must pay off no more than 3 of 10,000 insured risks against such accident in a given year? (Given $e^{-0.5} = 0.6065$)

6. A manufacturer of screws knows that 4% of his product is defective. If he sells the screws in boxes of 100 and guarantee that not more than 5 screws will be defective. What is the probability that box will fail to meet the guaranteed quality?

7. A manufacturer knows that the condensers he makes contain on the average 1% of the defectives. He packs them in boxes of 100. What is the probability that a box picked at random will contain 4 or more defective condensers?

8. Assume that the probability of an individual coal-miner being killed in a mine accident during a year is $\frac{1}{2400}$ Use Poisson distribution to calculate the probability that in a mine employing 200 miners, there will be at least one fatal accident in a year.

9. An insurance company founds that only 0.01% of the population is involved in a certain type of accident each year. If its 1000 policy holders were randomly selected from the population, then what is the probability that not more than 2 of its clients are involved in such an accident next year?

10. If *X* is a Poisson variate such that $P(X = 2) = 9P(X = 4) + 90P(X = 6)$, find the mean of *X*.

11. If *X* is a Poisson variate such that $P(X = 1) = 0.01487;\ P(X = 2) = 0.04461$, find $P(X = 3)$.

12. If *X* is a Poisson variate such that $P(X = 1) = P(X = 2)$, find

(*i*) mean of the distribution (*ii*) $P(X = 0)$

(*iii*) $P(X = 4)$ (Given $e^{-2} = 0.1353$)

13. The number of accidents in a year involving taxi drivers in a city follows a Poisson distribution with mean equal to 3. Out of 1000 taxi drivers, find approximately the number of drivers with:

(*i*) no accident in a year (*ii*) more than 3 accidents in a year.

14. In a certain factory turning out razor blades, there is a small chance $\frac{1}{500}$ for any blade to be defective. The blades are supplied in packets of 10. Using Poisson distribution, calculate the approximate number of packets containing:

(*i*) no defective (*ii*) one defective

(*iii*) two defective blades respectively in a consignment of 10,000 packets.

(Given $e^{-0.02} = 0.9802$)

15. The distribution of typing mistakes committed by a typist is as follows:

Mistakes per page	0	1	2	3	4	5
No. of pages	142	156	69	27	5	1

Assuming Poisson model, find out the expected frequencies.

16. Accidents per day were recorded in certain city for a period of 400 days, as follows:

No. of accidents	0	1	2	3	4	5
No. of days	213	128	37	18	3	1

Assuming Poisson model, find out the expected frequencies.

17. The first proof of 200 pages of a book containing 560 pages revealed the following distribution of the number of printing errors:

No. of errors in page	0	1	2	3	4	5
No. of pages	112	63	20	3	1	1

Fit a Poisson distribution corresponding to these data.

ANSWERS

1. 0.51 **2.** (*i*) 0.1 (*ii*) 0.122 **3.** 0.18

4. (*i*) 0.18 (*ii*) 0.325 (*iii*) 0.135 (*iv*) 0.59 **5.** 0.3235

6. $1-e^{-4}\left[5+\frac{4^2}{2!}+\frac{4^3}{3!}+\frac{4^4}{4!}+\frac{4^5}{5!}\right]$ 7. 0.019 **8.** 0.08

9. 0.9998 **10.** 1 **11.** 0.08922

12. (*i*) 2 (*ii*) 0.1353 (*iii*) 0.0902 **13.** (*i*) 50 (*ii*) 353

14. (*i*) 9802 (*ii*) 196 (*iii*) 2 **15.** 147, 147, 74, 25, 6, 1 pages

16. 202, 138, 47, 11, 2, 0 **17.** 109, 66, 20, 4, 1, 0

4.15 NORMAL DISTRIBUTION

A continuous random variable X, taking all real values in the range $(-\infty, \infty)$ is said to follow a normal distribution with parameters mean μ and standard deviation σ if its probability density function is given by

$$f(x) = \frac{1}{\sigma\sqrt{2\pi}}e^{-\frac{1}{2}\left(\frac{x-\mu}{\sigma}\right)^2}; \quad -\infty < x < \infty,\ -\infty < \mu < \infty,\ \sigma > 0$$

The normal distribution with mean μ and variance σ^2 is denoted by $N(\mu, \sigma^2)$.

The graph of the normal distribution is called the normal curve. It is a bell-shaped curve and symmetrical about the mean μ. The two tails of the curve extend to $+\infty$ and $-\infty$ towards the positive and negative directions of the x-axis respectively and gradually approach the x-axis without ever meeting it.

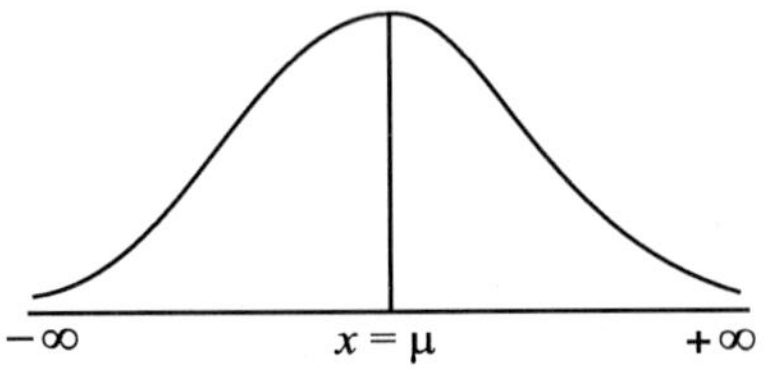

4.16 STANDARD FORM OF THE NORMAL DISTRIBUTION

A random variable Z which has a normal distribution with $\mu = 0$ and $\sigma = 1$ is said to have a standard normal distribution. The probability density function for the normal distribution in standard form is given by

$$f(Z) = \frac{1}{\sqrt{2\pi}} e^{-\frac{z^2}{2}}$$

where $Z = \dfrac{x-\mu}{\sigma}$; Z is called the standardized normal random variate and is denoted by $N(0, 1)$.

$$P(a \le Z \le b) = \text{Area under the standard normal curve between } Z = a \text{ and } Z = b.$$

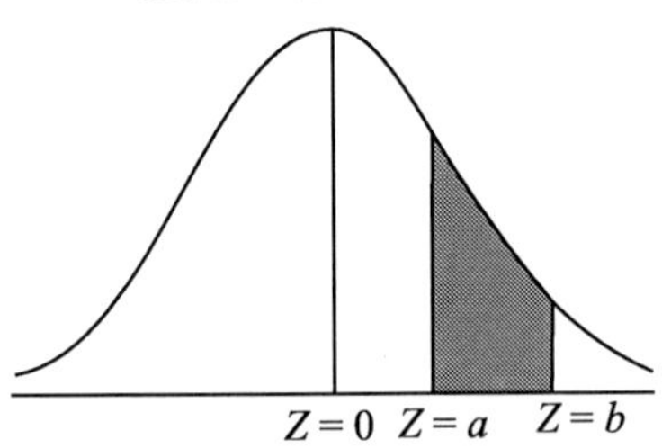

Note: The probabilities $P(z_1 \le Z \le z_2)$, $P(z_1 < Z \le z_2)$, $P(z_1 \le Z < z_2)$ and $p(z_1 < Z < z_2)$ are regarded to be the same.

4.17 PROPERTIES OF THE NORMAL DISTRIBUTION

A normal probability curve with mean μ and standard deviation σ is given by the equation

$$f(x) = \frac{1}{\sigma\sqrt{2\pi}} e^{-\frac{1}{2}\left(\frac{x-\mu}{\sigma}\right)^2}$$

has the following properties:

(*i*) $f(x) \geq 0$

(*ii*) $\int_{-\infty}^{+\infty} f(x)\,dx = 1$ *i.e.*, the total area under the normal curve above the x-axis is 1.

(*iii*) The normal curve is bell shaped and symmetrical about the line $x = \mu$, *i.e.*, mean.

(*iv*) It is a unimodal distribution *i.e.*, mean = median = mode.

(*v*) The height of the normal curve is maximum at the mean value. The maximum ordinate at $x = \mu$ is given by $y = \dfrac{1}{\sigma\sqrt{2\pi}}$.

(*vi*) $P(\mu - \sigma < x < \mu + \sigma) = 68\%$

$P(\mu - 2\sigma < x < \mu + 2\sigma) = 95.5\%$

$P(\mu - 3\sigma < x < \mu + 3\sigma) = 99.7\%$

ILLUSTRATIVE EXAMPLES

Example 1. The marks obtained by a group of students who appeared for a test were normally distributed with mean 80 and standard deviation 6. Find the standard scores for the student who scored

(*i*) 98 marks (*ii*) 58 marks (*iii*) 50 marks.

Solution. Suppose x is normally distributed with mean $\mu = 80$ and standard deviation $\sigma = 6$.

We know that the standard normal variate $Z = \dfrac{x - \mu}{\sigma}$

(*i*) When $x = 98$, $Z = \dfrac{98 - 80}{6} = \dfrac{18}{6} = 3$

(*ii*) When $x = 58$, $Z = \dfrac{58 - 80}{6} = \dfrac{-22}{6} = -3.67$

(*iii*) When $x = 50$, $Z = \dfrac{50 - 80}{6} = \dfrac{-30}{6} = -5$

Example 2. Find the area under the standard normal curve in each of the following:

(*i*) $P(0 \leq Z \leq 1.4)$ (*ii*) $P(-1.67 \leq Z < 0)$

(*iii*) $P(0.65 \leq Z < 2.35)$ (*iv*) $P(-3 \leq Z < 1.6)$

Solution. Using the table of area of standard normal curve, we have

(*i*) $P(0 \leq Z \leq 1.4) = 0.4192$

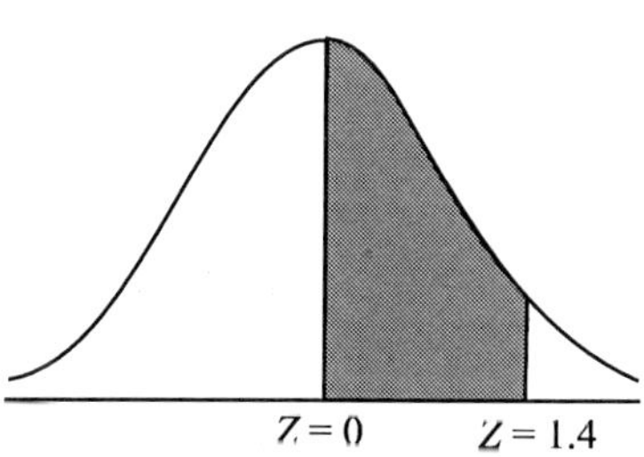

(*ii*) $P(-1.67 \leq Z < 0) = P(0 < Z \leq 1.67) = 0.4525$

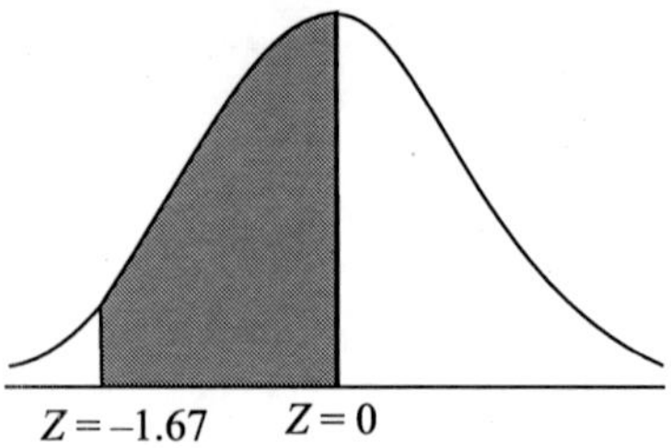

(*iii*) $P(0.65 \leq Z < 2.35) = P(0 < Z \leq 2.35) - P(0 < Z \leq 0.65)$ (by symmetry)

$$= 0.4906 - 0.2422 = 0.2484$$

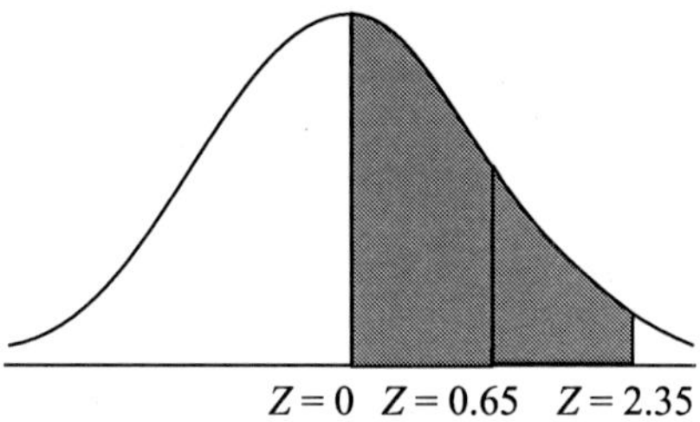

(*iv*) $P(-3 \leq Z < 1.6) = P(-3 \leq Z \leq 0) + P(0 \leq Z \leq 1.6)$

$$= P(0 \leq Z \leq 3) + P(0 \leq Z \leq 1.6) \qquad \text{(by symmetry)}$$

$$= 0.4987 + 0.4452 = 0.9439$$

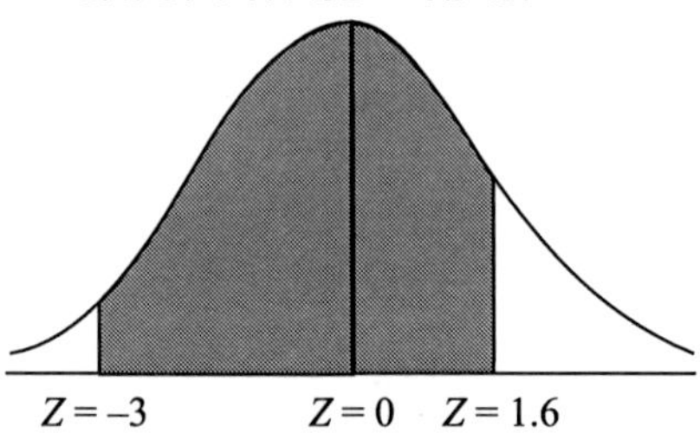

Example 3. Students of a class were given an aptitude test. Their marks were found to be normally distributed with mean 60 and standard deviation 5. What percentage of the students scored more than 60 marks?

Solution. Here $\mu = 60, \sigma = 5, x = 60$

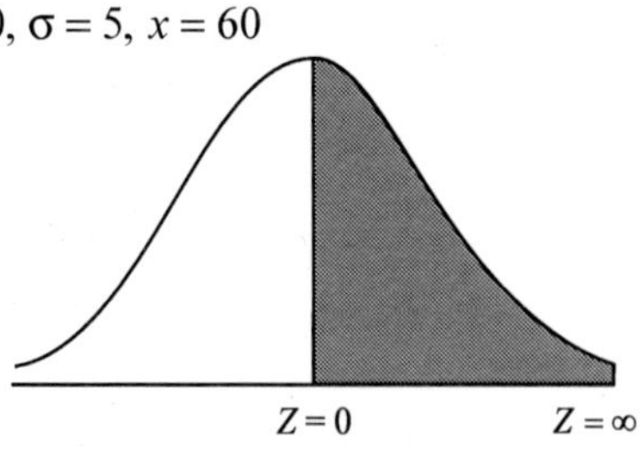

$$Z = \frac{x - \mu}{\sigma}$$

$$= \frac{60-60}{5} = 0$$

$$P(x > 60) = P(Z > 0)$$

$$= P(0 < Z < \infty)$$

$$= 0.5 = 50\%$$

Example 4. A sample of 100 dry battery cells tested to find the length of life produced the following results:

$$\mu = 12 \text{ hours}, \sigma = 3 \text{ hours}$$

Assuming data to be normally distributed, what percentage of battery cells are expected to have life:

(*i*) more than 15 hours (*ii*) less than 6 hours

(*iii*) between 10 and 14 hours?

Solution. Here, x denotes the length of life of dry battery cells.

We know that $Z = \frac{x-\mu}{\sigma}$

$$= \frac{x-12}{3}$$

(*i*) When $x = 15$, $Z = \frac{15-12}{3} = \frac{3}{3} = 1$

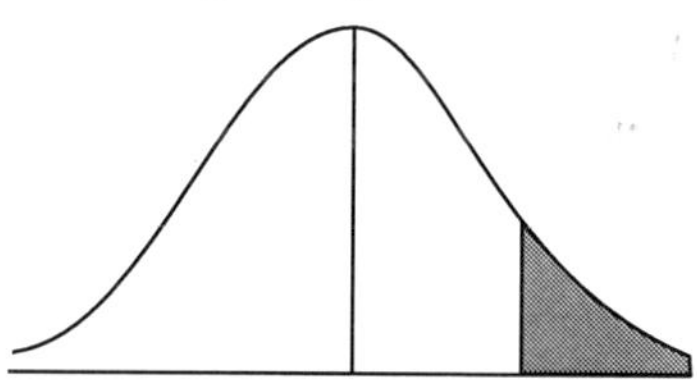

$Z=0$ $Z=1$

$$P(x > 15) = P(Z > 1)$$

$$= P(0 < Z < \infty) - P(0 < Z < 1)$$

$$= 0.5 - 0.3413$$

$$= 0.1587 = 15.87\%$$

(*ii*) When $x = 6$, $Z = \frac{6-12}{3} = \frac{-6}{3} = -2$

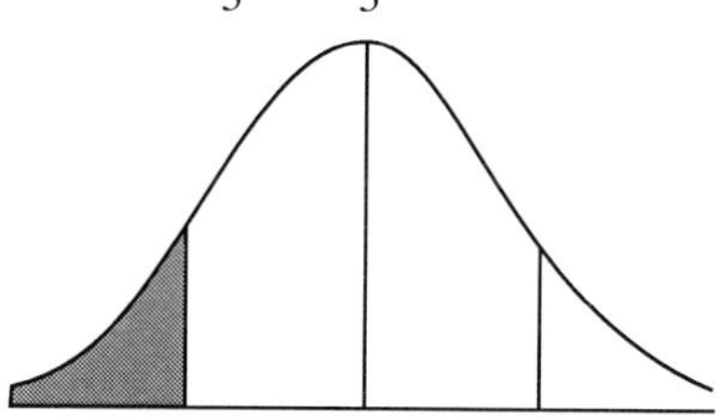

$Z=-2$ $Z=0$ $Z=2$

$$P(x < 6) = P(Z < -2)$$

$$= P(Z > 2)$$

$$= P(0 < Z < \infty) - P(0 < Z < 2)$$
$$= 0.5 - 0.4772$$
$$= 0.0228 = 2.28\%$$

(*iii*) When $x = 10$, $\quad Z = \frac{10-12}{3} = \frac{-2}{3} = -0.67$

When $x = 14$, $\quad Z = \frac{14-12}{3} = \frac{2}{3} = 0.67$

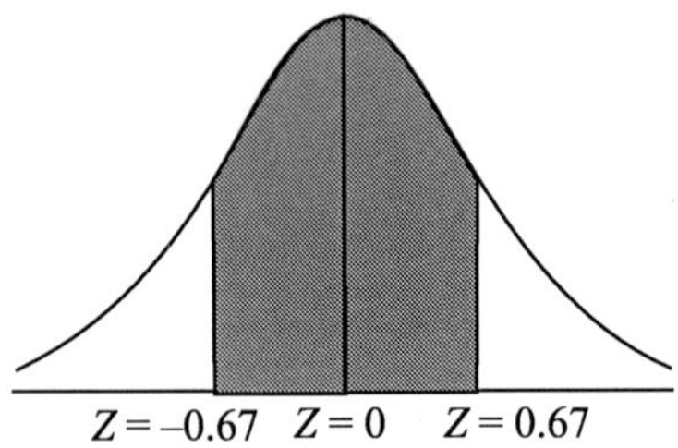

$$P(10 < x < 14) = P(-0.67 < Z < 0.67)$$
$$= P(-0.67 < Z < 0) + P(0 < Z < 0.67)$$
$$= P(0 < Z < 0.67) + P(0 < Z < 0.67)$$
$$= 2P(0 < Z < 0.67)$$
$$= 2 \times 0.2486$$
$$= 0.4972 = 49.72\%$$

Example 5. A normal distribution is given with mean 50 and standard deviation 8. Find the probability that x assumes a value between 38 and 72.

Solution. Here, $\quad \mu = 50, \ \sigma = 8$

$$Z = \frac{x-\mu}{\sigma} = \frac{x-50}{8}$$

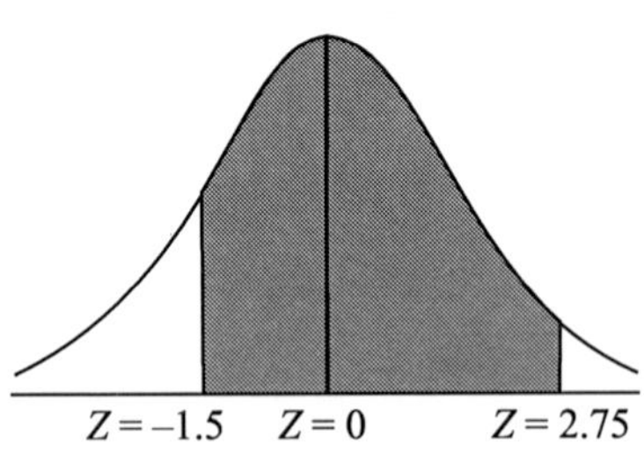

When $x = 38$, $\quad Z = \frac{38-50}{8} = \frac{-12}{8} = -1.5$

When $x = 72$, $\quad Z = \frac{72-50}{8} = \frac{22}{8} = 2.75$

$$\begin{aligned} P(38 < x < 72) &= P(-1.5 < Z < 2.75) \\ &= P(-1.5 < Z < 0) + P(0 < Z < 2.75) \\ &= P(0 < Z < 1.5) + P(0 < Z < 2.75) \\ &= 0.4332 + 0.4970 \\ &= 0.9302 \end{aligned}$$

Example 6. In a sample of 1000 cases, the mean of a certain test is 14 and standard deviation is 2.5. Assuming the distribution to be normal, find:

(*i*) how many students score between 12 and 15?
(*ii*) how many scores above 18?
(*iii*) how many scores below 8?
(*iv*) how many scores 16?

Solution. Here, $N = 1000$, $\mu = 14$ and $\sigma = 2.5$

$$Z = \frac{x - \mu}{\sigma} = \frac{x - 14}{2.5}$$

(*i*) When $x = 12$, $\quad Z = \dfrac{12 - 14}{2.5} = \dfrac{-2}{2.5} = -0.8$

When $x = 15$, $\quad Z = \dfrac{15 - 14}{2.5} = \dfrac{1}{2.5} = 0.4$

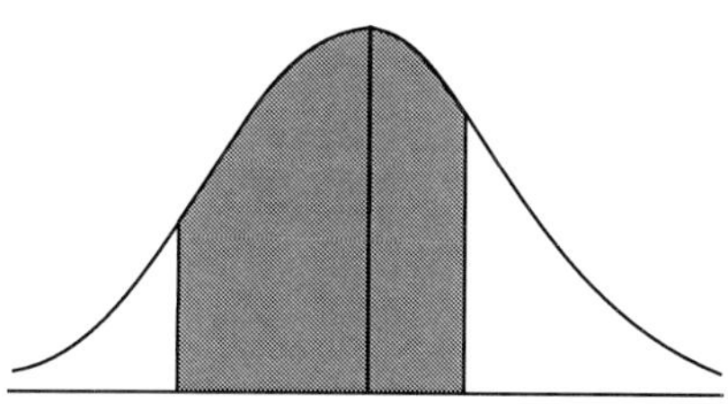

$$\begin{aligned} P(12 < x < 15) &= P(-0.8 < Z < 0.4) \\ &= P(-0.8 < Z < 0) + P(0 < Z < 0.4) \\ &= P(0 < Z < 0.8) + P(0 < Z < 0.4) \\ &= 0.2881 + 0.1554 = 0.4435 \end{aligned}$$

Number of students scoring between 12 and 15 $= 1000 \times 0.4435 = 443.5 \approx 444$ (approx.)

(*ii*) When $x = 8$, $\quad Z = \dfrac{18 - 14}{2.5} = \dfrac{4}{2.5} = 1.6$

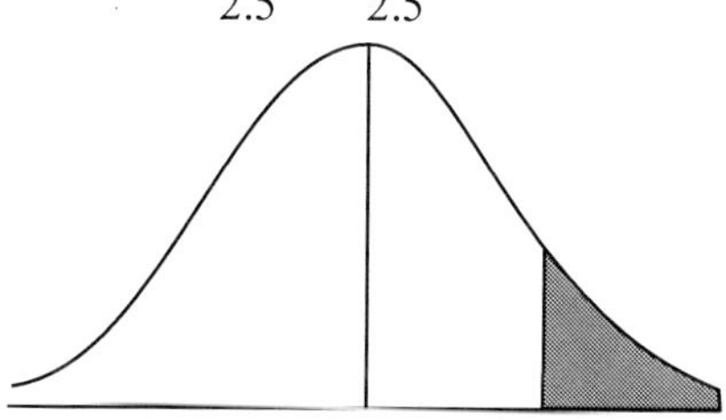

$$P(x > 18) = P(Z > 1.6)$$
$$= P(0 < Z < \infty) - P(0 < Z < 1.6)$$
$$= 0.5 - 0.4452$$
$$= 0.0548$$

Number of students scoring above 18 $= 1000 \times 0.0548 = 54.8 \approx 55$ (approx.)

(*iii*) When $x = 8$, $Z = \dfrac{8-14}{2.5}$

$$= \frac{-6}{2.5} = -2.4$$

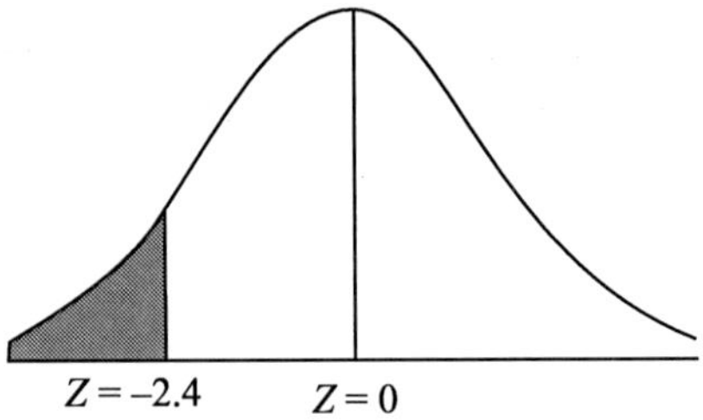

$$P(x < 8) = P(Z < -2.4)$$
$$= P(Z > 2.4)$$
$$= P(0 < Z < \infty) - P(0 < Z < 2.4)$$
$$= 0.5 - 0.4918$$
$$= 0.0082$$

Number of students scoring below 8 $= 1000 \times 0.0082 = 8.2 \approx 8$ (approx.)

(*iv*) Area between $x = 15.5$ and $x = 16.5$

When $x = 15.5$, $Z = \dfrac{15.5-14}{2.5} = \dfrac{1.5}{2.5} = 0.6$

When $x = 16.5$, $Z = \dfrac{16.5-14}{2.5} = \dfrac{2.5}{2.5} = 1$

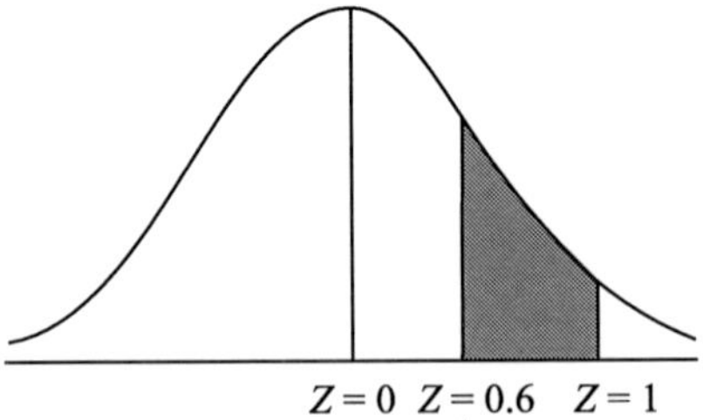

$$P(15.5 < x < 16.5) = P(0.6 < Z < 1)$$
$$= P(0 < Z < 1) - P(0 < Z < 0.6)$$
$$= 0.3413 - 0.2257 = 0.1156$$

Number of students scoring 16 $= 1000 \times 0.1156 = 115.6 \approx 116$ (approx.)

Example 7. The life of army shoes are normally distributed with mean 8 months and standard deviation 2 months. If 5000 pairs are issued, how many pairs would be expected to need replacement after 12 months?

Solution. Here $\mu = 8, \sigma = 2, N = 5000$

$$Z = \frac{x-\mu}{\sigma} = \frac{12-8}{2} = \frac{4}{2} = 2$$

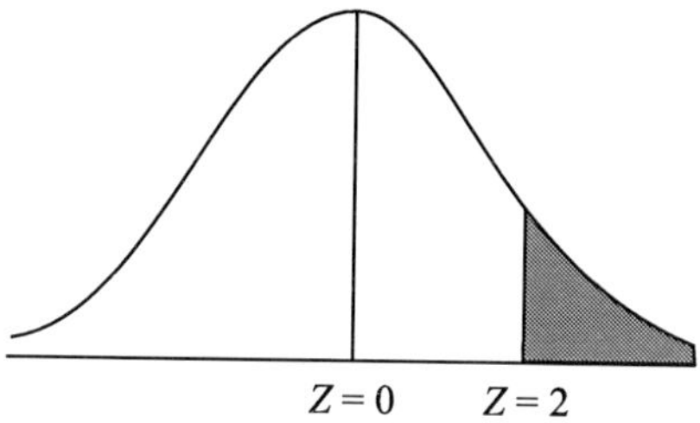

$$\begin{aligned} P(x > 12) &= P(Z > 2) \\ &= P(0 < Z < \infty) - P(0 < Z < 2) \\ &= 0.5 - 0.4772 = 0.0228 \end{aligned}$$

Example 8. Assuming that the diameters of 1000 brass plugs taken consecutively from a machine form a normal distribution with mean 0.7515 inches and standard deviation 0.0020 inches. How many of the plugs are likely to be rejected if the diameter is to be 0.752 ± 0.004 inches?

Solution. Here, $\mu = 0.7515, \sigma = 0.0020, N = 1000$

Least diameter of non-defective plug $= 0.752 - 0.004 = 0.748$ inches

Greatest diameter of non-defective plug $= 0.752 + 0.004 = 0.756$ inches

When x = 0.748, $\quad Z = \dfrac{0.748 - 0.7515}{0.0020} = \dfrac{-0.0035}{0.0020} = -1.75$

When x = 0.756, $\quad Z = \dfrac{0.756 - 0.7515}{0.0020} = \dfrac{0.0045}{0.0020} = 2.25$

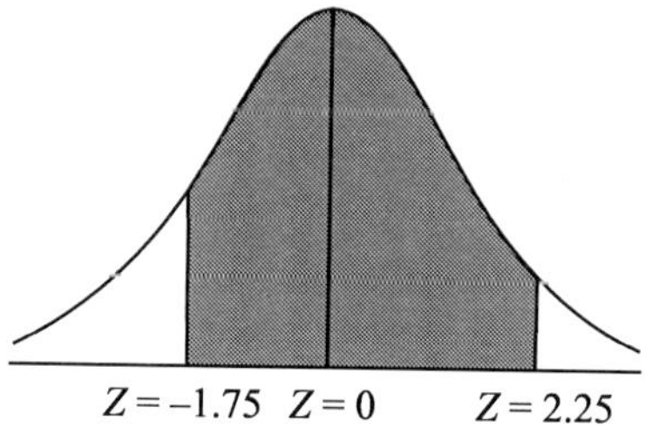

$$\begin{aligned} P(0.748 \le x \le 0.756) &= P(-1.75 \le Z \le 2.25) \\ &= P(-1.75 \le Z \le 0) + P(0 \le Z \le 2.25) \\ &= P(0 \le Z \le 1.75) + P(0 \le Z \le 2.25) \\ &= 0.4599 + 0.4878 = 0.9477 \end{aligned}$$

Number of plugs to be accepted $= 1000 \times 0.9477 = 947.7 \approx 948$ (approx.)

Number of plugs likely to be rejected $= 1000 - 948 = 52.$

Example 9. A manufacturer knows from the experience that the resistance of the resistors he produces is normal with mean 100 ohms and standard deviation 2 ohms. What percentage of resistors will have resistance between 98 ohms and 102 ohms?

Solution. Here $\mu = 100, \sigma = 2$

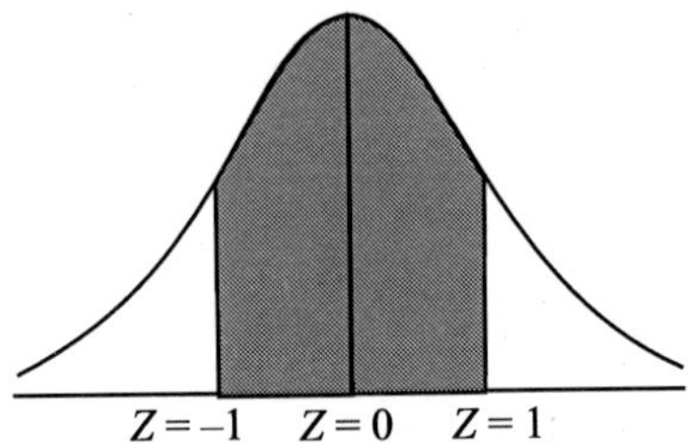

$$Z = \frac{x-\mu}{\sigma} = \frac{x-100}{2}$$

When $x = 98$, $$Z = \frac{98-100}{2} = \frac{-2}{2} = -1$$

When $x = 102$, $$Z = \frac{102-100}{2} = \frac{2}{2} = 1$$

$$\begin{aligned} P(98 < x < 102) &= P(-1 < Z < 1) \\ &= P(-1 < Z \leq 0) + P(0 \leq Z < 1) \\ &= P(0 \leq Z \leq 1) + P(0 \leq Z \leq 1) \\ &= 2P(0 \leq Z \leq 1) \\ &= 2 \times 0.3413 \\ &= 0.6826 \end{aligned}$$

Resistors having resistance between 98 ohms and 102 ohms = 68.26% (approx.)

Example 10. In a normal distribution, 31% of the items are under 45 and 8% are over 64. Find the mean and standard deviation of the distribution.

Solution. Let μ be the mean and σ be the standard deviation.

When $x = 45$, $$Z_1 = \frac{45-\mu}{\sigma}$$

When $x = 64$, $$Z_2 = \frac{64-\mu}{\sigma}$$

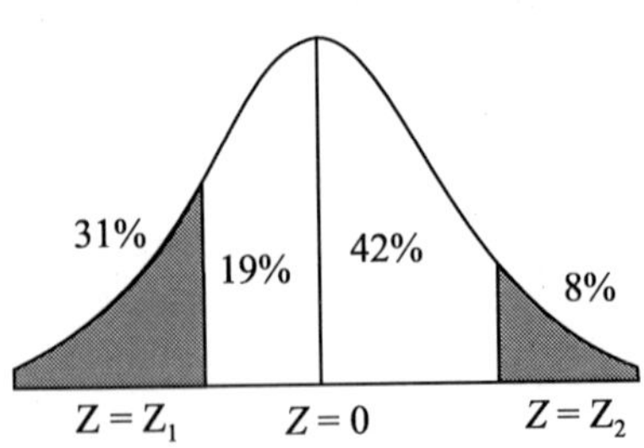

Area between 0 and $Z_1 = 0.50 - 0.31 = 0.19$

From the table, when area is $0.19,\ Z_1 = -0.496 \quad (Z_1 < 0)$

Area between 0 and $Z_2 = 0.50 - 0.08 = 0.42$

From the table, when area is $0.42,\ Z_2 = 1.405$

We have $-0.496 = \dfrac{45 - \mu}{\sigma}$ and $1.405 = \dfrac{64 - \mu}{\sigma}$

Now, on solving the above equations for μ and σ, we have

$$\mu = 50 \text{ and } \sigma = 10.$$

Example 11. Fit a normal curve to the following data:

Length of line (In Cm.)	4	6	8	10	12	14	16	18	20	22	24
Frequency	1	7	15	22	35	43	38	20	13	5	1

Solution. First find the mean μ and standard deviation σ as follows:

Let assumed mean $A = 14$

Length of line (*In Cm.*) x	*Frequency* f	$d = x - A$ $= x - 14$	fd	fd^2
4	1	– 10	– 10	100
6	7	– 8	– 56	448
8	15	– 6	– 90	540
10	22	– 4	– 88	352
12	35	– 2	– 70	140
14	43	0	0	0
16	38	2	76	152
18	20	4	80	320
20	13	6	78	468
22	5	8	40	320
24	1	10	10	100
Total	200		–30	2940

$$\text{Mean } (\mu) = A + \frac{\sum fd}{\sum f} = 14 + \frac{(-30)}{200} = 14 - 0.15 = 13.85$$

$$\text{Standard deviation } (\sigma) = \sqrt{\frac{\sum fd^2}{\sum f} - \left(\frac{\sum fd}{\sum f}\right)^2}$$

$$= \sqrt{\frac{2940}{200} - \left(\frac{-30}{200}\right)^2} = \sqrt{14.7 - 0.0225}$$

$$= \sqrt{14.6775} = 3.83$$

Hence, the equation of the normal curve is given by

$$f(x) = \frac{1}{\sigma\sqrt{2\pi}} e^{-\frac{1}{2}\left(\frac{x-\mu}{\sigma}\right)^2} = \frac{1}{(3.83)\sqrt{2\pi}} e^{-\frac{(x-13.85)^2}{29.355}}$$

EXERCISE 4.4

1. On a final examination in Statistics, the mean was 72, and the standard deviation was 15. Determine the standard scores of students receiving grades:

   (*i*) 60 (*ii*) 93 (*iii*) 72

2. If scores are normally distributed with a mean of 30 and a standard deviation of 5, what percent of the scores is:

   (*i*) greater than 30 (*ii*) greater than 37 (*iii*) between 28 and 34?

3. Find the area under the normal curve in each of the cases:

   (*i*) $Z = 0$ and $Z = 1.2$ (*ii*) $Z = -0.68$ and $Z = 0$

   (*iii*) $Z = -0.46$ and $Z = 2.21$ (*iv*) $Z = 0.81$ and $Z = 1.94$

   (*v*) To the left of $Z = -0.6$ (*vi*) Right of $Z = -1.28$.

4. Find the value of Z in each of the following cases :

   (*i*) area between 0 and Z is 0.3770

   (*ii*) area to the left of Z is 0.8621

5. If X is a normal variate with mean 12 and standard deviation 4, then find

   (*i*) $P(X \geq 20)$ (*ii*) $P(X \leq 20)$

6. The scores of candidates in a certain test are normally distributed with mean 500 and standard deviation 100. What percentage of candidates receives the scores between 350 and 550?

7. In a sample of 1000 items, the mean weight and standard deviation are 45 kgs and 15kgs respectively. Assuming the distribution to be normal, find the number of items weighing between 40 kgs and 60 kgs.

8. A workshop produces 2000 units per day. The average weight of units is 130 kgs with a standard deviation of 10 kgs. Assuming normal distribution, how many units are expected to weigh less than 142 kgs?

9. The mean of a normal distribution is 50 and 5% of the values are greater than 60. Find the standard deviation of the distribution.

10. The time taken to complete a particular type of job is distributed approximately normal with a mean of 1.8 hours and a standard deviation 0.1 hour. If 'Normal time work' finishes at 6:00 p.m. and a job is started at 4:00 p.m. then, what is the probability that the job will need overtime payments?

11. The marks of the students in a class are normally distributed with mean 70 and standard deviation 5. If the instructor decides to give 'A' grade to the top 15% students of the class, how many marks a student must get to be able to get 'A' grade?

12. Find the values of mean and standard deviation from the following data relating to a normal distribution?10% of the items are under 40, 95% of the items are under 75.

13. In a sample of 240 workers in a factory, the mean and standard deviation of wages were Rs. 113.50 and Rs. 30.30 respectively. Find the percentage of workers getting wages between Rs. 90 and Rs. 170 in the whole factory assuming that the wages are normally distributed.

14. In a distribution exactly normal, 7% of the items are under 35 and 89% are under 63. What are the mean and standard deviation of the distribution?

15. Fit a normal curve to the following data:

Variable (r)	0	1	2	3	4	5
Frequency (f)	10	14	19	8	5	4

ANSWERS

1. (*i*) -0.8 (*ii*) 1.4 (*iii*) 0 **2.** (*i*) 50% (*ii*) 8.08% (*iii*) 44.35

3. (*i*) 0.3849 (*ii*) 0.2518 (*iii*) 0.6637 (*iv*) 0.1828 (*v*) 0.2743 (*vi*) 0.8997

4. (*i*) $Z = \pm 1.16$ (*ii*) $Z = 1.09$ **5.** (*i*) 0.0228 (*ii*) 0.9772

6. 62.47% **7.** 471 (approx) **8.** 1770 **9.** 6.1

10. 0.0228 **11.** 75.18 **12.** 55.32; 11.97 **13.** 75%

14. 50.30; 10.33 **15.** $f(x) = \dfrac{1}{(1.40)\sqrt{2\pi}} e^{-\frac{(x-1.93)^2}{3.92}}$

■ ■ ■

UNIT – V

Measures of Central Tendency

5.1 INTRODUCTION

It is a single value within the range of data which represents a group of individual values in a simple manner. Since the value lies within the range of data, it is known as measures of central tendency. There are different types of measures of central values, here we are discussing only three measures as follows:

1. Arithmetic Mean
2. Median
3. Mode

5.2 ARITHMETIC MEAN

1. For Ungrouped Data

The arithmetic mean (A.M.) of the values $x_1, x_2, ..., x_n$ of a variable x is denoted by $\bar{x}$ and is defined as

$$\text{A.M. } (\bar{x}) = \frac{x_1 + x_2 + ... + x_n}{n} = \frac{1}{n}\sum_{i=1}^{n} x_i$$

where n is the number of observations.

As some the values of x will be less and some more than, in some sense, lies at the centre of all the values justifying that it is a measure of central tendency.

2. For Grouped Data

If the values $x_1, x_2, ..., x_n$ of a variable x, occurring with frequencies $f_1, f_2, ..., f_n$ respectively, then the arithmetic mean (A.M.) is defined as

$$\text{A.M. } (\bar{x}) = \frac{f_1x_1 + f_2x_2 + ... + f_nx_n}{f_1 + f_2 + ... + f_n} = \frac{\sum_{i=1}^{n} f_i x_i}{\sum_{i=1}^{n} f_i} = \frac{1}{N}\sum_{i=}^{n} f_i x_i$$

where $N = \sum_{i=1}^{n} f_i$ denotes the total frequency.

If the raw data is in the form of a frequency distribution in which each class interval contains one or more values of the variable (the class interval may or may not be the same) for all classes in such situation we find mid-points of each class interval by taking the mean of the upper and lower limits of the interval. These mid-points denoted by $x_1, x_2, ..., x_n$. Then A.M. is defined as

$$\text{A.M. } (\overline{x}) = \frac{f_1x_1 + f_2x_2 + ... + f_nx_n}{f_1 + f_2 + ... + f_n} = \frac{\sum_{i=1}^{n} f_i x_i}{\sum_{i=1}^{n} f_i} = \frac{1}{N}\sum_{i=1}^{n} f_i x_i$$

5.3 SHORT CUT METHOD (STEP DEVIATION METHOD) FOR COMPUTING ARITHMETIC MEAN

If the values of x_i (or the mid-values x_i in the different classes of a grouped data) are large in magnitude, then the computation of A.M. becomes quite lengthy and tedious. In this case the computation of A.M. may be considerably simplified arithmetically by shifting the origin of reference and at the same time changing the scale.

For example if we take A as the assumed mean (new origin) and take h units of the variate x_i equal to one unit of the new variate u_i then

$$u_i = \frac{x_i - A}{h} \quad or \quad x_i = A + hu_i$$

Then A.M. is calculated as

$$\overline{x} = \frac{1}{N}\sum_{i=1}^{n} f_i x_i = \frac{1}{N}\sum_{i=1}^{n} f_i (A + hu_i)$$

$$= \frac{1}{N}\left[\sum_{i=1}^{n} f_i A + \sum_{i=1}^{n} f_i hu_i\right]$$

$$= A.\frac{1}{N}\sum_{i=1}^{n} f_i + h.\frac{1}{N}\sum_{i=1}^{n} f_i u_i$$

$$= A + h\overline{u}, \quad \text{where } \overline{u} = \frac{1}{N}\sum_{i=1}^{n} f_i u_i$$

Note: In case of class intervals, if the upper class limit of any class interval is not same as the lower class limit of the succeeding class interval then we have to make it same by subtracting 0.5 from lower class limit and adding 0.5 to upper class limit.

5.4 COMBINED MEAN

If $\bar{x}_1$ and $\bar{x}_2$ are the mean of two groups computed from n_1 and n_2 observations respectively, then the mean $\bar{x}$ (combined mean) of all the groups taken is given by

$$\bar{x} = \frac{n_1\bar{x}_1 + n_2\bar{x}_2 + \ldots + n_k\bar{x}_k}{n_1 + n_2 + \ldots + n_k} = \frac{\sum_{i=1}^{k} n_i x_i}{\sum_{i=1}^{k} n_i}$$

5.5 WEIGHTED ARITHMETIC MEAN

If $x_1, x_2, \ldots, x_n$ be the n values of a variable x and for each there is associated a weight w_i, *i.e.*, $w_1, w_2, \ldots, w_n$, then their weighted arithmetic mean (weighted mean) is given by

$$\bar{x}_w = \frac{w_1x_1 + w_2x_2 + \ldots + w_nx_n}{w_1 + w_2 + \ldots + w_n} = \frac{\sum_{i=1}^{n} w_i x_i}{\sum_{i=1}^{n} w_i}$$

5.6 ADVANTAGES OF ARITHMETIC MEAN

1. It is based on all the observations.
2. It is easy to compute.
3. It is by far the most commonly used.
4. It is a calculated value and uniquely defined.
5. It is not based on position in the series.

5.7 DISADVANTAGES OF ARITHMETIC MEAN

1. Since it includes each and every item in a set of data, it is generally affected by extreme values. In some cases its value may be distorted by extreme values that it becomes undesirable as a measure of central tendency.
2. It can not be calculated if any value of the set of data is missing.
3. It can not provide the information about the individual values $x_1, x_2, \ldots x_n$.
4. In cases where the items can not be represented quantitatively, but can be ranked, the arithmetic mean is not an appropriate measure of central tendency.

5.8 PROPERTIES OF ARITHMETIC MEAN

1. The algebraic sum of deviations of a set of n values from their arithmetic mean is zero, *i.e.*, $\sum_{i=1}^{n}(x_i - \bar{x}) = 0$.

Proof : Let $x_1, x_2, \ldots, x_n$ be the set of n observations.

Then, $\sum_{i=1}^{n}(x_i - \overline{x}) = (x_1 - \overline{x}) + (x_2 - \overline{x}) + \ldots + (x_n - \overline{x})$

$$= (x_1 + x_2 + \ldots + x_n) - (\overline{x} + \overline{x} + \ldots + n \text{ times})$$

$$= n\overline{x} - n\overline{x} = 0$$

Similarly, $\sum_{i=1}^{n} f_i(x_i - \overline{x}) = 0$

2. If a constant amount is added or subtracted from each value of the set of data, mean is also added or subtracted by the same constant amount.

Proof : Let $x_1, x_2, \ldots, x_n$ be the set of n observations.

Then mean $(\overline{x}) = \dfrac{x_1 + x_2 + \ldots + x_n}{n} = \dfrac{1}{n}\sum_{i=1}^{n} x_i$

If we add or subtract constant amount a to each observations, the new observations will be $x_1 \pm a,\ x_2 \pm a,\ \ldots, x_n \pm a$.

Now the new mean,

$$(\overline{x}_{new}) = \frac{(x_1 \pm a) + (x_2 \pm a) + \ldots + (x_n \pm a)}{n}$$

$$= \frac{(x_1 + x_2 + \ldots + x_n) + (a \pm a \pm \ldots \pm n \text{ times})}{n}$$

$$= \frac{(x_1 + x_2 + \ldots + x_n)}{n} \pm \frac{na}{n}$$

$$= \overline{x} \pm a$$

i.e., the mean is also added or subtracted by a.

3. If each value of the set of data is multiplied or divided by a constant, the mean is also multiplied or divided by a constant.

Proof : Let $x_1, x_2, \ldots, x_n$ be the set of n observations.

Then mean $(\overline{x}) = \dfrac{x_1 + x_2 + \ldots + x_n}{n} = \dfrac{1}{n}\sum_{i=1}^{n} x_i$

If we multiply constant amount a to each observations, the new observations will be $ax_1,\ ax_2,\ \ldots, ax_n$.

Now the new mean,

$$(\overline{x}_{new}) = \frac{ax_1 + ax_2 + \ldots + ax_n}{n}$$

$$= \frac{a(x_1 + x_2 + \ldots + x_n)}{n} = a.\frac{1}{n}\sum_{i=1}^{n} x_i$$

$$= a\overline{x}$$

i.e., the mean is also multiplied by constant a.

ILLUSTRATIVE EXAMPLES

Example 1. Calculate the A.M. of the following observations:

$$30, 33, 34, 35, 37, 39, 41, 45, 48.$$

Solution. Here, $n = 9$

$$\text{A.M.}\ (\bar{x}) = \frac{1}{9}\sum_{i=1}^{9} x_i$$

$$= \frac{30+33+34+35+37+39+41+45+48}{9}$$

$$= \frac{342}{9} = 38$$

Example 2. Calculate the A.M. of the following data:

x	4	6	8	10	11
f	1	3	4	1	3

Solution.

x	f	fx
4	1	4
6	3	18
8	4	32
10	1	10
11	3	33
	$N = \Sigma f = 12$	$\Sigma fx = 97$

$$\text{A.M.}\ (\bar{x}) = \frac{\sum_{i=1}^{5} f_i x_i}{\sum_{i=1}^{5} f_i} = \frac{97}{12} = 8.083$$

Example 3. The arithmetic mean of the following data is 30.5 marks. Find the missing figure.

Marks (x)	10	20	?	40	50
Frequency (f)	8	10	20	15	7

Solution. Let the missing term be a.

Marks (x)	*Frequency* (f)	fx
10	8	80
20	10	200
a	20	$20a$
40	15	600
50	7	350
	$N = \Sigma f = 60$	$\Sigma f = 1230 + 20a$

$$\text{A.M. } (\bar{x}) = \frac{\sum_{i=1}^{5} f_i x_i}{N}$$

$$30.5 = \frac{1}{60} \times (1230 + 20a)$$

$$20a = 1830 - 1230 = 600$$

$$a = \frac{600}{20} = 30$$

Example 4. The marks obtained by 30 students in an examination in Mathematics are given in the following distribution table. Calculate the A.M.

Marks	0-10	10-20	20-30	30-40	40-50	50-60	60-70	70-80
Frequency	2	2	4	10	6	3	1	2

Solution.

Marks	*Frequency* (f)	*Mid-points* (x)	fx
0-10	2	5	10
10-20	2	15	30
20-30	4	25	100
30-40	10	35	350
40-50	6	45	270
50-60	3	55	165
60-70	1	65	65
70-80	2	75	150
	$N = \Sigma f = 30$		$\Sigma fx = 1140$

$$\text{A.M. } (\bar{x}) = \frac{\sum_{i=1}^{8} f_i x_i}{N}$$

$$= \frac{1140}{30} = 38$$

Example 5. The arithmetic mean of the following frequency distribution is 33. Find the missing frequency.

Class interval	0-10	10-20	20-30	30-40	40-50	50-60
Frequency	10	15	30	—	25	20

Solution. Let the missing frequency be a.

Class interval	*Frequency* (f)	*Mid Points* (x)	fx
0-10	10	5	50
10-20	15	15	225

20-30	30	25	750
30-40	a	35	$35a$
40-50	25	45	1125
50-60	20	55	1100
	$N=\Sigma f=100+a$		$\Sigma fx=3250+35a$

$$\text{A.M.}\ (\bar{x}) = \frac{\sum_{i=1}^{6} f_i x_i}{N}$$

$$33 = \frac{1}{100+a} \times (3250+35a)$$

$$3300+33a = 3250+35a$$

$$a = \frac{50}{2} = 25$$

Example 6. Calculate the A.M. of the following data by using short cut method :

x	4	6	8	10	11
f	1	3	4	1	3

Solution. Using $u_i = \frac{x_i - A}{h}$, where $A = 8, h = 1$

x	f	$u_i = x_i - 8$	fu
4	1	–4	–4
6	3	–2	–6
8	4	0	0
10	1	2	2
11	3	3	9
	$N=\Sigma f=12$		$\Sigma fu=1$

$$\text{A.M.}\ (\bar{x}) = A + h\bar{u}$$

$$= A + \frac{h}{N}\sum_{i=1}^{5} f_i u_i$$

$$= 8 + \frac{1}{12} \times 1 = 8 + 0.083 = 8.083$$

Example 7. Calculate the A.M. of the following data by using short cut method :

x	61	64	67	70	73
f	5	18	42	27	8

Solution. Using $u_i = \frac{x_i - A}{h}$, where $A = 67, h = 3$

x_i	f_i	$u_i = \frac{x_i - 67}{3}$	$f_i u_i$
61	5	–2	–10
64	18	–1	–18
67	42	0	0
70	27	1	27
73	8	2	16
	$N = \Sigma f = 100$		$\Sigma fu = 15$

$$\text{A.M. } (\overline{x}) = A + h\overline{u}$$

$$= A + \frac{h}{N}\sum_{i=1}^{5} f_i u_i$$

$$= 67 + \frac{3}{100} \times 15 = 67 + 0.45 = 67.45$$

Example 8. Calculate the A.M. of the following data by using short cut method:

Marks	0-10	10-20	20-30	30-40	40-50	50-60	60-70	70-80
Frequency	2	2	4	10	6	3	1	2

Soluton. Using $u_i = \frac{x_i - A}{h}$, where $A = 45, h = 10$

Marks	*Frequency* (f_i)	*Mid-points* (x_i)	$u_i = \frac{x_i - 45}{10}$	$f_i u_i$
0-10	2	5	–4	–8
10-20	2	15	–3	–6
20-30	4	25	–2	–8
30-40	10	35	–1	–10
40-50	6	45	0	0
50-60	3	55	1	3
60-70	1	65	2	2
70-80	2	75	3	6
	$N = \Sigma f = 30$			$\Sigma f = -21$

$$\text{A.M. } (\overline{x}) = A + h\overline{u}$$

$$= A + \frac{h}{N}\sum_{i=1}^{8} f_i u_i$$

$$= 45 + \frac{10}{30} \times (-21) = 45 - 7 = 38$$

Example 9. Calculate the A.M. of the following data by using short cut method:

Class interval	0-9	10-19	20-29	30-39	40-49
Frequency	3	5	2	4	1

Solution. Here, in each class interval upper class limit is not same as the lower class limit of succeeding one, to do so subtract 0.5 from lower class limits and adding 0.5 to upper class limits.

Class interval	*Frequency (f)*	*Mid-points (x)*	*fx*
–0.5-9.5	3	4.5	13.5
9.5-19.5	5	14.5	72.5
19.5-29.5	2	24.5	49.5
29.5-39.5	4	34.5	138.0
39.5-49.5	1	44.5	44.5
	$N = \Sigma f = 15$		$\Sigma fx = 317.5$

$$\text{A.M. } (\bar{x}) = \frac{\sum_{i=1}^{5} f_i x_i}{N}$$

$$= \frac{317.5}{15} = 21.167$$

Example 10. The three parts of a distribution are as follows:

Part	*Mean*	*Number (frequency)*
I	68	20
II	75	18
III	86	12

Find the mean of the entire distribution.

Solution. Here, $n_1 = 20,\ \bar{x}_1 = 68,\ n_2 = 18,\ \bar{x}_2 = 75,\ n_3 = 12,\ \bar{x}_3 = 86$

$$\text{Combined arithmetic mean } \bar{x} = \frac{n_1\bar{x}_1 + n_2\bar{x}_2 + n_3\bar{x}_3}{n_1 + n_2 + n_3}$$

$$= \frac{20\times 68 + 18\times 75 + 12\times 86}{20 + 18 + 12}$$

$$= \frac{3742}{50} = 74.84$$

Example 11. A distribution consists of three components with frequencies 30, 20 and 60 having their means 16, 8 and 4 respectively. Find the mean of the combined distribution.

Solution. Here, $n_1 = 30, n_2 = 20, n_3 = 60,\ \bar{x}_1 = 16,\ \bar{x}_2 = 8,\ \bar{x}_3 = 4$

$$\text{Combined arithmetic mean } \bar{x} = \frac{n_1\bar{x}_1 + n_2\bar{x}_2 + n_3\bar{x}_3}{n_1 + n_2 + n_3}$$

$$= \frac{30 \times 16 + 20 \times 8 + 60 \times 4}{30 + 20 + 60}$$

$$= \frac{880}{110} = 8$$

Example 12. Mean marks in a subject of 80 students in two classes *A* and *B* is 50. There are 45 students in class *A*. The mean marks of the students in class *B* is 48. Find the mean marks of the students in class *A*.

Solution. Here, $n_1 = 45, n_2 = 80 - 45 = 35, \bar{x} = 50, \bar{x}_2 = 48$

We have to find the value of $\bar{x}_1$

Combined arithmetic mean $\bar{x} = \dfrac{n_1\bar{x}_1 + n_2\bar{x}_2}{n_1 + n_2}$

$$50 = \frac{45 \times \bar{x}_1 + 35 \times 48}{45 + 35}$$

$$45\bar{x}_1 = 4000 - 1680 = 2320$$

$$\bar{x}_1 = \frac{2320}{45} = 51.56 \text{ marks}$$

Mean marks of the students in class $A = 51.56$

Example 13. The mean of 20 numbers is 50. By mistake marks of the two students were taken as 64 and 67 instead of 46 and 76. Find the corrected mean.

Solution. Total marks of the 20 students = 20 × 50 = 1000

Corrected total = Total – (sum of wrong marks) + (sum of correct marks)

$= 1000 - (64 + 67) + (46 + 76) = 991$

$$\text{Corrected mean} = \frac{\text{Corrected total}}{\text{No. of students}} = \frac{991}{20} = 49.55$$

Corrected mean = 49.55 marks.

Example 14. The mean of seven observations is 8. A new observation 16 is added. Find the mean of eight observations.

Solution. Total of seven observations = 7 × 8 = 56

Total of eight observations = 56 + 16 = 72

$$\text{Mean of eight observations} = \frac{72}{8} = 9$$

Example 15. The mean weight of 150 students in a class is 60 kg. The mean weight of boy students is 70 kg. and that of girl students is 55 kg. Find the number of boys and girls in the class.

Solution. Let the number of boys be n_1 and that of girls be n_2. We have $n_1 + n_2 = 150$,

$\bar{x} = 60, \bar{x}_1 = 70, \bar{x}_2 = 55$

Combined arithmetic mean $\bar{x} = \dfrac{n_1\bar{x}_1 + n_2\bar{x}_2}{n_1 + n_2}$

$$60 = \frac{n_1 \times 70 + (150 - n_1) \times 55}{150}$$

$$70n_1 - 55n_1 = 9000 - 8250$$

$$15n_2 = 750 \Rightarrow n_1 = 50$$

Now, $n_2 = 150 - n_1 = 150 - 50 = 100$

Number of boys in the class = 50

Number of girls in the class = 100.

Example 16. A candidate obtains the following percentage of marks in an examination. English 60, Hindi 75, Mathematics 63, Physics 55, Chemistry 62. Find the weighted mean if weights 1, 2, 1, 3, 3 respectively are allotted to the subjects.

Solution. Here, $n = 5, x_1 = 60, x_2 = 75, x_3 = 63, x_4 = 55, x_5 = 62$

$$w_1 = 1, w_2 = 2, w_3 = 1, w_4 = 3, w_5 = 3$$

Weighted mean $\bar{x}_w = \dfrac{\sum_{i=1}^{5} w_i x_i}{\sum_{i=1}^{5} w_i}$

$$= \frac{1\times60 + 2\times75 + 1\times63 + 3\times55 + 3\times62}{1+2+1+3+3}$$

$$= \frac{624}{10} = 62.4$$

Example 17. Find the weighted arithmetic mean of first n natural numbers whose weights are equal to the corresponding numbers.

Solution. Weighted arithmetic mean

$$\bar{x}_w = \frac{\sum_{i=1}^{n} w_i x_i}{\sum_{i=1}^{n} w_i}$$

$$= \frac{1\times1 + 2\times2 + 3\times3 + \ldots + n\times n}{1+2+3+\ldots+n}$$

$$= \frac{1^2 + 2^2 + 3^2 + \ldots + n^2}{1+2+3+\ldots+n}$$

$$= \frac{\dfrac{n(n+1)(2n+1)}{6}}{\dfrac{n(n+1)}{2}} = \frac{(2n+1)}{3}$$

Note: Sum of the squares of first n natural numbers $= \dfrac{n(n+1)(2n+1)}{6}$ and sum of the first n natural numbers $= \dfrac{n(n+1)}{2}$.

Example 18. A person invests Rs. 1000 at 7%, Rs. 3000 at 8% and Rs. 16000 at 9%. What is the overall percentage yield of these investments?

Solution. Here, $w_1 = 1000, w_2 = 3000, w_3 = 16000, x_1 = 7, x_2 = 8, x_3 = 9$

$$\text{Weighted mean } \bar{x}_w = \frac{\sum_{i=1}^{3} w_i x_i}{\sum_{i=1}^{3} w_i}$$

$$= \frac{1000 \times 7 + 3000 \times 8 + 16000 \times 9}{1000 + 3000 + 16000}$$

$$= \frac{175000}{20000} = 8.75\%$$

5.9. MEDIAN

The median is a positional measure of a data set. It is most preferable measure of location for asymmetric distributions. Median is the value of the variable which divides the total frequency into two equal parts. Fifty per cent values are to be the left of the median and fifty per cent are to the right of the median. The median of a series is the middle value when the items are arranged in the ascending or descending order.

Median From Ungrouped Data : If the values x_i in the series are arranged in ascending or descending order of their magnitude, then the middle most value in this arrangement is called the median.

When n is odd : If the number of values, say n, in the series is odd then the value corresponding to $\left(\dfrac{n+1}{2}\right)^{th}$ item is the median.

When n is even: If the number of values, say n, in the series is even then the value corresponding to the average of $\left(\dfrac{n}{2}\right)^{th}$ and $\left(\dfrac{n}{2}+1\right)^{th}$ item is the median.

i.e.,

$$\text{Median} = \frac{\left(\frac{n}{2}\right)^{th} item + \left(\frac{n}{2}+1\right)^{th} item}{2}$$

Median From Grouped Data : In grouped data there may be two series, discrete series or continuous series.

(*a*) Discrete Series: The steps involved are

(*i*) Arrange the items in ascending or descending order of their magnitude.

(*ii*) Compute the cumulative frequency.

(*iii*) Find the value of the $\left(\frac{N+1}{2}\right)^{th}$ item.

(*iv*) The size of the item for which C.F. is just greater them $\frac{N+1}{2}$ value is the median.

Note: $$N = \sum_{i=1}^{n} f_i$$

(*b*) Continuous Series: The steps involved are

(*i*) Arrange the items in ascending or descending order of their magnitude.

(*ii*) Compute the cumulative frequency.

(*iii*) Find the value of the $\left(\frac{N}{2}\right)^{th}$ item.

(*iv*) After finding the class in which the median value lies, the exact value has to be interpolated from the relevant class.

To calculate the median value the following formula is used:

$$\text{Median } (M_d) = L_1 + \frac{\frac{N}{2} - c.f.}{f} \times h$$

where L_1 = real lower limit of the median class

$$N = \sum_{i=1}^{n} f_i$$

$\frac{N}{2}$ = item whose value has to be interpolated

$c.f.$ = cumulative frequency of class preceding to the median class

f = frequency of the median class

h – size of the class interval in the median class.

Note: The class whose cumulative frequency is the first to exceed is called the median class.

5.10. GRAPHICAL METHOD FOR FINDING THE MEDIAN

The steps involved are

(*i*) Arrange the items in ascending or descending order of their magnitude.

(*ii*) Compute the cumulative frequency.

(*iii*) Draw the cumulative curve (Ogive). Plot the variable values on the x-axis and the cumulative frequencies on the y-axis.

(*iv*) Locate the point $\frac{N}{2}$ on the y-axis and draw a horizontal line from this point which cuts the Ogive at a point, say, M.

(*v*) From this point M, draw a perpendicular to the x-axis.

(*vi*) The point where the perpendicular touches the x-axis is the median.

5.11. ADVANTAGES OF MEDIAN

1. Median is not affected by extreme values because it is a positional average.
2. Median can be calculated in case of distribution with open ends intervals.
3. Median can be located even if the data are incomplete.
4. Median is applicable to cases which are not capable of precise quantitative studies.

5.12. DISADVANTAGES OF MEDIAN

1. A slight change in the series may bring drastic change in the value of median.
2. Median may not be representative value of a given series when the distribution is irregular.
3. In case of even number of observations or continuous series, median is an estimated value other than any value in the series.
4. Median is not suitable for further mathematical treatment except its use in mean deviation.

ILLUSTRATIVE EXAMPLES

Example 1. Find the median of the observations:

$$6, 14, 9, 17, 11, 16, 18, 23, 5.$$

Solution. Arranging the data in ascending order, we get

$$5, 6, 9, 11, 14, 16, 17, 18, 23$$

Number of items $n = 9$

Since the number of items is odd, so $\left(\frac{n}{2}+1\right)^{th}$ item *i.e.*, $\frac{9+1}{2} = 5^{th}$ item is the median.

$$\text{Median} = 14$$

Example 2. Find the median of the observations:

$$16, 14, 12, 9, 11, 8, 7, 18.$$

Solution. Arranging the data in ascending order, we get

$$7, 8, 9, 11, 12, 14, 16, 18$$

Number of items $n = 8$

Since the number of items is even, so

$$\text{Median} = \frac{\left(\frac{n}{2}\right)^{\text{th}} \text{item} + \left(\frac{n}{2}+1\right)^{\text{th}} \text{item}}{2}$$

$$= \frac{4^{\text{th}} \text{item} + 5^{\text{th}} \text{item}}{2} = \frac{11+12}{2} = 11.5$$

Example 3. Find the median of the observations:

$$23, 22, 21, 30, 38, 25, 28, 23, 18, 8, 13, 43.$$

Solution. Arranging the data in ascending order, we get

$$8, 13, 18, 21, 22, 23, 25, 28, 30, 38, 43$$

Number of items $\quad n = 12$

Since the number of items is even, so

$$\text{Median} = \frac{\left(\frac{n}{2}\right)^{\text{th}} \text{item} + \left(\frac{n}{2}+1\right)^{\text{th}} \text{item}}{2}$$

$$= \frac{6^{\text{th}} \text{item} + 7^{\text{th}} \text{item}}{2} = \frac{23+23}{2} = 23$$

Example 4. The median of the following observations arranged in ascending order is 40.

$$20, 22, 31, 35, x-1, x+1, 42, 45, 49, 56$$

Find the value of x.

Solution. Since the number of items $n = 10$ (even)

So, $$\text{Median} = \frac{\left(\frac{n}{2}\right)^{\text{th}} \text{item} + \left(\frac{n}{2}+1\right)^{\text{th}} \text{item}}{2}$$

$$= \frac{5^{\text{th}} \text{item} + 6^{\text{th}} \text{item}}{2}$$

$$40 = \frac{(x-1)+(x+1)}{2}$$

$$2x = 80 \Rightarrow x = 40$$

Example 5. The median of eight observations

33, 50, 39, 36, 47, 38, 43 and x is 40, where $39 < x < 43$.

Find the value of x.

Solution.

Arranging the data in ascending order, we get

$$33, 36, 38, 39, x, 43, 47, 50$$

Number of items $n = 8$ (even)

So, $$\text{Median} = \frac{\left(\frac{n}{2}\right)^{\text{th}} \text{item} + \left(\frac{n}{2}+1\right)^{\text{th}} \text{item}}{2}$$

$$= \frac{4^{\text{th}} \text{ item} + 5^{\text{th}} \text{ item}}{2}$$

$$40 = \frac{39 + x}{2}$$

$$x = 80 - 39 \Rightarrow x = 41$$

Example 6. Find the median for the following frequency distribution:

x	3	5	6	9	10	15
f	5	9	18	13	8	7

Solution. First we make the cumulative frequency table as:

x	f	*Cumulative frequency (c.f.)*
3	5	5
5	9	14
(6)	18	(32)
9	13	45
10	8	53
15	7	60
Total	$N = 60$	

Here, $N = 60, \frac{N+1}{2} = \frac{60+1}{2} = 30.5$

The value of x for which *c.f.* is just greater than 30.5 is 6.

So median = 6.

Example 7. Find the median for the following frequency distribution:

x	60	50	52	61	56
f	5	8	6	3	7

Solution. First arranging the data in ascending order and then we make the cumulative frequency table as:

x	f	*Cumulative frequency (c.f.)*
50	8	8
52	6	14
(56)	7	(21)
60	5	26
61	3	29
Total	$N = 29$	

Here, $N = 29, \frac{N+1}{2} = \frac{29+1}{2} = 15$

The value of x for which *c.f.* is just greater than 15 is 56.

So median = 56.

Example 8. Find the median for the following frequency distribution:

Class (Marks)	0-10	10-20	20-30	30-40	40-50
No. of candidates	5	8	20	10	7

Solution. First we make the cumulative frequency table as:

Class (Marks)	*No. of candidates*	*Cumulative frequency (c.f.)*
0-10	5	5
10-20	8	13 (*c.f.*)
(L_1) 20-30	20 (*f*)	33
30-40	10	43
40-50	7	43
Total	$N = 50$	

Here, $\frac{N}{2} = \frac{50}{2} = 25$. So median class is 20-30 and $h = 10$

$$\text{Median} = L_1 + \frac{\frac{N}{2} - c.f.}{f} \times h$$

$$= 20 + \frac{25 - 13}{20} \times 10 = 20 + 6 = 26$$

Example 9. Find the median for the following data:

Marks below	10	20	30	40	50	60	70	80
No. of students	15	35	60	84	96	127	198	250

Solution. The data is given in cumulative frequency form. First we have to find frequencies as follows:

Class (Marks)	*No. of students*	*Cumulative frequency (c.f.)*
0-10	15	15
10-20	$35 - 15 = 20$	35
20-30	$60 - 35 = 25$	60
30-40	$84 - 60 = 24$	84
40-50	$96 - 84 = 12$	96 (*c.f.*)
(L_1) 50-60	$127 - 96 = 31$ (*f*)	127
60-70	$198 - 127 = 71$	198
70-80	$250 - 198 = 52$	250
Total	$N = 250$	

Here, $\frac{N}{2} = \frac{250}{2} = 125$. So median class is 50-60 and $h = 10$

$$\text{Median} = L_1 + \frac{\frac{N}{2} - c.f.}{f} \times h$$

$$= 50 + \frac{125 - 96}{31} \times 10 = 50 + 9.355 = 59.355$$

Example 10. Find the missing frequencies f_1 and f_2 in the following distribution. It is given that median of the distribution is 26 and the total number of candidates is 50.

Marks	0-10	10-20	20-30	30-40	40-50
No. of candidates	5	f_1	20	f_2	7

Solution. First we make the cumulative frequency table as:

Class (Marks)	*No. of candidates*	*Cumulative frequency (c.f.)*
0-10	5	5
10-20	f_1	$5+f_1$ (c.f.)
(L_1)20-30	20 (f)	$25+f_1$
30-40	f_2	$25+f_1+f_2$
40-50	7	$32+f_1+f_2$

Here, $\frac{N}{2}=\frac{50}{2}=25$. So median class is 20-30 and $h=10$

$$\text{Median} = L_1+\frac{\frac{N}{2}-c.f.}{f}\times h$$

$$26 = 20+\frac{25-(5+f_1)}{20}\times 10$$

$$6 = 20+\frac{20-f_1}{2}$$

$$\Rightarrow \qquad f_1 = 20-12=8$$

Total number of candidates = 50

$$5+f_1+20+f_2+7 = 50 \Rightarrow f_1+f_2=50-32=18$$

$$f_2 = 18-f_1=18-8$$

Hence, $f_1=8$ and $f_2=10$.

Example 11. Draw the less than ogive for the following distribution and find median graphically:

Marks	*No. of students*
0-30	3
30-40	14
40-50	30
50-60	34
60-70	28
70-80	23
80-90	15
90-100	3
Total	$N=150$

Solution. First we make the cumulative frequency table as:

Marks	*No. of students*	*Cumulative frequency (c.f.)*
0-30	3	3
30-40	14	17
40-50	30	47
50-60	34	81
60-70	28	109
70-80	23	132
80-90	15	147
90-100	3	150
Total	$N = 150$	

Now we draw the less than cumulative frequency curve (Ogive) against the upper limit of the corresponding class in the above table as

$$\frac{N}{2} = \frac{150}{2} = 75$$

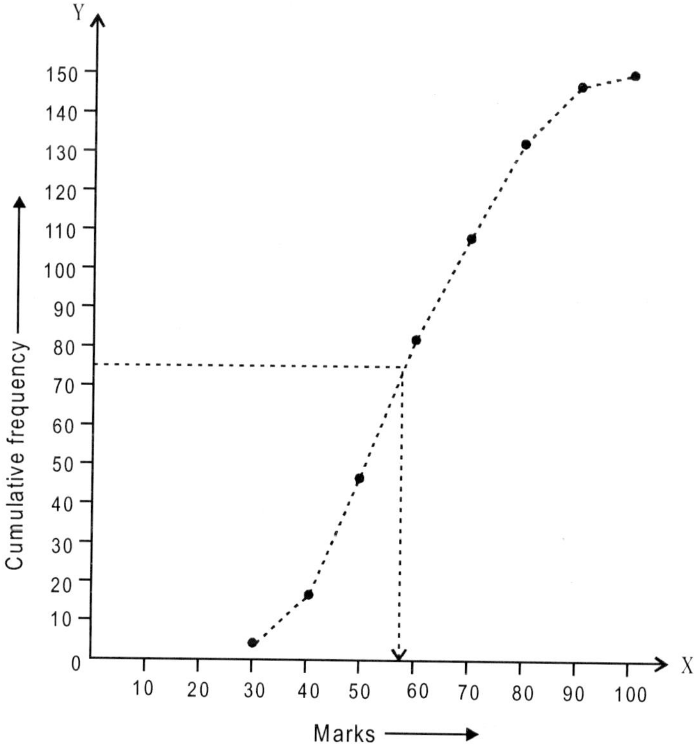

Median = 58 marks

5.13. MODE

The variate value x having maximum frequency in a distribution is known as its mode. In other words, the mode is that value of a series which appears most frequently than any other. Mode is not necessarily unique, there can be more than one mode of a distribution.

Mode From Grouped Data : In grouped data there may be two series, discrete series or continuous series.

(*a*) ***Discrete Series :*** Mode of a discrete frequency distribution can be determined by inspection. The value having maximum frequency is the mode of the distribution.

(*b*) ***Continuous Series:*** Mode of a continuous frequency distribution with classes in ascending order can be calculated by using the following formula:

$$\text{Mode } (M_o) = L_0 + \frac{f_0 - f_1}{2f_0 - f_1 - f_2} \times h$$

where

L_0 = lower limit of the modal class

f_0 = frequency of the modal class

f_1 = frequency of the preceding class to the modal class

f_2 = frequency of the succeeding class to the modal class

h = size of class interval

Note:

1. While calculating value of mode it should be seen that the class intervals of the different classes are equal otherwise the above formula will not give the correct answer. In this case mode can be find by using an empirical relationship between mean, median and mode as

   Mean – Mode = 3 (Mean – Median) or Mode = 3 Median – 2 Mean

2. A frequency distribution having more than one mode is called multimodal distribution. If there is only one mode it is called unimodal distribution.
3. The class having the maximum frequency is called the modal class.

5.14. GRAPHICAL METHOD FOR FINDING THE MODE

The steps involved are :

(*i*) Draw a histogram of the given distribution.

(*ii*) Join the top right corner of the highest rectangle by a straight line to the top right corner of the preceding rectangle. Similarly, the top left corner of the highest rectangle is joined to the left corner of the rectangle of the right.

(*iii*) From the point of intersection of these two diagonal lines, draw a perpendicular to the x-axis.

(*vi*) The point where the perpendicular touches the x-axis is the mode.

5.15. GROUPING METHOD FOR FINDING THE MODE

Grouping method is used when there exists one or more of the following cases:

(*i*) If the maximum frequency is repeated.

(*ii*) If the maximum frequency occurs in the very beginning or at the end of the distribution.

(*iii*) If there are irregularities in the distribution.

In grouping method, the values are first arranged in ascending order of their magnitude, then grouping table is prepared. Normally the grouping table has six columns.

Column 1: It contains the original frequencies.

Column 2: The frequencies are grouped in twos, the total are indicated against each pair.

Column 3: Leave the first frequency and remaining frequencies are grouped in twos.

Column 4: The frequencies are grouped in threes.

Column 5: Leave the first frequency and remaining frequencies are grouped in threes.

Column 6: Leaving the first two frequencies, the remaining frequencies are grouped in threes.

Note: Underline the highest value of each column.

Now an analysis table is formed using grouping table for finding the value which is repeated the maximum number of times. This value is the mode.

5.16. ADVANTAGES OF MODE

1. Mode can be found by mere inspection in some cases.
2. Mode is not at all affected by the extreme values.
3. Mode is of great importance in manufacturing of garments, shoes etc. due to most common size.
4. Mode is one of the value in the series.
5. Mode is not affected by open end classes.

5.17. DISADVANTAGES OF MODE

1. Mode is not based on all observations.
2. Mode is not capable of further mathematical treatment.
3. Mode is not rigidly defined. Different methods can give different values of mode.
4. Mode is not unique.
5. Mode is based only on concentrated values.

ILLUSTRATIVE EXAMPLES

Example 1. Find the mode from the following data:

$$38, 26, 28, 23, 25, 26, 36, 25, 26, 28.$$

Solution. Arranging the data as frequency distribution, we have

Value	*Frequency*
23	1
25	2
26	3
28	2
36	1
38	1

It is clear from the table that the highest frequency is for the value 26.

$$\text{Mode} = 26$$

Example 2. Find the mode from the following frequency distribution:

x	8	10	12	14	16	18	20
f	3	5	9	10	20	9	4

Solution. In the given problem mode can be find by inspection.

The variate value 16 has the maximum frequency 20.

$$\text{Mode} = 16$$

Example 3. Find mode from the following frequency distribution:

Class	52-55	55-58	58-61	61-64
Frequency	15	20	25	10

Solution.

Class	*Frequency*
52-55	15
55-58	20 (f_1)
(L_0) 58-61	25 (f_0)
61-64	10 (f_2)

Here, maximum frequency 25 occurs in the class 58-61, so it is the modal class.

$$L_0 = 58, f_0 = 25, f_1 = 20, f_2 = 10 \text{ and } h = 3$$

$$\text{Mode } (M_o) = L_0 + \frac{f_0 - f_1}{2f_0 - f_1 - f_2} \times h$$

$$= 58 + \frac{25 - 20}{(2 \times 25) - 20 - 10} \times 3$$

$$= 58 + 0.75 = 58.75$$

Example 4. Find mode from the following frequency distribution:

Class interval	0-10	10-20	20-30	30-40	40-50	50-60
Frequency	8	7	16	24	15	7

Solution.

Class interval	*Frequency*
0-10	8
10-20	7
20-30	16 (f_1)
(L_0) 30-40	24 (f_0)
40-50	15 (f_2)
50-60	7

Here, maximum frequency 24 occurs in the class 30-40, so it is the modal class.

$$L_0 = 30, f_0 = 24, f_1 = 16, f_2 = 15 \text{ and } h = 10$$

$$\text{Mode } (M_o) = L_0 + \frac{f_0 - f_1}{2f_0 - f_1 - f_2} \times h$$

$$= 30 + \frac{24 - 16}{(2 \times 24) - 16 - 15} \times 10$$

$$= 30 + 4.706 = 3.4.706$$

Example 5. The expenditure of 100 families is given below:

Expenditure	0-10	10-20	20-30	30-40	40-50
No. of families	14	?	27	?	15

Given that mode is 24. Determine the missing frequencies.

Solution. Let the missing frequencies be a and b.

Expenditure	*No. of families*
0-10	14
10-20	a (f_1)
(L_0) 20-30	27 (f_0)
30-40	b (f_2)
40-50	15

Here, mode is 24, so modal class is 20-30.

$$L_0 = 20, f_0 = 27, f_1 = a, f_2 = b \text{ and } h = 10$$

$$\text{Mode } (M_o) = L_0 + \frac{f_0 - f_1}{2f_0 - f_1 - f_2} \times h$$

$$24 = 20 + \frac{27 - a}{(2 \times 27) - a - b} \times 10$$

$$4 = \frac{27-a}{(2\times 27)-a-b}\times 10$$

$$3a - 2b = 27 \qquad ...(1)$$

Since $N = 100$, we have $56 + a + b = 100$

$$a + b = 44 \qquad ...(2)$$

On solving equations (1) and (2), we get $a = 23$ and $b = 21$.

Hence, missing frequencies are 23 and 21.

Example 6. Find the mode of the following frequency distribution:

Size (x)	11	12	13	14	15	16	17	18	19	20	21	22
Frequency	3	8	15	23	35	40	32	28	20	45	14	6

Solution. Here the given frequency distribution is not regular since the frequencies are increasing steadily upto 40 and then decreasing upto 20 but the frequency 45 after 20 does not seem to be consistent with the distribution. In this case we use the grouping method.

Grouping Table

Size (x)	*Frequency*					
	(1)	(2)	(3)	(4)	(5)	(6)
11	3	11		26		
12	8		23		46	
13	15	38				73
14	23		58	98		
15	35	75			107	
16	40		72			100
17	32	60		80		
18	28				93	
19	20	65				79
20	45		59	65		
21	14	20				
22	6					

To find the mode we use the following analysis table:

Analysis Table

Column No. (a)	*Maximum frequency* (b)	*Value or combination of values of x giving maximum frequency in* (b) (c)
1	45	20
2	75	15, 16
3	72	16, 17

4	98	14, 15, 16
5	107	15, 16, 17
6	100	16, 17, 18

On examining the values in column (*c*) of analysis table we find that the value 16 is repeated the maximum number of times. Hence, the value of the mode is 16.

Example 7. Draw a histogram for the following frequency distribution and determine mode graphically. Verify this value by direct calculation.

Wages (Rs.) less than	50	100	150	200	250	300	350	400	450
No. of workers	29	224	465	582	634	644	650	653	655

Solution. Here the data is given in the form of cumulative frequency. First of all we have to convert it into a normal continuous frequency distribution as follows:

Wages (Rs.)	*No. of workers (f)*
0-50	29
50-100	$224-29=195\ (f_1)$
(L_0) 100-150	$465-224=241\ (f_0)$
150-200	$582-465=117\ (f_2)$
200-250	$634-582=52$
250-300	$644-634=10$
300-350	$650-644=6$
350-400	$653-650=3$
400-450	$655-653=2$

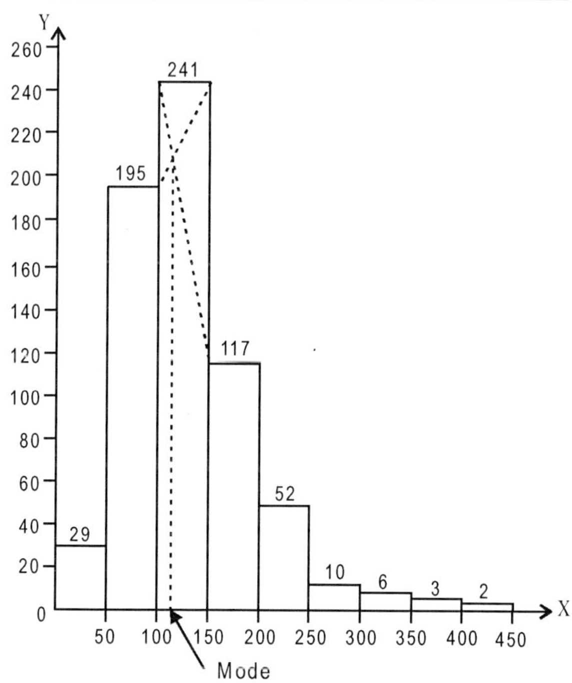

Mode = 113.5

Direct Calculation : Here maximum frequency 241 occurs in the class 100-150 which is the modal class.

$$L_0 = 100, f_0 = 241, f_1 = 195, f_2 = 117 \text{ and } h = 50$$

$$\text{Mode } (M_o) = L_0 + \frac{f_0 - f_1}{2f_0 - f_1 - f_2} \times h$$

$$= 100 + \frac{241 - 195}{(2 \times 241) - 195 - 117} \times 50$$

$$= 100 + 13.5 = 113.5$$

Example 8. Find the mode of the following frequency distribution:

Class interval	0-10	10-20	20-30	30-40	40-50	50-60	60-70	70-80
Frequency	9	12	15	16	17	15	10	13

Solution. Here the given frequency distribution is not regular we use the grouping method.

Grouping Table

Size (x)	*Frequency*					
	(1)	(2)	(3)	(4)	(5)	(6)
0-10	9	21				
10-20	12		27	36		
20-30	15	31			<u>43</u>	
30-40	16		<u>33</u>	<u>48</u>		<u>48</u>
40-50	17	<u>32</u>				
50-60	15		25		42	
60-70	10	23				38
70-80	13					

To find the mode we use the following analysis table:

Analysis Table

Column No. (a)	*Maximum frequency* (b)	*Size of the group of containing maximum frequency in* (b) (c)				
1	17				40-50	
2	32				40-50	50-60
3	33			30-40	40-50	
4	48			30-40	40-50	50-60
5	43	10-20	20-30	30-40		
6	48		20-30	30-40	40-50	
No. of times size groups occurs		1	2	4	5	2

Here, the size group 40-50 is occurring maximum number of times (*i.e.*, 5 times) and therefore 40-50 is the modal class.

$$L_0 = 40, f_0 = 17, f_1 = 16, f_2 = 15 \text{ and } h = 10$$

$$\text{Mode } (M_o) = L_0 + \frac{f_0 - f_1}{2f_0 - f_1 - f_2} \times h$$

$$= 40 + \frac{17 - 16}{(2 \times 17) - 16 - 15} \times 10$$

$$= 40 + 3.33 = 43.33$$

EXERCISE 5.1

1. Calculate the arithmetic mean of the following observations:
   55, 30, 28, 20, 22, 25
2. Calculate the arithmetic mean of the following observations:
   48, 35, 65, 40, 43, 52, 57, 40, 70
3. Calculate the A.M. of the following data:

x	65	66	67	68	69	70	71	72	73
f	1	4	5	7	11	10	6	4	2

4. Calculate the A.M. of the following data by step deviation method:

x	25	35	45	55	65	75	85
f	3	61	132	153	140	51	2

5. The A.M. of the following data is 30.5 marks. Find the missing frequency.

Marks (x)	10	20	30	40	50
Frequency	8	10	--	15	7

6. Calculate the A.M. of the following frequency distribution:

Class	0-10	10-20	20-30	30-40	40-50
Frequency	12	8	15	10	5

7. The A.M. of the following frequency distribution is 33. Find the missing frequency.

Class	0-10	10-20	20-30	30-40	40-50	50-60
Frequency	10	15	--	25	25	20

8. Find the mean marks of students from the following data by using short cut method:

Marks	0-10	10-20	20-30	30-40	40-50
Frequency	4	5	7	4	10

9. A distribution consists of three components with frequencies 150, 100 and 300 having their means 16, 8 and 4 respectively. Find the mean of combined distribution.

10. The mean of 100 numbers is 60. By mistake two numbers were taken as 38 and 84 instead of 83 and 48. Find the correct mean.

11. A candidate obtained the following percentages of marks in an examination. English 45%, Statistics 70%, Economics 58%, Income Tax 55%, Cost accountancy 74%

Find the weighted mean if weights 1, 3, 1, 2, 1 respectively are allotted to the subjects.

12. A person invests Rs. 2000 at 5%, Rs. 3000 at 6% and Rs. 10000 at 9%. What is the overall percentage yield of these investments?

13. Find the median of the observations:

$49, 67, 56, 54, 68, 57, 61, 71, 54, 81, 84.$

14. Find the median of the observations:

$33, 75, 84, 50, 43, 72, 68, 40, 90, 61.$

15. The median of 8 observations 31, 48, 37, 34, 45, 36, 41 and x is 38, where $37 < x < 41$. Find the value of x.

16. Find the median of the following data:

Income (Rs.)	100	150	80	200	250	180
No. of persons	24	26	16	20	6	30

17. Find the median of the following frequency distribution:

Marks	10-20	20-30	30-40	40-50	50-60	60-70	70-80
No.of candidates	12	30	34	65	45	25	18

18. Find the missing frequencies f_1 and f_2 in the following distribution. It is given that median of the distribution is 26 and the total number of candidates is 50.

Marks	0-10	10-20	20-30	30-40	40-50
No. of candidates	5	8	f_1	10	f_2

19. Find the mode from the following data:

$$27, 28, 26, 35, 40, 37, 38, 28, 25, 27, 40, 28, 30.$$

20. Find mode from the following frequency distribution:

x	12	14	16	18	20	22	24
f	5	7	11	12	22	11	6

21. Find mode from the following frequency distribution:

Class	10-15	15-20	20-25	25-30	30-35	35-40	40-45	45-50
Frequency	22	45	67	73	85	190	64	55

22. The expenditure of 100 families is given below:

Expenditure	0-10	10-20	20-30	30-40	40-50
No. of families	14	23	?	21	?

Given that mode is 24. Determine the missing frequencies.

23. Find the mode of the following frequency distribution:

Size (x)	32	33	34	35	36	37	38	39	40	41
Frequency	7	14	30	28	35	34	16	14	36	16

ANSWERS

1. 30	**2.** 50	**3.** 69.18	**4.** 54.72
5. 20	**6.** 22.6	**7.** 30	**8.** 28.67
9. 8	**10.** 60.09	**11.** 62.125%	**12.** 7.86%
13. 61	**14.** 64.5	**15.** 39	**16.** Rs. 150
17. 46	**18.** 20,7	**19.** 28	**20.** 20
21. 37.27	**22.** 27,15	**23.** 36	

5.18. STANDARD DEVIATION

The idea of standard deviation (S.D.) was given by Karl Pearson in 1893. Standard deviation is defined as the square root of the arithmetic mean of the square of the deviation from the mean. It is denoted by σ (sigma). Standard deviation is also known as mean error, mean square error or root mean square deviation from mean.

5.19. STANDARD DEVIATION FROM UNGROUPED DATA

1. **Direct Method:** The formula for calculating standard deviation (S.D.) of n values $x_1, x_2, ..., x_n$ is

$$\text{S.D.}\,\sigma = \sqrt{\frac{1}{n}\sum_{i=1}^{n}(x_i - \bar{x})^2}$$

When mean is not a whole number but a fractional value then the computation is quite complicated. In this case, we use the following formula:

$$\text{S.D.}\,(\sigma) = \sqrt{\frac{1}{n}\sum_{i=1}^{n}x_i^2 - \bar{x}^2}$$

or

$$\text{S.D.}\,(\sigma) = \sqrt{\frac{1}{n}\sum_{i=1}^{n}x_i^2 - \left(\frac{\sum_{i=1}^{n}x_i}{n}\right)^2}$$

2. **Short-cut Method:** The formula for calculating S.D. is

$$\text{S.D.}\,(\sigma) = \sqrt{\frac{1}{n}\sum_{i=1}^{n}\frac{dx_i^2}{n} - \left(\frac{dx_i}{n}\right)^2}$$

or

$$\text{S.D.}\,(\sigma) = \sqrt{\frac{1}{n}\sum_{i=1}^{n}\frac{dx_i^2}{n} - (\bar{x} - A)^2}$$

where $dx_i = x_i - A (i = 1, 2, ..., k)$; A = assumed mean

5.20 STANDARD DEVIATION FROM GROUPED DATA

1. **Direct Method:** For a frequency distribution in which the variate value x_i occurs f_i time ($i = 1, 2, ..., k$) the formula is

$$\text{S.D.}(\sigma) = \sqrt{\frac{\sum_{i=1}^{k} f_i (x_i - \bar{x})^2}{N}}$$

or
$$\text{S.D.}(\sigma) = \sqrt{\frac{\sum_{i=1}^{k} f_i x_i^2}{N} - \bar{x}^2}$$

$$\sigma = \sqrt{\frac{\sum_{i=1}^{k} f_i x_i^2}{N} - \left(\frac{\sum_{i=1}^{k} f_i x_i}{N}\right)^2}$$

where
$$N = \sum_{i=1}^{k} f_i$$

2. **Short-cut Method :** The formula for calculating S.D. is

$$\text{S.D.}(\sigma) = \sqrt{\frac{\sum_{i=1}^{k} f_i dx_i^2}{N} - \left(\frac{\sum_{i=1}^{k} f_i dx_i}{N}\right)^2}$$

or
$$\text{S.D.}(\sigma) = \sqrt{\frac{1}{n}\sum_{i=1}^{n} \frac{dx_i^2}{n} - (\bar{x} - A)^2}$$

where $dx_i = x - A$ ($i = 1, 2, ..., k$); A = assumed mean

5.21. STANDARD DEVIATION BY STEP DEVIATION METHOD

In order to simplify further the calculation work, some common factor may be taken out from all the deviations. The formula is

$$\text{S.D.}(\sigma) = C \times \sqrt{\frac{\sum_{i=1}^{k} f_i dx_i'^2}{N} - \left(\frac{\sum_{i=1}^{k} f_i dx_i'}{N}\right)^2}$$

where $dx_i = \dfrac{x_i - A}{C}$ ($i = 1, 2, ..., k$); A = assumed mean and C = common factor

5.22. COMBINED STANDARD DEVIATION

If two sets contain n_1 and n_2 observations with means $\overline{x}_1$ and $\overline{x}_2$ and standard deviations σ_1 and σ_2 respectively, then the standard deviation of the combined set is given by

$$\text{Combined standard deviation } \sigma_{12} = \sqrt{\frac{n_1(\sigma_1^2 + d_1^2) + n_2(\sigma_2^2 + d_2^2)}{n_1 + n_2}}$$

Where $d_1 = \overline{x}_{12} - \overline{x}_1$, $d_2 = \overline{x}_{12} - \overline{x}_2$, and $\overline{x}_{12} = \dfrac{n_1\overline{x}_1 + n_2\overline{x}_2}{n_1 + n_2}$

Note:

1. Standard deviation is independent of change of origin and dependent of change of scale.
2. Standard deviation is always positive.

5.23. ADVANTAGES OF STANDARD DEVIATION

1. It is rigidly defined.
2. It is based on all the values.
3. It is based on arithmetic mean, therefore, it has all the advantages of that measure.
4. It is used for further mathematical treatments.

5.24. DISADVANTAGES OF STANDARD DEVIATION

1. It is not easy to understand and not easy to calculate.
2. It is affected by extreme values.
3. It can not be used for comparing the dispersion of two or more series of observations given in different units.

5.25. VARIANCE

The square of standard deviation *i.e.* σ_2 is known as variance.

$$\text{Variance} = (\text{S.D.})^2$$

For finding the value of variance just takes the square of S.D. in all above formulas.

ILLUSTRATIVE EXAMPLES

Example 1. Find the standard deviation of the data11,14,15,17,18.

Solution. Here, $n = 5$

$$\text{Mean } (\overline{x}) = \frac{\sum x}{n} = \frac{1}{5} \times (11 + 14 + 15 + 17 + 18) = \frac{75}{5} = 15$$

x_i	$(x_i - \overline{x})$	$(x_i - \overline{x})^2$
11	–4	16
14	–1	1
15	0	0

17	2	4
18	3	9
		$\Sigma(x_i - \bar{x})^2 = 30$

$$\text{Standard deviation } (\sigma) = \sqrt{\frac{1}{n}\sum(x_i - \bar{x})^2}$$

$$= \sqrt{\frac{1}{5} \times 30} = \sqrt{6} = 2.45$$

Example 2. Find the variance and standard deviation from the following frequency distribution:

x	2	4	6	8	10	12	14	16
f	4	4	5	15	8	5	4	6

Solution.

x_i	f_i	$x_i f_i$	$(x_i - \bar{x})$	$(x_i - \bar{x})^2$	$f_i(x_i - \bar{x})^2$
2	4	8	−7	49	196
4	4	16	−5	25	100
6	5	30	−3	9	45
8	15	120	−1	1	15
10	8	80	1	1	8
12	5	60	3	9	45
14	4	56	5	25	100
16	6	80	7	49	245
	$N = \Sigma f_i = 50$				$\Sigma f_i(x_i - \bar{x})^2 = 754$

$$N = \Sigma f_i = 50$$

$$\text{Mean } (\bar{x}) = \frac{\sum fx}{\sum f} = \frac{450}{50} = 9$$

$$\text{Variance } (\sigma^2) = \frac{\sum f_i(x_i - \bar{x})^2}{N} = \frac{754}{50} = 15.0$$

$$\text{Standard deviation } (\sigma) = \sqrt{\text{Variance}} = \sqrt{15.08} = 3.88$$

Example 3. Find the standard deviation if the sum of squares of deviations taken from mean 40 of 10 values is 360.

Solution. Here, mean $\bar{x} = 40, n = 10$ and $\sum_{i=1}^{10}(x_i - \bar{x})^2 = 360$

$$\text{Standard deviation } (\sigma) = \sqrt{\frac{1}{n}\sum_{i=1}^{n}(x_i - \overline{x})^2}$$

$$= \sqrt{\frac{1}{10} \times 360} = \sqrt{36} = 6$$

Example 4. Find the standard deviation and variance from the following record of goals scored by a team in a football season:

No. of goals in match	0	1	2	3	4
No. of matches	1	9	7	5	3

Solution.

No. of goals in a match (x_i)	*No. of* (f_i)	$f_i x_i$	$(x_i - \overline{x})$	$(x_i - \overline{x})^2$	$f_i(x_i - \overline{x})^2$
0	1	0	–2	4	4
1	9	9	–1	1	9
2	7	14	0	0	0
3	5	15	1	1	5
4	3	12	2	4	12
	$N = \Sigma f_i = 25$	$\Sigma f_i x_i = 50$			$\Sigma f_i(x_i - \overline{x})^2 = 30$

$$\text{Mean } (\overline{x}) = \frac{\sum f_i x_i}{\sum f_i} = \frac{50}{25} = 2$$

$$\text{Standard deviation } (\sigma) = \sqrt{\frac{1}{N}\sum_{i=1}^{n} f_i(x_i - \overline{x})^2}$$

$$= \sqrt{\frac{1}{25} \times 30} = \sqrt{1.2} = 1.095$$

$$\text{Variance } (\sigma^2) = (1.095)^2 = 1.2$$

Example 5. Prices of a particular commodity in five years in two cities are as follows:

Prices in city A (in Rs.)	*Prices in city B (in Rs.)*
16	10
23	12
19	18
22	15
20	10

Find the city which has less variance.

Solution.

For City A

Prices in city A (x_i)	$(x_i - \bar{x})$	$(x_i - \bar{x})^2$
16	–4	16
23	3	9
19	–1	1
22	2	4
20	0	0
$\Sigma x_i = 100$		$\Sigma(x_i - \bar{x})^2 = 30$

$$\text{Mean} = \frac{\sum x_i}{n} = \frac{100}{5} = 20$$

$$\text{Standard deviation } (\sigma) = \sqrt{\frac{1}{n}\sum_{i=1}^{n}(x_i - \bar{x})^2}$$

$$= \sqrt{\frac{1}{5} \times 30} = \sqrt{6} = 2.45$$

$$\text{Variance } (\sigma^2) = (2.45)^2 = 6$$

For City B

Prices in city *B* (y_i)	$(y_i - \bar{y})$	$(y_i - \bar{y})^2$
10	–5	25
12	–3	9
18	3	9
15	0	0
20	5	25
$\Sigma y_i = 75$		$(y_i - \bar{y})^2 = 68$

$$\text{Mean } (\bar{y}) = \frac{\sum y_i}{n} = \frac{75}{5} = 15$$

$$\text{Standard deviation } (\sigma) = \sqrt{\frac{1}{n}\sum_{i=1}^{n}(y_i - \bar{y})^2}$$

$$= \sqrt{\frac{1}{5} \times 68} = \sqrt{13.6} = 3.69$$

$$\text{Variance } (\sigma^2) = (3.69)^2 = 13.60$$

City A has less variance than city B.

Example 6. Find the standard deviation from the following frequency distribution by short cut method:

Size of item	6	7	8	9	10	11
Frequency	3	6	9	13	8	5

Solution.

Let $A = 8$ (assumed mean)

x_i	f_i	$dx_i = x_i - A$	$f_i dx_i$	$f_i dx_i^2$
6	3	–2	–6	12
7	6	–1	–6	6
8	9	0	0	0
9	13	1	13	13
10	8	2	16	32
11	5	3	15	45
	$N = \Sigma f_i = 44$		$\Sigma f_i dx_i = 32$	$\Sigma f_i dx_i^2 = 108$

$$\text{Standard deviation } (\sigma) = \sqrt{\frac{\sum f_i dx_i^2}{N} - \left(\frac{\sum f_i dx_i}{N}\right)^2}$$

$$= \sqrt{\frac{108}{44} - \left(\frac{32}{44}\right)^2} = \sqrt{2.454 - 0.529} = \sqrt{1.925} = 1.387$$

Example 7. Find the standard deviation and variance from the following frequency distribution:

Class interval	0-8	8-16	16-24	24-32
Frequency	4	8	2	1

Solution.

Let $A = 12$ (assumed mean)

Class interval	f_i	x_i	$dx_i = x_i - A$	$f_i dx_i$	$f_i dx_i^2$
0-8	4	4	–8	–32	256
8-16	8	12(A)	0	0	0
16-24	2	20	8	16	128
24-32	1	28	16	16	256
	$N = \Sigma f_i = 15$			$\Sigma f_i dx_i = 0$	$\Sigma f_i dx_i^2 = 640$

$$\text{Standard deviation } (\sigma) = \sqrt{\frac{\sum f_i dx_i^2}{N} - \left(\frac{\sum f_i dx_i}{N}\right)^2}$$

$$= \sqrt{\frac{640}{15} - 0} = \sqrt{42.667} = 6.53$$

$$\text{Variance } (\sigma)^2 = (6.53)^2 = 42.667$$

Example 8. Find the standard deviation from the following frequency distribution using step deviation method:

Class interval	20-30	30-40	40-50	50-60	60-70	70-80	80-90
Frequency	3	6	13	15	14	5	4

Solution.

Let $A = 55$ (assumed mean) and $C = 10$

Class interval	f_i	x_i	$dx_i' = \frac{x_i - A}{C}$	$f_i dx_i'$	$f_i dx_i'^2$
20-30	3	25	−3	−9	27
30-40	6	35	−2	−12	24
40-50	13	45	−1	−13	13
50-60	15	55 (A)	0	0	0
60-70	14	65	1	14	14
70-80	5	75	2	10	20
80-90	3	85	3	12	36
	$N = \Sigma f_i = 60$			$\Sigma f_i dx_i' = 2$	$\Sigma f_i dx_i'^2 = 134$

$$\text{Standard deviation } (\sigma) = C \times \sqrt{\frac{\sum f_i dx_i'^2}{N} - \left(\frac{\sum f_i dx_i'}{N}\right)^2}$$

$$= 10 \times \sqrt{\frac{134}{60} - \left(\frac{2}{60}\right)^2}$$

$$= 10 \times \sqrt{2.232} = 14.94$$

Example 9. The means of two samples of sizes 500 and 600 were 186 and 175 respectively. The standard deviations for the two samples were 9 and 10 respectively. Find the mean and variance of the combined sample.

Solution. Here, $n_1 = 500, n_2 = 600, \bar{x}_1 = 186, \bar{x}_2 = 175, \sigma_1 = 9$ and $\sigma_2 = 10$

$$\text{Combined mean } (\bar{x}_{12}) = \frac{n_1\bar{x}_1 + n_2\bar{x}_2}{n_1 + n_2}$$

$$= \frac{500 \times 186 + 600 \times 175}{500 + 600} = \frac{198000}{1100} = 180$$

Now,

$$d_1 = \bar{x}_{12} - \bar{x}_1 = 180 - 186 = -6$$

$$d_2 = \bar{x}_{12} - \bar{x}_2 = 180 - 175 = 5$$

Combined standard deviation $(\sigma_{12}) = \sqrt{\dfrac{n_1(\sigma_1^2 + d_1^2) + n_2(\sigma_2^2 + d_2^2)}{n_1 + n_2}}$

$$= \sqrt{\frac{500\times[(9)^2 + (-6)^2] + 600\times[(10)^2 + (5)^2]}{500 + 600}}$$

$$= \sqrt{\frac{585 + 750}{11}} = \sqrt{121.363} = 11.016$$

Example 10. Two samples of sizes 40 and 50 have the same mean 53 but different standard deviations 19 and 8 respectively. Find the standard deviation of the combined sample of size 90.

Solution. Here, $n_1 = 40, n_2 = 50,\ \bar{x}_1 = \bar{x}_2 = 53, \sigma_1 = 19$ and $\sigma_2 = 8$

Combined mean $(\sigma_{12}) = \dfrac{n_1\bar{x}_1 + n_2\bar{x}_2}{n_1 + n_2}$

$$= \frac{40\times 53 + 50\times 53}{40 + 50} = \frac{53\times 90}{90} = 53$$

Now, $d_1 = \bar{x}_{12} - \bar{x}_1 = 53 - 53 = 0$

$d_2 = \bar{x}_{12} - \bar{x}_2 = 53 - 53 = 0$

Combined standard deviation $(\sigma_{12}) = \sqrt{\dfrac{n_1(\sigma_1^2 + d_1^2) + n_2(\sigma_2^2 + d_2^2)}{n_1 + n_2}}$

$$= \sqrt{\frac{40\times[(19)^2 + (0)^2] + 50\times[(8)^2 + (0)^2]}{40 + 50}}$$

$$= \sqrt{\frac{17640}{90}} = \sqrt{196} = 14$$

EXERCISE 5.2

1. Find the standard deviation and variance of the following data:
   19, 22, 27, 28, 28, 34, 35, 36, 41, 48
2. Find the variance and standard deviation of the following data:
   44, 54, 46, 63, 55, 42, 34, 48, 70, 38
3. Find the standard deviation of the following data:

x	6	7	8	9	10	11	12
f	3	6	9	13	8	5	4

4. Find the standard deviation and variance of the following data:

x	10	15	18	20	25
f	3	2	5	8	2

5. Find the standard deviation and variance of the following data:

Class interval	0-4	4-8	8-12	12-16	16-20
Frequency	5	8	10	8	5

6. Find the standard deviation and variance of the following data by using step deviation method:

Wages upto (in Rs.)	15	30	45	60	75	90	105	120
No. of workers	12	30	65	107	157	202	222	230

7. Calculate the standard deviation and variance of the following data by using step deviation method:

Marks	20-29	30-39	40-49	50-59	60-69	70-79	80-89	90-99
No. of students	5	12	15	20	18	10	6	4

8. There are 60 male and 40 female workers in a factory. The standard deviation of their wages was computed as Rs. 8 and Rs. 11 respectively. The mean wages of the two groups were found to be equal. Compute the combined standard deviation of the wages of all the workers.

9. The mean weight of 150 students is 60 kg. The mean weight of boys is 70 kg with a standard deviation of 10 kg and for the girls the mean weight is 55 kg and the standard deviation is 15 kg. Find the number of boys and combined standard deviation.

ANSWERS

1. 8.316
2. 112.572; 10.61
3. 1.607
4. 4.1; 16.81
5. 4.9; 24.01
6. 25.88; 669.7744
7. 17.65; 311.5225
8. 9.316
9. 50; 15.275

5.26 CORRELATION AND REGRESSION

5.27 CORRELATION ANALYSIS

In a bivariate distribution, if the change in one variable is accompanied by a change in the other variable in such a way that an increase in one variable results in an increase or decrease in the other, then the two variables are said to be correlated. For example, income and expenditure, heights and weights of students in a class, price and demand of certain commodities.

If the increase (or decrease) in one variable results in a corresponding increase (or decrease) in the other, correlation is said to be direct or positive. But if the increase (or decrease) in one variable results in a corresponding decrease (or increase), in the other, correlation is said to be negative. If two variables vary in such a way that their ratio is always constant, then the correlation is said to be perfect.

When we plot the corresponding values of two variables, taking one on X-axis and the other along Y-axis, it shows a collection of dots. This collection of dots is called a dot diagram or a scatter diagram.

If all the plotted points lie in a straight line and show an upward trend, then the correlation is perfect positive. If all the plotted points lie in a straight line and show a downward trend, then the correlation is perfect negative.

If the plotted points are not on a straight line but seem to be scattered around a straight line, the variables are correlated. Closer the scatter of points around a line, higher is the degree of correlation. If the plotted points are not clustered around a straight line but are widely scattered over the diagram, then there is a very low degree of correlation between the variables. If the plotted points show no trend at all, then the variables are independent and are not correlated.

5.28 KARL PEARSON'S COEFFICIENT OF CORRELATION

The correlation coefficient $r(x, y)$ between two variables x and y is given by

$$r(x,y) = \frac{\operatorname{cov}(x,y)}{\sqrt{\operatorname{var} iance(x)}\sqrt{\operatorname{var} iance(y)}}$$

$$= \frac{\operatorname{cov}(x,y)}{\sigma_x \sigma_y}$$

$r(x,y)$ is also denoted by $\rho(x,y)$ or r_{xy} or simply by r.

$$r(x,y) = \frac{\frac{1}{n}\sum_{i=1}^{n}(x_i-\bar{x})(y_i-\bar{y})}{\sqrt{\frac{1}{n}\sum_{i-1}^{n}(x_i-\bar{x})^2}\sqrt{\frac{1}{n}\sum_{i=1}^{n}(y_i-\bar{y})^2}}$$

or
$$r(x,y) = \frac{n\sum_{i=1}^{n}x_iy_i - \sum_{i=1}^{n}x_i\sum_{i=1}^{n}y_i}{\sqrt{n\sum_{i=1}^{n}x_i^2-(\sum_{i=1}^{n}x_i)^2}\sqrt{n\sum_{i=1}^{n}y_i^2-(\sum_{i=1}^{n}y_i)^2}}$$

If the values of x_i and y_i's are large or involve fractions, then define

$$u_i = \frac{x_i - a}{h} \quad \text{and} \quad v_i = \frac{y_i - b}{k}$$

where a and b arc assumed means of x-series and y-series respectively, h and k are constants. This property is known as change of origin and scale. Correlation coefficient is independent of change of origin and scale. In this $r(x, y)$ is given by the formula:

$$r(x,y)=r(u,v) = \frac{n\sum_{i=1}^{n} u_i v_i - \sum_{i=1}^{n} u_i \sum_{i=1}^{n} v_i}{\sqrt{n\sum_{i=1}^{n} u_i^2 - (\sum_{i=1}^{n} u_i)^2}\sqrt{n\sum_{i=1}^{n} v_i^2 - (\sum_{i=1}^{n} v_i)^2}}$$

5.29 PROPERTIES OF THE CORRELATION COEFFICIENT

(*i*) $-1 \le r \le +1$

(*ii*) If $r = -1$, then there is perfect negative correlation between x and y.

(*iii*) If $r = 1$, then there is perfect positive correlation between x and y.

(*iv*) If $r = 0$, then there is no correlation between x and y.

(*v*) If $-1 \le r < 0$, then there is negative correlation between x and y.

(*vi*) If $0 < r \le 1$, then there is positive correlation between x and y

5.30 SPEARMAN'S RANK CORRELATION

Sometimes we have to deal with problems in which data cannot be measured quantitatively but qualitative assessment is possible, *e.g.*, beauty, honesty, morality etc. In such cases we assign ranks to the individuals possessing these attributes or characteristics. Ranks are assigned starting from lowest to highest value or from highest to lowest value.

The coefficient of rank correlation r is given by

$$r(x,y) = 1 - \frac{6\sum_{i=1}^{n} d_i^2}{n(n^2-1)}$$

where d_i is the difference of corresponding rank and n is the number of pairs of observations.

5.31 CORRELATION IN CASE OF REPEATED RANKS

When two or more items have same values, it is difficult to assign different ranks to them. In such a case average ranks (same ranks) are assigned to each of them. In this case the following modified formula is used to find the rank coefficient of correlation:

$$r(x,y) = 1 - \frac{6\left[\sum_{i=1}^{n} d_i^2 + \frac{m_1(m_1^2-1)}{12} + \frac{m_2(m_2^2-1)}{12} + \ldots\right]}{n(n^2-1)}$$

where $m_1, m_2, \ldots$ are the number of times the item is repeated in a series.

ILLUSTRATIVE EXAMPLES

Example 1. Find the coefficient of correlation for the following data:

$$n=10,\quad \Sigma x=50,\ \Sigma y=-30,\ \Sigma x^2=290,\ \Sigma y^2=300,\ \Sigma xy=-115$$

Solution.

$$r(x,y)=\frac{n\sum xy-\sum x\sum y}{\sqrt{n\sum x^2-\left(\sum x\right)^2}\sqrt{n\sum y^2-\left(\sum y\right)^2}}$$

$$=\frac{10\times(-115)-(50)\times(-30)}{\sqrt{10\times 290-(50)^2}\sqrt{10\times 300-(-30)^2}}$$

$$=\frac{-1150+1500}{\sqrt{400}\sqrt{2100}}=\frac{350}{200\sqrt{21}}=0.3819$$

Example 2. A computer while calculating correlation coefficient between two variables x and y from 25 pairs of observations obtained the following results:

$$n=25,\quad \Sigma x=125,\quad \Sigma y=100,\quad \Sigma x^2=650,\quad \Sigma y^2=460,\quad \Sigma xy=508.$$

It was, however later, discovered at the time of checking that he had copied down two pairs as:

x	6	8
y	14	6

while the correct values were

x	8	6
y	12	8

Obtain the correct value of the correlation coefficient.

Solution.

Corrected Σx = given Σx – (sum of incorrect values) + (sum of correct values)

Corrected $\Sigma x = 125-(6+8)+(8+6)=125$

Corrected $\Sigma y = 100-(14+6)+(12+8)=100$

Corrected $\Sigma x^2 = 650-(6^2+8^2)+(8^2+6^2)=650$

Corrected $\Sigma y^2 = 460-(14^2+6^2)+(12^2+8^2)=436$

Corrected $\Sigma xy = 508-(6\times 14+8\times 6)+(8\times 12+6\times 8)=520$

$$\text{Corrected } r(x,y)=\frac{n\Sigma xy-\Sigma x\Sigma y}{\sqrt{n\Sigma x^2-(\Sigma x)^2}\sqrt{n\Sigma y^2-(\Sigma y)^2}}$$

$$=\frac{25\times 520-125\times 100}{\sqrt{25\times 650-(125)^2}\sqrt{25\times 436-(100)^2}}$$

$$=\frac{500}{25\times 30}=0.66$$

Example 3. Calculate the Karl Pearson's coefficient of correlation for the following data:

x	2	4	6	8	10
y	20	12	18	10	40

Solution.

x	y	x^2	y^2	xy
2	20	4	400	40
4	12	16	144	48
6	18	36	324	108
8	10	64	100	80
10	40	100	1600	400
$\Sigma x = 30$	$\Sigma y = 100$	$\Sigma x^2 = 220$	$\Sigma y^2 = 2568$	$\Sigma xy = 676$

Here, $n = 5$

$$r(x, y) = \frac{n\sum xy - \sum x \sum y}{\sqrt{n\sum x^2 - \left(\sum x\right)^2}\sqrt{n\sum y^2 - \left(\sum y\right)^2}}$$

$$= \frac{5 \times 676 - 30 \times 100}{\sqrt{5 \times 220 - (30)^2}\sqrt{5 \times 2568 - (100)^2}}$$

$$= \frac{3380 - 3000}{\sqrt{1100 - 900}\sqrt{12840 - 10000}}$$

$$= \frac{380}{\sqrt{200}\sqrt{2840}} = 0.5042$$

Example 4. Calculate the Karl Pearson's coefficient of correlation between x and y for the following data:

x	150	153	154	155	157	160	163	164
y	65	66	67	70	68	53	70	63

Solution. Let $u_i = x_i - 155$ and $v_i = y_i - 68$

x	y	u	v	u^2	v^2	uv
150	65	–5	–3	25	9	15
153	66	–2	–2	4	4	4
154	67	–1	–1	1	1	1
155	70	0	2	0	4	0
157	68	2	0	4	0	0
160	53	5	–15	25	225	–75
163	70	8	2	64	4	16
164	63	9	–5	81	25	–45
Total		16	–22	204	272	–84

Here, $n = 8$

$$r(x,y) = r(u,v) = \frac{n\sum uv - \sum u \sum v}{\sqrt{n\sum u^2 - \left(\sum u\right)^2}\sqrt{n\sum v^2 - \left(\sum v\right)^2}}$$

$$= \frac{8\times(-84) - 16\times(-22)}{\sqrt{8\times 204 - (16)^2}\sqrt{8\times 272 - (-22)^2}}$$

$$= \frac{-672 + 352}{\sqrt{1376}\sqrt{1692}} = -0.2097$$

Example 5. Calculate the rank correlation coefficient for the following data:

Student	A	B	C	D	E	F	G	H	I	J
Rank in Maths	9	10	6	5	7	2	4	8	1	3
Rank in Stats	1	2	3	4	5	6	7	8	9	10

Solution. Here the ranks are given and $n = 10$

Student	R_1	R_2	$d = R_1 - R_2$	d^2
A	9	1	8	64
B	10	2	8	64
C	6	3	3	9
D	5	4	1	1
E	7	5	2	4
F	2	6	–4	16
G	4	7	–3	9
H	8	8	0	0
I	1	9	–8	64
J	3	10	–7	49
				$\Sigma d^2 = 280$

$$r = 1 - \frac{6\sum d^2}{n(n^2-1)} = 1 - \frac{6\times 280}{10(10^2-1)}$$

$$= 1 - \frac{1680}{990} = 1 - 1.697 = -0.697$$

Example 6. Ten competitors in a beauty contest are ranked by three judges in the following order:

First Judge	1	6	5	10	3	2	4	9	7	8
Second Judge	3	5	8	4	7	10	2	1	6	9
Third Judge	6	4	9	8	1	2	3	10	5	7

Using the rank correlation method, discuss which pair of judges has the nearest approach to common taste in beauty?

Solution. Let R_1, R_2 and R_3 be the ranks given by three judges.

R_1	R_2	R_3	$d_{12}=R_1-R_2$	$d_{13}=R_1-R_3$	$d_{23}=R_2-R_3$	d_{12}^2	d_{13}^2	d_{23}^2
1	3	6	–2	–5	–3	4	25	9
6	5	4	1	2	1	1	4	1
5	8	9	–3	–4	–1	9	16	1
10	4	8	6	2	–4	36	4	16
3	7	1	–4	2	6	16	4	36
2	10	2	–8	0	8	64	0	64
4	2	3	2	1	–1	4	1	1
9	1	10	8	–1	–9	64	1	81
7	6	5	1	2	1	1	4	1
8	9	7	–1	1	2	1	1	4
					Total	200	60	214

Here, $n = 10$

Rank correlation coefficient between first and second judges

$$r_{12} = 1-\frac{6\sum d_{12}^2}{n(n^2-1)} = 1-\frac{6\times 200}{10(10^2-1)} = 1-\frac{40}{33} = -0.212$$

Rank correlation coefficient between first and third judges

$$r_{13} = 1-\frac{6\sum d_{13}^2}{n(n^2-1)} = 1-\frac{6\times 60}{10(10^2-1)} = 1-\frac{4}{11} = 0.636$$

Rank correlation coefficient between second and third judges

$$r_{23} = 1-\frac{6\sum d_{23}^2}{n(n^2-1)} = 1-\frac{6\times 214}{10(10^2-1)} = 1-\frac{214}{165} = -0.297$$

Since r_{13} is maximum, therefore, the pair of judges first and third has the nearest approach to common tastes in beauty.

Example 7. The marks obtained by 9 students in Statistics and Mathematics are given below:

Marks in Statistics	35	23	47	17	10	43	9	6	28
Marks in Mathematics	30	33	45	23	8	49	12	4	31

Compute the rank correlation coefficient.

Solution. Here the marks are given. First find the ranks and then differences.

Marks in Statistics (X)	*Marks in Mathematics (Y)*	*Ranks in X (R_1)*	*Ranks in Y (R_2)*	$d = R_1 - R_2$	d^2
35	30	3	5	–2	4
23	33	5	3	2	4
47	45	1	2	–1	1
17	23	6	6	0	0

10	8	7	8	–1	1
43	49	2	1	1	1
9	12	8	7	1	1
6	4	9	9	0	0
28	31	4	4	0	0
				Total	12

Here, $n = 9$

$$r = 1 - \frac{6\sum d^2}{n(n^2-1)} = 1 - \frac{6\times 12}{9(9^2-1)}$$

$$= 1 - \frac{72}{720} = 1 - 0.1 = 0.9$$

Example 8. Calculate the rank correlation coefficient from the following data:

x	48	33	40	9	16	16	65	24	16	57
y	13	13	24	6	15	4	20	9	6	19

Solution. Here, the values of x and y are given and n=10. First find the ranks and then differences.

x	y	*Ranks in x* (R_1)	*Ranks in y* (R_2)	$d = R_1 - R_2$	d^2
48	13	3	5.5	–2.5	6.25
33	13	5	5.5	–0.5	0.25
40	24	4	1	3	9.00
9	6	10	8.5	1.5	2.25
16	15	8	4	4	16.00
16	4	8	10	–2	4.00
65	20	1	2	–1	1.00
24	9	6	7	–1	1.00
16	6	8	8.5	–0.5	0.25
57	19	2	3	–1	1.00
				Total	41

In series x, 16 is repeated 3 times, hence $m_1 = 3$. In series y, 13 and 6 repeated 2 times each, hence $m_2 = 2$ and $m_3 = 2$.

Therefore, the correlation coefficient for repeated rank is given by

$$r(x,y) = 1 - \frac{6\left[\sum_{i=1}^{n} d_i^2 + \frac{m_1(m_1^2-1)}{12} + \frac{m_2(m_2^2-1)}{12} + \frac{m_3(m_3^2-1)}{12}\right]}{n(n^2-1)}$$

$$= 1 - \frac{6\left[41 + \frac{3(3^2-1)}{12} + \frac{2(2^2-1)}{12} + \frac{2(2^2-1)}{12}\right]}{10(10^2-1)}$$

$$= 1 - \frac{6[41+2+0.5+0.5]}{990}$$

$$= 1 - \frac{264}{990} = 0.733$$

Example 9. Following are the scores of ten students in a class and their IQ:

Students	1	2	3	4	5	6	7	8	9	10
Scores	35	40	25	55	85	90	65	55	45	50
I.Q.	100	100	110	140	150	130	100	120	140	110

Calculate the Spearman's rank correlation coefficient.

Solution. Here, the values of x and y are given and n=10. First find the ranks and then differences.

Students	*Scores*	*I.Q.*	*Ranks for Scores* (R_1)	*Ranks for I.Q.* (R_2)	$d = R_1 - R_2$	d^2
1	35	100	9	9	0	0.00
2	40	100	8	9	–1	1.00
3	25	110	10	6.5	3.5	12.25
4	55	140	4.5	2.5	2	4.00
5	85	150	2	1	1	1.00
6	90	130	1	4	–3	9.00
7	65	100	3	9	–6	36.00
8	55	120	4.5	5	–0.5	0.25
9	45	140	7	2.5	4.5	20.25
10	50	110	6	6.5	–0.5	0.25
					Total	84.00

In scores, 55 is repeated 2 times, hence $m_1 = 2$. In *I.Q.*, 140 and 110 repeated 2 times each, hence $m_2 = 2$ and $m_3 = 2$, and 100 is repeated 3 times, hence $m_4 = 3$.

Therefore, the correlation coefficient for repeated rank is given by

$$r(x,y) = 1 - \frac{6\left[\sum_{i=1}^{n} d_i^2 + \frac{m_1(m_1^2-1)}{12} + \frac{m_2(m_2^2-1)}{12} + \frac{m_3(m_3^2-1)}{12} + \frac{m_4(m_4^2-1)}{12}\right]}{n(n^2-1)}$$

$$= 1 - \frac{6\left[84 + \frac{2(2^2-1)}{12} + \frac{2(2^2-1)}{12} + \frac{2(2^2-1)}{12} + \frac{3(3^2-1)}{12}\right]}{10(10^2-1)}$$

$$= 1 - \frac{6[84+0.5+0.5+0.5+2.0]}{990} = 1 - \frac{525}{990} = 0.47$$

Example 10. The following table gives the scores obtained by 11 students in Statistics and Mathematics:

Scores in Statistics	40	46	54	60	70	80	82	85	85	90	95
Scores in Mathematics	45	45	50	43	40	75	55	72	65	42	70

Calculate the Spearman's rank correlation coefficient.

Solution. Here, the values of x and y are given and $n = 10$. First find the ranks and then differences.

Students	*Scores in Statistics* (x)	*Scores in Mathematics* (y)	*Ranks of* x (R_1)	*Ranks of* y (R_2)	$d = R_1 - R_2$	d^2
1	40	45	11	7.5	3.5	12.25
2	46	45	10	7.5	2.5	6.25
3	54	50	9	6	3	9.00
4	60	43	8	9	–1	1.00
5	70	40	7	11	–4	16
6	80	75	6	1	5	25.00
7	82	55	5	5	0	0.00
8	85	72	3.5	2	1.5	2.25
9	85	65	3.5	4	–0.5	0.25
10	90	42	2	10	–8	64.00
11	95	70	1	3	–2	4.00
					Total	140.00

In series x, 85 is repeated 2 times, hence $m_1 = 5$. In series y, 45 is repeated 2 times, hence $m_2 = 2$.

Therefore, the correlation coefficient for repeated rank is given by

$$r(x, y) = 1 - \frac{6\left[\sum_{i=1}^{n} d_i^2 + \frac{m_1(m_1^2 - 1)}{12} + \frac{m_2(m_2^2 - 1)}{12}\right]}{n(n^2 - 1)}$$

$$= 1 - \frac{6\left[140 + \frac{2(2^2 - 1)}{12} + \frac{2(2^2 - 1)}{12}\right]}{11(11^2 - 1)}$$

$$= 1 - \frac{6[140 + 0.5 + 0.5]}{1320}$$

$$= 1 - \frac{846}{1320} = 0.359$$

EXERCISE 5.3

1. Calculate the Karl Pearson's correlation coefficient from the following data:

x	80	90	100	110	120	130	140	150	160
y	15	19	16	19	17	18	16	18	15

2. Calculate the Karl Pearson's correlation coefficient between height of father and height of son from the following data:

Height of father (in inches)	64	65	66	67	68	69	70
Height of son (in inches)	66	67	65	68	70	68	72

3. Calculate the Karl Pearson's correlation coefficient from the following data using 20 as the working mean for price and 70 as the working mean for demand:

Price	14	16	17	18	19	20	21	22	23
Demand	84	78	70	75	66	67	62	58	60

4. Calculate the correlation coefficient from the following results:

$$n = 10,\ \Sigma x = 140,\ \Sigma y = 150, \Sigma(x-10)^2 = 180,$$

$$\Sigma(y-15)^2 = 215, \Sigma(x-10)(y-15) = 60$$

5. The correlation coefficient between x and y for 20 items is 0.3. Mean of x is 15 and mean of y is 20 while standard deviations for x and y are 4 and 5 respectively. At the time of calculations one item 27 has wrongly been taken as 17 in case of x series and 35 instead of 30 in case of y series. Find the corrected coefficient of correlation.

6. Calculate the coefficient of rank correlation from the following data:

x	10	12	8	15	20	25	40
y	15	10	6	25	16	12	18

7. Calculate the coefficient of rank correlation from the following data:

x	4	6	8	10	12	14	16	18
y	10	15	20	25	30	35	40	45

8. In a beauty contest two judges rank the ten competitors in the following order:

Competitors	*A*	*B*	*C*	*D*	*E*	*F*	*G*	*H*	*I*	*J*
Rank by Judge I	6	4	3	1	2	7	9	8	10	5
Rank by Judge II	4	1	6	7	5	8	10	9	3	2

Determine if the two judges have the same taste in beauty?

9. Calculate the coefficient of rank correlation from the following data:

x	68	64	75	50	64	80	75	40	55	64
y	62	58	68	45	81	60	68	48	50	70

10. The coefficient of rank correlation of the marks obtained by 10 students in Statistics and Mathematics was found to be 0.5. It was then detected that the difference in ranks in the two subjects for one particular student was wrongly taken to be 3 in place of 7. What should be the corrected rank correlation coefficient?

11. The coefficient of rank correlation between marks in Statistics and marks in Mathematics obtained by certain group of students is 0.8. If the sum of squares of the differences in marks is 33, find the number of students in the group.

ANSWERS

1. –0.11547 **2.** 0.81 **3.** –0.9542 **4.** 0.9151
5. 0.515 **6.** 0.57 **7.** 1.00 **8.** Yes
9. 0.545 **10.** 0.2576 **11.** 10

5.32 REGRESSION

Regression analysis attempts to establish the nature of relationship between the variables. It also helps to determine the functional relationship between the variables so that one can predict or estimate the value of one variable for the given value of the other variable. Regression measures the nature and extent of correlation.

5.33 LINEAR REGRESSION

If the variable in a bivariate distribution are correlated, then points in scatter diagram will be more or less concentrated round a curve. This curve is called the curve of regression. If the curve is a straight line, it is called a line of regression and the regression is said to be linear. Since the line of regression gives the best estimate to the value of dependent variable for any given value of the independent variable, therefore, it is called the line of best fit which is obtained by the method of least squares. Since any one of the two variables x and y can be taken as the independent variable and the other as a dependent variable. Therefore, there are two regression lines, one as the line of regression of y on x and the other as the line of regression of x on y.

5.34 LINES OF REGRESSION

Let the equation of line of regression of y on x be

$$y = a + bx \quad \ldots (1)$$

then

$$\bar{y} = a + b\bar{x} \quad \ldots (2)$$

Now subtracting equation (2) from (1), we have

$$y - \bar{y} = b(x - \bar{x}) \quad \ldots (3)$$

The normal equations for the equation (1) are

$$\Sigma y = na + b\Sigma x$$

$$\Sigma xy = a\Sigma x + b\Sigma x^2 \qquad \text{... (4)}$$

Shifting the origin to $(\bar{x}, \bar{y})$, (4) becomes

$$\Sigma(x-\bar{x})(y-\bar{y}) = a\Sigma(x-\bar{x}) + b\Sigma(x-\bar{x})^2 \qquad \text{... (5)}$$

We know that

$$r = \frac{Cov.(x, y)}{\sigma_x \sigma_y} = \frac{\frac{1}{n}\Sigma(x-\bar{x})(y-\bar{y})}{\sigma_x \sigma_y}, \ \Sigma(x-\bar{x}) = 0 \text{ and}$$

$$\sigma_x^2 = \frac{1}{n}\Sigma(x-\bar{x})^2$$

From equation (5), we have

$$nr\sigma_x\sigma_y = a.0 + b.n\sigma_x^2 \text{ or } b = r\frac{\sigma_y}{\sigma_x}$$

Hence, from equation (3), the line of regression of y on x is given by

$$y - \bar{y} = r\frac{\sigma_y}{\sigma_x}(x - \bar{x})$$

Similarly, the line of regression of x on y is given by

$$x - \bar{x} = r\frac{\sigma_x}{\sigma_y}(y - \bar{y})$$

$r\frac{\sigma_y}{\sigma_x}$ is called the regression coefficient of y on x and is denoted by b_{yx}

$$b_{yx} = r\frac{\sigma_y}{\sigma_x} = \frac{Cov(x, y)}{\sigma_x^2} = \frac{n\Sigma xy - \Sigma x\Sigma y}{n\Sigma x^2 - (\Sigma x)^2}$$

$r\frac{\sigma_x}{\sigma_y}$ is called the regression coefficient of x on y and is denoted by b_{xy}

$$b_{xy} = r\frac{\sigma_x}{\sigma_y} = \frac{Cov(x, y)}{\sigma_y^2} = \frac{n\Sigma xy - \Sigma x\Sigma y}{n\Sigma y^2 - (\Sigma y)^2}$$

Note: The line of regression of y on x is used to estimate the value of y for given value of x. The line of regression of x on y is used to estimate the value of x for given value of y.

5.35 PROPERTIES OF REGRESSION COEFFICIENTS

(*i*) The geometric mean of the two regression coefficients is the coefficient of correlation *i.e.*,

$$r = \sqrt{b_{yx} \times b_{xy}}$$

Proof: We know that the two regression coefficients are

$$b_{yx} = r\frac{\sigma_y}{\sigma_x} \text{ and } b_{xy} = r\frac{\sigma_x}{\sigma_y}$$

Now, Geometric mean $= \sqrt{b_{yx} \times b_{xy}}$

$$= \sqrt{r\frac{\sigma_y}{\sigma_x} \times r\frac{\sigma_x}{\sigma_y}}$$

$$= \sqrt{r^2} = \pm r$$

Note: r will be positive (+) if b_{yx} and b_{xy} are positive, r will be negative (–) if b_{yx} and b_{xy} are negative.

(*ii*) The arithmetic mean of the regression coefficients is greater than or equal to the coefficient of correlation *i.e.*,

$$\frac{b_{yx} + b_{xy}}{2} \geq r$$

Proof: We know that the two regression coefficients are

$$b_{yx} = r\frac{\sigma_y}{\sigma_x} \text{ and } b_{xy} = r\frac{\sigma_x}{\sigma_y}$$

Now, Arithmetic mean $= \frac{b_{yx} + b_{xy}}{2} \geq r$

or $$\frac{r\frac{\sigma_y}{\sigma_x} + r\frac{\sigma_x}{\sigma_y}}{2} \geq r \text{ or } \frac{\sigma_y}{\sigma_x} + \frac{\sigma_x}{\sigma_y} - 2 \geq 0$$

or $\frac{1}{\sigma_x \sigma_y}[\sigma_x^2 + \sigma_y^2 - 2\sigma_x\sigma_y] \geq 0$

or $$\frac{1}{\sigma_x \sigma_y}[\sigma_x - \sigma_y]^2 \geq 0 \text{ which is true}$$

$\because$ $$\sigma_x, \sigma_y \geq 0$$

$\therefore$ $$\frac{b_{yx} + b_{xy}}{2} \geq r$$

(*iii*) If one of the regression coefficient is greater than unity, the other must be less than unity *i.e.*, If $b_{yx} > 1$ then $b_{xy} < 1$.

Proof: The two regression coefficients are b_{yx} and b_{xy}.

Let $b_{yx} > 1$ then $\frac{1}{b_{yx}} < 1$

We know that $b_{yx} \times b_{xy} = r^2 \leq 1$ or $b_{yx} \times b_{xy} \leq 1$ or $b_{xy} \leq \dfrac{1}{b_{yx}} < 1$

Similarly, if $b_{xy} > 1$ then $b_{yx} < 1$.

(*iv*) Regression coefficients are independent of the origin but not of scale.

Proof: Let $u = \dfrac{x-a}{h}$ and $v = \dfrac{y-b}{k}$ where *a*, *b*, *h* and *k* are constants.

Now,
$$b_{yx} = r\frac{\sigma_y}{\sigma_x} = r\frac{k\sigma_v}{h\sigma_u} = \frac{k}{h}\left(r\frac{\sigma_v}{\sigma_u}\right) = \frac{k}{h}b_{vu}$$
$$(\because \sigma_x^2 = h^2\sigma_u^2,\ \sigma_y^2 = k^2\sigma_v^2)$$

Similarly,
$$b_{xy} = \frac{h}{k}b_{uv}$$

5.36 ANGLE BETWEEN TWO LINES OF REGRESSION

If θ be the acute angle between the two lines of regression in case of two variables *x* and *y*, then

$$\tan\theta = \frac{1-r^2}{r}\cdot\frac{\sigma_x\sigma_y}{\sigma_x^2+\sigma_y^2}$$

where r, σ_x and σ_y have their usual meanings.

Explain the significance when $r = 0$ and $r = \pm 1$.

Proof: Equation of the line of regression of *y* on *x* is

$$y - \bar{y} = r\frac{\sigma_y}{\sigma_x}(x - \bar{x})$$

and the equation of the line of regression of *x* on *y* is

$$x - \bar{x} = r\frac{\sigma_x}{\sigma_y}(y - \bar{y})$$

Their slopes are $m_1 = r\dfrac{\sigma_y}{\sigma_x}$ and $m_2 = \dfrac{\sigma_y}{r\sigma_x}$

$\therefore$
$$\tan\theta = \pm\frac{m_2 - m_1}{1 + m_1 m_2}$$

$$= \pm\frac{\dfrac{\sigma_y}{r\sigma_x} - r\dfrac{\sigma_y}{\sigma_x}}{1 + \dfrac{\sigma_y^2}{\sigma_x^2}} = \pm\frac{1-r^2}{r}\cdot\frac{\sigma_y}{\sigma_x}\cdot\frac{\sigma_x^2}{\sigma_x^2+\sigma_y^2}$$

$$= \pm\frac{1-r^2}{r}\cdot\frac{\sigma_x\sigma_y}{\sigma_x^2+\sigma_y^2}$$

Since $r^2 \leq 1$ and σ_x, σ_y are positive.

$\therefore$ Positive sign gives the acute angle between the lines.

Hence, $$\tan\theta = \frac{1-r^2}{r} \cdot \frac{\sigma_x \sigma_y}{\sigma_x^2 + \sigma_y^2}$$

When $r = 0$, $\tan\theta = \infty$ so that $\theta = \frac{\pi}{2}$

So the two lines of regression are perpendicular to each other.

When $r = \pm 1$, $\tan\theta = 0$ so that $\theta = 0$ or π

So the two lines of regression coincide and there is perfect correlation between the two variables x and y.

ILLUSTRATIVE EXAMPLES

Example 1. Find the equation of two lines of regression for the following data:

x	1	2	3	4	5
y	7	6	5	4	3

and hence find an estimate of y for $x = 3.5$ from the appropriate line of regression.

Solution.

x	Y	x^2	y^2	xy
1	7	1	49	7
2	6	4	36	12
3	5	9	25	15
4	4	16	16	16
5	3	25	9	15
$\Sigma x = 15$	$\Sigma y = 25$	$\Sigma x^2 = 55$	$\Sigma y^2 = 135$	$\Sigma xy = 65$

Here, $$n = 5$$

$$\bar{x} = \frac{1}{n}\Sigma x = \frac{15}{5} = 3, \quad \bar{y} = \frac{1}{n}\Sigma y = \frac{25}{5} = 5$$

Now, $$b_{yx} = \frac{n\Sigma xy - \Sigma x \Sigma y}{n\Sigma x^2 - (\Sigma x)^2} = \frac{5\times65 - 15\times25}{5\times55 - (15)^2} = \frac{13-15}{11-9} = 1$$

$$b_{xy} = \frac{n\Sigma xy - \Sigma x \Sigma y}{n\Sigma y^2 - (\Sigma y)^2} = \frac{5\times65 - 15\times25}{5\times135 - (25)^2} = \frac{13-15}{27-25} = -1$$

So, the line of regression of y on x is

$$y - \bar{y} = b_{yx}(x - \bar{x})$$

$$y - 5 = -1(x - 3)$$

$$y = -x + 8$$

and the line of regression of x on y is

$$x-\bar{x} = b_{xy}(y-\bar{y})$$
$$x-3 = -1(y-5)$$
$$x = -y+8$$

To estimate the value of y when x is given, we use the line of regression of y on x, *i.e.*

$$y = -x+8$$

Now, substitute $x = 3.5$, we have

$$y = -3.5+8 = 4.5$$

Example 2. The following table gives age (x) in years of cars and annual maintenance cost (y) in hundred rupees:

x	1	3	5	7	9
y	15	18	21	23	22

Estimate the maintenance cost for a 4 year old car after finding the appropriate line of regression.

Solution.

x	y	x^2	xy
1	15	1	15
3	18	9	54
5	21	25	105
7	23	49	161
9	22	81	198
$\Sigma x = 25$	$\Sigma y = 99$	$\Sigma x^2 = 165$	$\Sigma xy = 533$

Here, $n = 5$

$$\bar{x} = \frac{1}{n}\Sigma x = \frac{25}{5} = 5, \quad \bar{y} = \frac{1}{n}\Sigma y = \frac{99}{5} = 19.8$$

Now, $$b_{yx} = \frac{n\Sigma xy - \Sigma x\Sigma y}{n\Sigma x^2 - (\Sigma x)^2} = \frac{5\times 533 - 25\times 99}{5\times 165 - (25)^2}$$

$$= \frac{2665-2475}{825-625} = \frac{190}{200} = 0.95$$

The line of regression of y on x is given by

$$y-\bar{y} = b_{yx}(x-\bar{x})$$
$$y-19.8 = 0.95(x-5)$$
$$y = 0.95x+15.05$$

When $x = 4$ years, we have

$$y = 0.95\times 4+15.05 = 18.85 \text{ hundred rupees} = \text{Rs. } 1885$$

Example 3. From the following information on values of two variables x and y, find the two regression lines and the correlation coefficient between x and y.

$$n = 10,\ \Sigma x = 20,\ \Sigma y = 40,\ \Sigma x^2 240,\ \Sigma y^2 = 410,\ \Sigma xy = 200.$$

Solution. We know that

$$b_{yx} = \frac{n\Sigma xy - \Sigma x\Sigma y}{n\Sigma x^2 - (\Sigma x)^2} = \frac{10\times 200 - 20\times 40}{10\times 240 - (20)^2} = \frac{20-8}{24-4} = \frac{3}{5}$$

and

$$b_{xy} = \frac{n\Sigma xy - \Sigma x\Sigma y}{n\Sigma y^2 - (\Sigma y)^2} = \frac{10\times 200 - 20\times 40}{10\times 410 - (40)^2} = \frac{20-8}{41-16} = \frac{12}{25}$$

$$\bar{x} = \frac{1}{n}\Sigma x = \frac{20}{10} = 2,\quad \bar{y} = \frac{1}{n}\Sigma y = \frac{40}{10} = 4$$

The two regression lines are

$$y - \bar{y} = b_{yx}(x - \bar{x})$$

$$y - 4 = \frac{3}{5}(x-2)$$

$$y = 0.6x + 2.8$$

and

$$x - \bar{x} = b_{xy}(y - \bar{y})$$

$$x - 2 = \frac{12}{25}(y-4)$$

$$x = 0.48y + 0.08$$

We know that

$$r = \pm\sqrt{b_{yx}\times b_{xy}}$$

$$r = \sqrt{\frac{3}{5}\times\frac{12}{25}} = \sqrt{\frac{36}{125}} = 0.536$$

Example 4. For 100 students of a class, the regression equation of marks in Statistics (x) on the marks in Mathematics (y) is $3y - 5x + 180 = 0$. The mean marks in Mathematics is 50 and variance of marks in Statistics is $\frac{4}{9}th$ of the variance of marks in Mathematics. Find the mean marks in Statistics and the coefficient of correlation between marks in the two subjects.

Solution. Since the given line of regression is x on y so we have

$$x = \frac{3}{5}y + \frac{180}{5} = \frac{3}{5}y + 36$$

We have

$$b_{xy} = \frac{3}{5} = r\frac{\sigma_x}{\sigma_y}$$

Given $\qquad$ variance of $x = \frac{4}{9}$ variance of y

$$\frac{\text{variance of } x\ (\sigma_x^2)}{\text{variance of } y\ (\sigma_y^2)} = \frac{4}{9} \Rightarrow \frac{\sigma_x}{\sigma_y} = \frac{2}{3}$$

So,
$$\frac{3}{5} = r \times \frac{2}{3} \Rightarrow r = \frac{9}{10} = 0.9$$

($\because b_{xy}$ is positive, r is positive)

Since the mean of x and mean of y lie on the regression lines, we have

$$\bar{x} = \frac{3}{5}\bar{y} + 36 \Rightarrow \bar{x} = \frac{3}{5} \times 50 + 36 = 66 \qquad (\because \bar{y} = 50)$$

Example 5. The lines of regression of y on x and x on y are $y = x + 5$ and $16x - 9y = 94$ respectively. Find the variance of x if the variance of y is 16. Also find the covariance of x and y.

Solution. Regression equation of y on x is

$$y = x + 5 \Rightarrow b_{yx} = 1 \qquad \text{(coefficient of } x)$$

Regression equation of x on y is $16x - 9y = 94$ *i.e.*

$$x = \frac{9}{16}y + \frac{94}{16} \Rightarrow b_{xy} = \frac{9}{16} \qquad \text{(coefficient of } y)$$

We know that
$$r = \pm\sqrt{b_{yx} \times b_{xy}}$$

$$r = \sqrt{1 \times \frac{9}{16}} = \frac{3}{4} = 0.75$$

(r is positive, since b_{yx} and b_{xy} are positive)

$$b_{xy} = r\frac{\sigma_x}{\sigma_y} \Rightarrow \sigma_x = \frac{b_{xy} \times \sigma_y}{r} = \frac{\left(\frac{9}{16}\right) \times 4}{\left(\frac{3}{4}\right)} = 3$$

$(\because \sigma_y^2 = 16)$

$$r = \frac{Cov.(x, y)}{\sigma_x \sigma_y}$$

$$\Rightarrow \qquad Cov.(x, y) = r\sigma_x\sigma_y = \frac{3}{4} \times 3 \times 4 = 9$$

Example 6. The equations of two lines of regression are $4x + 3y + 7 = 0$ and $3x + 4y + 8 = 0$.

Find

(*i*) the mean values of x and y

(*ii*) the regression coefficients b_{yx} and b_{xy}

(*iii*) the correlation coefficient between x and y

(*iv*) the standard deviation of y, if the variance of x is 4

(*v*) the value of y for $x = 5$.

Solution.

(*i*) Since the mean of x and mean of y lie on the regression lines, we have

$$4\bar{x}+3\bar{y}+7 = 0 \quad \text{or} \quad 4\bar{x}+3\bar{y} = -7$$

and
$$3\bar{x}+4\bar{y}+8 = 0 \quad \text{or} \quad 3\bar{x}+4\bar{y} = -8$$

Now, on solving the above equations for $\bar{x}$ and $\bar{y}$, we have

$$\bar{x} = -\frac{4}{7} \text{ and } \bar{y} = -\frac{11}{7}$$

$$\text{Mean of } x = -\frac{4}{7} \text{ and mean of } y = -\frac{11}{7}$$

(*ii*) Let the regression line of y on x be

$$3x+4y+8 = 0 \text{ or } y = -\frac{3}{4}x-2$$

$$\therefore \qquad b_{yx} = -\frac{3}{4} \qquad \text{(coefficient of } x\text{)}$$

and the regression line of x on y be

$$4x+3y+7 = 0 \text{ or } x = -\frac{3}{4}y-\frac{7}{4}$$

$$\therefore \qquad b_{xy} = -\frac{3}{4} \qquad \text{(coefficient of } y\text{)}$$

Since
$$b_{yx} \times b_{xy} = -\frac{3}{4}\times-\frac{3}{4} = \frac{9}{16} < 1$$

Hence, the choice of regression lines is correct.

So
$$b_{yx} = -\frac{3}{4} \text{ and } b_{xy} = -\frac{3}{4}$$

(*iii*) We know that
$$r = \pm\sqrt{b_{yx} \times b_{xy}}$$

$$\therefore \qquad r = \pm\sqrt{\left(\frac{-3}{4}\right)\times\left(\frac{-3}{4}\right)} = \pm\frac{3}{4} = -0.75$$

(r is negative, sinceand b_{yx} are b_{xy} negative)

(*iv*) We have
$$\sigma_x^2 = 4 \Rightarrow \sigma_x = 2$$

Now,
$$b_{yx} = -\frac{3}{4} \text{ or } r\frac{\sigma_y}{\sigma_x} = -\frac{3}{4}$$

$$\left(-\frac{3}{4}\right)\times\frac{\sigma_y}{2} = -\frac{3}{4} \Rightarrow \sigma_y = 2$$

(*v*) Since we have to find y when x is given, we use line of regression of y on x

$$y = -\frac{3}{4}x-2$$

Putting $x = 5$, we have

$$y = -\frac{3}{4}\times 5 - 2 = -\frac{23}{4} = -5.75$$

Example 7. The two regression lines of the variables x and y are:

$$x = 19.13 - 0.87y \text{ and } y = 11.64 - 0.50x$$

Find

(*i*) mean of x and mean of y

(*ii*) correlation coefficient between x and y (*I.U. 2007, 08*)

Solution. (*i*) Since the mean of x and mean of y lie on the regression lines, we have

$$\bar{x} = 19.13 - 0.87\bar{y} \text{ or } \bar{x} + 0.87\bar{y} = 19.13$$

and $$\bar{y} = 11.64 - 0.50\bar{x} \text{ or } 0.50\bar{x} + \bar{y} = 11.64$$

Now, on solving the above equations for $\bar{x}$ and $\bar{y}$, we have

$$\bar{x} = 15.935 \text{ and } \bar{y} = 3.67$$

Mean of x = 15.935 and mean of y = 3.67

(*ii*) Let the regression line of y on x be

$$y = 11.64 - 0.50x$$

$\therefore$ $$b_{yx} = -0.50 \qquad \text{(coefficient of } x)$$

and the regression line of x on y be

$$x = 19.13 - 0.87y$$

$\therefore$ $$b_{xy} = -0.87 \qquad \text{(coefficient of } y)$$

We know that $$r = \pm\sqrt{b_{yx}\times b_{xy}}$$

$\therefore$ $$r = \pm\sqrt{-0.50\times -0.87} = -0.66$$

(r is negative, since b_{yx} and b_{xy} are negative)

Example 8. For two random variables, x and y with the same mean, the two regression equations are $y = ax + b$ and $x = cy + d$. Show that

(*i*) $\bar{x} = \dfrac{bc+d}{1-ac}$ and $\bar{y} = \dfrac{ad+b}{1-ac}$

(*ii*) $\dfrac{\sigma_x}{\sigma_y} = \sqrt{\dfrac{c}{a}}$ (*iii*) $r(x, y) = \sqrt{a\ \ c}$

Solution. (*i*) Since the mean of x and mean of y lie on the regression lines, we have

$$\bar{y} = a\bar{x} + b \text{ and } \bar{x} = c\bar{y} + d$$

Now, on solving the above equations for $\bar{x}$ and $\bar{y}$, we have

$$\bar{x} = \frac{bc+d}{1-ac} \text{ and } \bar{y} = \frac{ad+b}{1-ac}$$

(*ii*) Here $b_{yx} = \frac{r\sigma_y}{\sigma_x} = a$ and $b_{xy} = \frac{r\sigma_x}{\sigma_y} = c$

Now on dividing b_{xy} by b_{yx} we have

$$\frac{r\sigma_x}{\sigma_y} \times \frac{\sigma_x}{r\sigma_y} = \frac{c}{a} \text{ or } \frac{\sigma_x^2}{\sigma_y^2} = \frac{c}{a} \text{ or } \frac{\sigma_x}{\sigma_y} = \sqrt{\frac{c}{a}}$$

(*iii*)
$$r = \sqrt{b_{yx} \times b_{xy}} = \sqrt{ac}$$

EXERCISE 5.4

1. Find the regression coefficient b_{xy} for the following data:

   $n = 6, \Sigma x = 30, \Sigma y = 42, \Sigma x^2 = 184, \Sigma y^2 = 318, \Sigma xy = 199.$

2. Find the regression coefficient b_{yx} and b_{xy} between x and y for the following data:

   $n = 7, \quad \Sigma x = 24, \quad \Sigma y = 12, \quad \Sigma x^2 = 374, \quad \Sigma y^2 = 97, \quad \Sigma xy = 157.$

   Also, find the coefficient of correlation between x and y.

3. Find the regression line of x on y and estimate the value of x, when $y = 5$ from the following data:

   $n = 25, \quad \Sigma x = 125, \quad \Sigma y = 100, \quad \Sigma x^2 = 1650, \quad \Sigma y^2 = 1500, \quad \Sigma xy = 50.$

4. Find the line of regression of y on x for the following data:

x	10	9	8	7	6	4	3
y	8	12	7	10	8	9	6

5. Find the line of regression of y on x for the following data:

x	1	3	4	6	8	9	11	14
y	1	2	4	4	5	7	8	9

   Estimate the value of y, when $x = 10$.

6. Find the regression lines for the following data:

x	6	2	10	4	8
y	9	11	5	8	7

7. Find the regression lines for the following data:

x	1	2	3	4	5
y	2	5	3	8	7

   Estimate the value of y, when $x = 10$.

8. The equations of two lines of regression are $8x - 10y + 66 = 0$ and $40x - 18y = 214$.

   Find

   (*i*) the mean values of x and y

(*ii*) the correlation coefficient between x and y

(*iii*) the standard deviation of y, if the variance of x is 9.

9. Two random variables have the regression lines $3x+2y-26=0$ and $6x+y-31=0$. Find the mean values of x and y and the coefficient of correlation. If the variance of x is 25, find standard deviation of y from the data given.

ANSWERS

1. –0.46

2. $b_{yx}=0.397;\ b_{xy}=1.516;\ r=0.776$

3. $x=-\frac{9}{22}y+\frac{146}{22};\ 4.591$

4. $y=\frac{1}{3}x+\frac{133}{21}$

5. $y=\frac{7}{11}x+\frac{6}{11};\ 6.91$

6. $y=11.9-0.65x;\ x=16.4-1.3y$

7. $y=1.1+1.3x;\ x=0.5+0.5y;\ y=14.1$

8. (*i*) $\bar{x}=13,\ \bar{y}=17$ (*ii*) 0.6 (*iii*) 4

9. $\bar{x}=4,\ \bar{y}=7,\ \sigma_y=15,\ r=-0.5$

B. C. A.

SECOND SEMESTER EXAMINATION, 2007-08

MATHEMATICS -II

Time : 3 Hours | Total Marks : 100

Note : (*i*) Attempt any Five questions.

(*ii*) Marks are indicated against each question.

1. (*a*) If $u = \tan^{-1}\dfrac{xy}{\sqrt{1+x^2+y^2}}$, show that $\dfrac{\partial^2 u}{\partial x\,\partial y} = \dfrac{1}{(1+x^2+y^2)^{3/2}}$ **10**

(*b*) If $y_1 = \dfrac{x_2x_3}{x_1}, y_2 = \dfrac{x_3x_1}{x_2}, y_3 = \dfrac{x_1x_2}{x_3}$, show that Jacobian of y_1, y_2, y_3 with respect to x_1, x_2, x_3, is 4 **10**

2. (*a*) Use variation of parameters method to solve $y'' + y = \sec x$. **10**

(*b*) Solve the following differential equation : **10**

$(ax + hy + g)dx + (hx + by + f)dy = 0$

3. (*a*) Solve : $4\dfrac{\partial^2 z}{\partial x^2} - 4\dfrac{\partial^2 z}{dx\,\partial y} + \dfrac{\partial^2 z}{\partial y^2} = e^{x+2y}$ **10**

(*b*) Find the value of p for which the line $x \cos a + y \sin a = p$ may touch the circle $x^2 + y^2 - 2ax \cos \alpha - 2\ ay \sin \alpha = 0$. **10**

4. (*a*) An Urn I contains 3 white and 4 red balls and Urn II contains 5 white and 6 red balls. One ball is drawn at random from one of the Urns and is bound to be white. Find the probability that it was drawn from Urn I.

(*b*) Fit Poisson's distribution to the following and calculate theoretical frequencies : **10**

Deaths	0	1	2	3	4
Feequencies	122	60	15	2	1

5. (*a*) The two regression lines of the variables x any y are : **10**

$x = 19.13 - 0.87$ andy $y = 11.64 - 0.50\,x$

Find :

(*i*) mean of x's

(*ii*) mean of y's

(*iii*) correlation of coefficient between x and y.

(*b*) Calculate the mode of the following data :

Wages (Rs.)	Below 100	100-200	200-300
No. of weeks	8	12	25

Wages (Rs.)	300-400	400-500	Above 500
No. of weeks	15	10	5

6. (*a*) Expand $\log x$ in powers of $(x - 1)$, by Taylor's theorem. **10**

(*b*) Six dice are thrown 729 times. How many times do you expect atleast three dice to show a five or a six ? **10**

7. (*a*) Find equations of the tangents to the hyperbola $2x^2 - 3y^2 = 6$, which is parallel to the line $y = 3x + 4$. **10**

(*b*) For two random variables x any y with the same mean the two regression equations are $y = ax + b$ and $x = ay + b$. Show that : **10**

$\frac{b}{\beta} = \frac{1-a}{1-\alpha}$. Also find the common mean.